WELDING
AND
WELDING TECHNOLOGY

WELDING AND WELDING TECHNOLOGY

RICHARD L. LITTLE

Associate Dean of Career Education
Central Arizona College

MCGRAW-HILL BOOK COMPANY

New York San Francisco St. Louis
Düsseldorf Johannesburg Kuala Lumpur
London Mexico Montreal
New Delhi Panama Rio de Janeiro
Singapore Sydney Toronto

This book was set in Baskerville by Black Dot, Inc. The editors were Cary F. Baker, Jr., and Marge Woodhurst, the designer was Janet Bollow, and the production supervisor was Charles A. Goehring. The drawings were done by John F. Foster.

WELDING
AND
WELDING TECHNOLOGY

Printed in the United States of America.

Library of Congress Cataloging in Publication Data

Little, Richard L.
  Welding and welding technology.

  Includes bibliographies.
  1. Welding.   I. Title.
TS227.L658   671.5'2   72–3652
ISBN 0–07–038095–3

456789–MAMM–7654

CONTENTS

PREFACE

The basic purpose of this text is to present welding and welding technology in such a manner that the beginning college student can understand the underlying theories of welding, apply these theories to the process, and then apply the process to the fabrication of goods. The theories and principles of welding technology, while based on mathematics and science, are presented so that a scientific background is not mandatory although the student will learn to apply an amount of mathematics and science in welding technology.

The text follows the same basic format in all its five major sections. Each section is divided into specific technical units. The first unit for most of the sections is an introduction which presents the operating theories and principles, and the remaining units explain how to apply the theory to the fabrication of goods.

A set of questions to reinforce the information within the unit and suggested further readings are also included at the end of each unit. Where applicable, each unit has sample projects or exercises that emphasize the manipulative aspects of welding technology. Many of the suggested activities are ideas to instructors for possible demonstrations.

The text provides the basic information to explain the "how as well as the why" of the welding processes. The material contained in the introductory unit of Sections I, II, and III is technical in nature and could be omitted if the educational program is not technically oriented. Thus the text will appeal to the more general educational approach to the welding processes as well as to the technical.

The last two sections are more general in nature than the first three. Section IV on special welding processes is an overview of many welding processes which are either too expensive, too new, or too hazardous to be included in a welding curriculum but with which welding students should be familiar. The last section is a basic study of metals; it presents the basis for a better understanding of how welding affects metal, how to relieve entrapped stresses, and how to test the effectiveness of the weld joint.

The text is not machine oriented; that is, step-by-step directions for operating particular machines are not emphasized. Instead, concepts of welding processes are explained so that a student will be able to operate any machine, whether he has ever operated it before or not, and will know not only how to operate it but also why it does what it does.

RICHARD L. LITTLE

GAS WELDING

1 INTRODUCTION

Oxyacetylene welding is a fusion-welding process; the coalescence of metals is produced by an oxygen-acetylene flame. Extreme heat is concentrated on the edges or on the edge and surface of the pieces of metal being joined until the molten metal flows together. The type of joint design determines whether a filler metal should be used to complete the weldment. Filler metal is added by inserting it into the molten puddle of the base metals. The puddle then solidifies making the weld bead (Figure 1–1).

The extremely high heat depends on the mixture of two types of gaseous substances: oxygen and acetylene. The oxygen supports higher combustion; the acetylene is the fuel for the combustion.

Oxygen is a nonmetallic chemical element, designated by the symbol O, which can be found in a free state or in

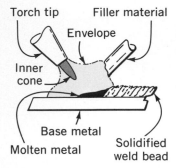

Fig. 1-1 Oxygen-acetylene welding.

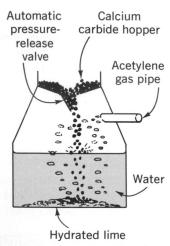

Fig. 1-2 Carbide-to-water acetylene generator.

combination with the other elements in nature. The atmosphere is composed of about one-fifth (20.95 percent) oxygen by volume. The element was first isolated in 1774 by Joseph Priestley when red mercury oxide was heated in a confined heated chamber in an experiment. Oxygen is now produced commercially in the almost pure state by two methods: electrolysis and the liquid-air process. In the electrolytic method water is broken down into hydrogen and oxygen by sending an electrical current through a solution of caustic soda and water. Oxygen is released at one terminal and hydrogen at the other one. Because of the high cost, this method of generating pure oxygen is not used to any great extent.

The liquid-air process is the prime method used in the production of commercial-quality oxygen. By this method, the air is first washed with caustic soda; then the temperature is lowered to $-317°F$. This low temperature liquifies or vaporizes all of the material within the chamber. The basic principle behind this method is that all gases vaporize at different temperatures. Oxygen, being one of the chief constituents of air and having a higher vaporization point than nitrogen and argon, vaporizes last. When the liquid air is allowed to evaporate slowly, the nitrogen and the argon boil off more rapidly than the oxygen until finally almost pure oxygen remains. The gas from the boiling liquid is compressed into steel cylinders specially designed to withstand pressure of 2200 psi (pounds per square inch) at a room temperature of 70°F. The oxygen is then ready to be transported for use with oxyacetylene welding and cutting equipment.

Acetylene gas—a hydrocarbon with the chemical formula C_2H_2—is not found in a free state in nature. This gas was first discovered in 1815 by Sir Humphry Davy, but it was of little commercial use until 1892 when it was produced by means of calcium carbide. Acetylene gas is produced in this country by the carbide-to-water method, which allows water to react upon calcium carbide in a carbide-to-water generator (Figure 1-2). The reaction between water and carbide is instantaneous; the carbon in the carbide combines with the hydrogen in the water, forming acetylene gas. This chemical reaction is

$$\underset{\substack{\text{Calcium}\\\text{carbide}}}{CaC_2} + \underset{\text{Water}}{2H_2O} = \underset{\text{Acetylene}}{C_2H_2} + \underset{\substack{\text{Hydrated}\\\text{lime}}}{Ca(OH)_2}$$

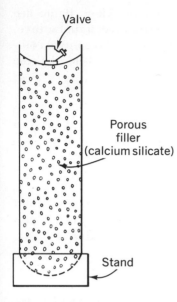

Fig. 1-3 Acetylene cylinder.

A carbide-to-water generator operates on a ratio of 1 gal (gallon) of water to 1 lb (pound) of carbide. The rate at which these parts are mixed, in some generators, is 1 lb carbide hopper capacity/hr to produce 1 ft³ (cubic foot) acetylene gas.

Acetylene gas under pressure becomes very unstable and in the free state will explode before reaching a pressure of 30 psi. This instability places special requirements on the storage of acetylene. A storage cylinder is filled with a mixture of calcium silicate, a material that is 92 percent porous. The cylinder is then filled with acetone, which is the solvent agent of acetylene gas and which has an absorptive capacity of up to 35 volumes of acetylene per volume of acetone per atmosphere of pressure. This enables about 420 volumes of acetylene to be compressed at 250 psi. Under these conditions, the gas is present in the form in which it is to be used. Acetylene comes out of the acetone solution at a slow constant rate as the pressure in the cylinder is released. The rate, however, depends on the temperature of the gas (Figure 1-3).

The burning of oxygen and acetylene is accomplished in five steps. First, the cylinder pressure is released to a working-line pressure by means of special regulators. The gases are transported, by hoses usually, to the torch body. These gases are mixed in the mixing chamber, and then this mixture is ejected into the atmosphere (Figure 1-4). All that is needed now to supply combustion is for this mixture to be ignited. When the spark of ignition is supplied, the triangle of combustion is then complete.

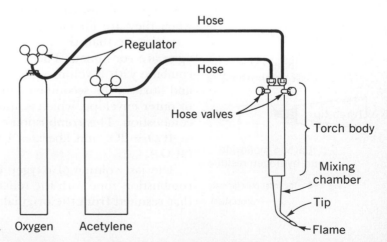

Fig. 1-4 Gas welding system.

The three requirements for oxygen-acetylene flame are those for any combustion: an oxygen source, a fuel source, and a kindling (or ignition) temperature. To control combustion, one need only to adjust or remove one of these three elements (Figure 1-5).

THE OXYACETYLENE FLAME   The combination of oxygen (O) and acetylene (C_2H_2) to support maximum combustion has been found to be

$$2C_2H_2 + 5O_2$$

That is, 2 volumes of acetylene combine with 5 volumes of oxygen for the complete burning of the two gases. Upon the complete burning or combustion the residue is

$$4CO_2 + 2H_2O$$

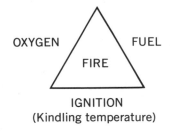

OXYGEN   FUEL

FIRE

IGNITION
(Kindling temperature)

Fig. 1-5   Combustion triangle.

or 4 volumes of carbon dioxide and 2 volumes of water vapor. However, this burn does not all happen at one time. Normally, it occurs in two stages. First, equal volumes of oxygen and acetylene are ejected from the torch tip to burn in the atmosphere ($2C_2H_2 + 2O_2$—2 volumes of acetylene and 2 volumes of oxygen). When these equal amounts of the gases go through combustion, the inner cone is created and is readily visible at the end of the torch tip (Figure 1-6). The inner cone is referred to as the primary combustion zone or the primary stage of combustion. This first combustion has a residue of 4 volumes of carbon monoxide and 2 volumes of hydrogen. Since both these residual gases are capable of supporting combustion, when they are liberated, they ignite and burn at a lower temperature than the primary burn of acetylene and oxygen are consumed. The 4 volumes of carbon monoxide combine with 2 volumes of oxygen from the atmosphere and burn in the secondary combustion zone (Figure 1-7) or outer envelope, which is also called the second stage of combustion. This combustion can be expressed chemically as $4CO + 2O_2$ and liberates 4 volumes of carbon dioxide ($4CO_2$).

The last volume of oxygen combines in the secondary combustion zone with the remaining 2 hydrogen volumes that resulted from the original combustion release. These

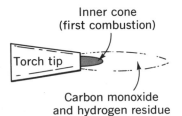

Inner cone
(first combustion)

Torch tip

Carbon monoxide
and hydrogen residue

Fig. 1-6   The inner cone (primary
combustion).

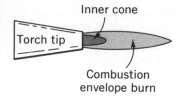

Fig. 1-7 The combustion envelope (secondary combustion).

two gases react and liberate 2 volumes of water vapor ($2H_2O$).

The complete combustion of oxyacetylene could be represented by the following formula:

$$2C_2H_2 + 5O_2 = 4CO_2 + 2H_2O$$

where $4CO_2 = 4CO + 2O_2$
$2H_2O = 2H_2 + O_2$

The approximate temperature range that can be attained by the oxyacetylene flame is from 5000 to 6300°F at the inner cone, around 3800°F in the middle of the envelope, and approximately 2300°F at the extreme end of the secondary combustion envelope (Figure 1-8). These temperatures can be varied to some extent by changing the mixture of gases or by causing an improper balance in the volumes of oxygen and acetylene ejected from the tip of the torch. This imbalance is usually accomplished by adjusting the needle valves on the blowpipe or torch body.

Flame adjustment Oxygen and acetylene can be ejected from the torch tip in three possible gas mixtures: an excess of acetylene, an equal mixture, or an excess of oxygen. These mixtures can be identified by the appearance of the flame at the tip of the torch. The heat liberated by the combustion also is dependent upon the type of mixture. The lower inner-cone temperature is associated with carbon-rich mixtures (excess of acetylene in mixture), and the highest cone temperature is associated with the oxygen-rich mixture (excess oxygen in the mixture).

The first step in igniting the flame is to open the acetylene valve on the torch and ignite the acetylene gas coming out of the tip. Enough oxygen will be drawn in from the atmosphere to burn the fuel gas partially. The needle valve should be opened until the flame separates from the tip and then closed just enough for the flame to join the tip. This is a method of estimating proper acetylene flow. The main characteristic of this flame is the abundance of free carbon released into the air. In fact, this flame is sometimes used to apply carbon to mold faces in the foundry, for the carbon acts as an insulator between the molten metal and the mold face.

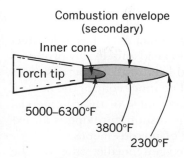

Fig. 1-8 Burn temperatures.

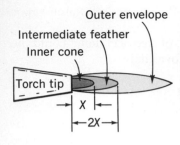

Outer envelope
Intermediate feather
Inner cone

Torch tip

X

2X

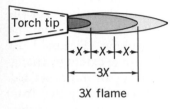

Torch tip

X X X

3X

3X flame

Fig. 1–9 Carburizing-reducing flame.

There are three types of flames: the carburizing or reducing, the balanced or neutral, and the oxidizing. The carburizing or reducing flame has an excess of acetylene (Figure 1–9) and is characterized by three stages of combustion instead of two as in the other two types of flames. The extra combustion stage, called the intermediate feather, can be adjusted by the amount of acetylene imbalance induced at the torch body needle valve. The length of the intermediate feather is usually measured, by eye, in terms of the inner-cone length. A 2× carburizing flame, then, would be a feather approximately 2 times as long as the inner cone (Figure 1–9). The carburizing flame does not completely consume the available carbon; therefore, its burning temperature is lower and the leftover carbon is forced into the metal. This action is characterized by the molten weld puddle appearing to boil. After this carbon-rich weld bead solidifies, the weld will have a pitted surface; these pits can extend through the entire weld bead. The bead will also be hardened and extremely brittle because of the excessive amount of carbon that was injected into the molten base metal. This excess of carbon, however, is an ideal condition when welding high-carbon steel.

Upon further adjustment of the torch needle valves, the intermediate feather can be drawn into the inner cone. The oxygen is usually increased in order to maintain the minimal amount of acetylene flow. Increase of the oxygen flow causes the intermediate feather to recede into the inner cone.

The instant that the feather disappears into the cone, the oxygen-acetylene mixture produces a balanced or neutral flame. This two-stage neutral flame, which should have a symmetrical inner cone and should make a hissing sound (Figure 1–8), is the most-used flame in both the welding and the flame cutting of metals. It has little effect upon the base metal or the weld bead and usually produces a clean-appearing, sound weld bead with properties equal to the base-metal properties. The inner cone of the neutral flame is not hot enough to melt most of the commercial metals, and the secondary combustion envelope is an excellent cleanser and protector of ferrous metals.

Further opening of the oxygen needle valve on the torch body shortens the inner cone approximately two-tenths of its original length. The inner cone then loses its symmetrical shape, and the flame gives off a loud roar. These three actions are the characteristics of the oxidizing flame, which is also the hottest flame that can be produced by any oxygen-fuel source. The oxidizing flame injects oxygen into the molten metal of the weld puddle, causing the metal to oxidize or burn too quickly, as demonstrated by the bright sparks that are thrown off the weld puddle. Also, the excess of oxygen causes the weld bead and the surrounding area to have a scummy or dirty appearance. However, a slightly oxidizing flame is helpful when welding most copper-base metals, zinc-base metals, and a few types of ferrous metals, such as manganese steel and cast iron. The oxidizing atmosphere creates a base-metal oxide that protects the base metal. For example, in welding brass, the zinc alloyed in the copper has a tendency to separate and fume away. The formation of a covering copper oxide prevents the zinc from dissipating.

OTHER HEAT SOURCES Many fuels can be used to weld metals, but the oxyacetylene flame is the most versatile and the hottest of all the flames produced by the various gas heat sources. Some of the more common fuels in use are natural gas, propane, Mapp, hydrogen, and butane (Chart 1-1). Because natural gas and Mapp gas are relatively inexpensive, they are becoming more popular in industries that use flame cutting. The only differences in equipment for the oxygen-natural gas or oxygen-Mapp cutting systems are the fuel-gas regulators. Other than this difference in regulators, the systems are operated in the same manner as the oxyacetylene flame cutting system.

Some heat sources that are used do not rely upon pure oxygen to support combustion. Such heat sources generally aspirate or draw in the available oxygen from the sur-

CHART 1-1 Fuel gas temperature

FUEL	AIR, °F	OXYGEN,* °F
Acetylene (C_2H_2)	4800	6300
Hydrogen (H_2)	4000	5400
Propane (C_3H_8)	3800	5300
Butane (C_4H_{10})	3900	5400
Mapp (——)	2680	5300
Natural gas (CH_4 and H_2)	3800	5025

*Oxidizing flame temperature

rounding air and operate on the bunsen burner principle. The bunsen burner consists typically of a straight tube with a series of small holes to allow the atmosphere to be drawn into the tube and mixed with the fuel gas. The gas is then ejected and ignited, producing an air-fuel flame. The common fuels used in air-fuel welding are acetylene, natural gas, propane, and butane. The air-fuel welding processes are used to solder and braze many low-melting-temperature metals.

The oxygen-hydrogen process was once used extensively to weld low-temperature metals such as aluminum, lead, and magnesium; but it is not as popular today because more versatile and faster welding processes such as TIG (tungsten inert gas) and MIG (metal inert gas) have replaced the oxygen-hydrogen flame. The oxyhydrogen welding process, however, is similar to the oxygen-acetylene system, with the only difference being a special regulator used in metering the hydrogen gas.

QUESTIONS

1. What are the characteristics of a neutral flame? Oxidizing flame? Carburizing flame?

2. What ratio or mixture should be supplied through the oxyacetylene torch tip?

3. How is acetylene produced?

4. How is oxygen produced?

5. What is the SAFE working pressure limit of acetylene?

6. What type of oxyacetylene flame should be used to weld carbon steel?

7. What type of oxyacetylene should be used to weld copper-base metals?

8. What are the basic elements of fire?

9. How is acetylene stored? Oxygen?

10. What are the components of a gas welding system?

11. What is the maximum temperature attainable with the oxyacetylene flame? Oxygen-hydrogen flame? Mapp gas?

12. What is the function of the secondary combustion envelope?

13. What is the bunsen burner principle?

PROBLEMS AND DEMONSTRATIONS

1. Adjust the basic oxyacetylene flame. Begin with the acetylene flame. Add oxygen until a carburizing flame is achieved. Add more oxygen until a neutral flame appears; then add more oxygen until the oxidizing flame appears.

   NOTE: Always shut off gas supply to extinguish any gas fire.

2. Select a piece of ferrous metal about $3 \times 3 \times \frac{1}{8}$ in. and weld three beads or puddles without a filler rod. Weld one bead with a carburizing flame, one bead with a neutral flame, and one bead with an oxidizing flame. Contrast the reaction of the weld puddles during welding and contrast the weld beads after they have cooled.

SUGGESTED FURTHER READINGS

Althouse, Andrew D., Carl H. Turnquist, and William A. Bowditch: "Modern Welding," The Goodheart-Willcox Company, Homewood, Ill., 1967.

Babor, Joseph A., and Alexander Lehrman: "Introductory College Chemistry," Thomas Y. Crowell Company, New York, 1946.

Flame Cutting Facts, Smith Welding Equipment, Minneapolis, Minn., 1966.

Giachino, J. W., W. Weeks, and E. Brunes: *Welding Skills and Practices,* American Technical Society, Chicago, 1965.

Harris, Norman C.: "Experiments in Applied Physics," 2d ed., McGraw-Hill Book Company, New York, 1972.

Jefferson, T. B.: "The Welding Encyclopedia," 16th ed., Monticello Books, Morton Grove, Ill., 1968.

Lapp, Ralph E., and the editors of *Life*: "Matter," Time Incorporated, New York, 1963.

The Oxy-Acetylene Handbook, 2d ed., Union Carbide Corporation, New York, 1960.

Rossi, B. E.: "Welding Engineering," McGraw-Hill Book Company, New York, 1954.

2 EQUIPMENT

The basic oxyacetylene torch consists of a cylinder of oxygen, a cylinder of acetylene, an oxygen regulator, an acetylene regulator, a hose system to carry the gases, and the torch (Figure 2–1).

OXYGEN Oxygen is available commercially in two forms: gaseous and liquid. The gaseous oxygen is compressed into steel cylinders for delivery to the consumer. It can be stored in this manner almost indefinitely and can be ready for use in an instant. These two qualities make the cylinder the most popular storage medium for gaseous oxygen among small- and medium-sized consumers of oxygen.

Because of the high pressures in the steel cylinders and the possibility of cylinder-wall deterioration, cylinders

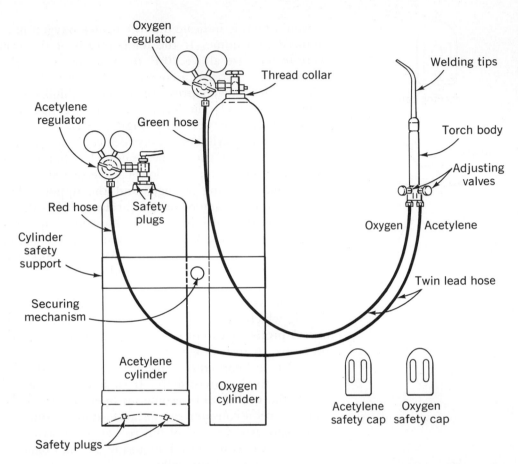

Fig. 2-1 Oxygen-acetylene welding system.

always remain the property of the company selling the compressed gases. The Interstate Commerce Commission (ICC) has established standards for compressed-gas cylinders because they are continually moved about, sometimes across state lines. These standards require that oxygen cylinders be tested periodically with water pressure to 3360 psi.

The three common sizes for cylinders or tanks are 80, 122, and 244 ft³. Even though cylinders may vary in size, they are all filled to 2200 psi at 70°F. Outside temperature changes will, of course, change the pressure within the tank. For example, if the outside temperature were to drop to a few degrees below freezing, the pressure within the tank would drop to approximately 2000 psi. However, this drop in pressure would not indicate any loss of oxygen.

Protector cap

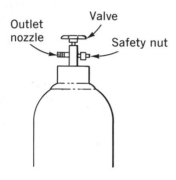

Fig. 2-2 Oxygen-cylinder valve.

It would merely indicate that the cooled oxygen has been reduced in volume. If the temperature of the cylinder were to increase above 70°F, the pressure inside the cylinder would increase accordingly. Because of the possibility of the oxygen pressure becoming high enough to rupture the steel cylinder, a safety device has been designed into the oxygen valve (Figure 2-2).

The oxygen-cylinder safety device is a special two-seated valve that should be operated either completely off or completely on. The lower valve seat shuts or seals the cylinder during storage, and the upper valve seat prevents the leakage of oxygen around the valve mechanism when the valve is opened completely. Operation of the valve in a partially open position results in the escape of oxygen into the atmosphere. An important point to remember is that the valve should be either open completely or closed completely.

In the use or handling of oxygen cylinders, the time during which they are being transported is most hazardous. A protector cap or safety cap should be screwed over the valve to insure against the possibility of damage to the valve and consequent instant release of high-pressure oxygen into the atmosphere. If such an accident should occur, the tank could become an operational rocket and could travel a considerable distance, doing damage to all in its path. An example of this rocket action would be the release of the compressed air inside a child's balloon and the erratic flight of the balloon.

LIQUID OXYGEN Oxyacetylene installations or other large consumers may find it more economical to purchase oxygen in the liquid form, which is delivered and stored at −297.3°F. Liquid oxygen (LOX) can be supplied in large storage tanks that can be transported only by large semi-tractor-trailer trucks or in 160-liter (42.4-gal) containers, which are approximately 62 in. in height and 22 in. in diameter and hold 4850 ft³ of oxygen. One of these containers would replace approximately sixteen 244-ft³ oxygen cylinders. The major drawback of these larger containers, however, is that the average evaporation loss is 1.5 percent per 24 hr of storage. Therefore, they are economical only for those consumers that use large amounts of oxygen daily. This system converts liquid oxygen to the gaseous state at 150 psi at 350 ft³/hr (cubic feet per hour) and gives intermittent service up to 1000 ft³/hr (Figure 2-3).

The liquid-oxygen system is used by all major distributors of commercial oxygen to charge the cylinders that are used by the consumer.

ACETYLENE Acetylene gas can be manufactured either by the water-to-carbide method or the carbide-to-water method. The carbide-to-water method is the primary method of production in the United States. The water-to-carbide system may still exist but only in a few special instances, such as in miners' lamps, which, however, are seldom used now because of the possibility that they might ignite gas fumes in a mine.

There are two types of carbide-to-water generators: low-pressure and medium-pressure. The low-pressure generator produces less than 1 psi, and the medium-pressure acetylene generator produces from 1 to 15 psi. The ICC prohibits the production of acetylene gas at any pressure higher than 15 psi. The 15-psi limit has a built-in safety margin to insure against any free combustion.

Either type of generator can be portable or stationary. However, generally the low-pressure generator is considered portable, while the medium-pressure generator is considered stationary. Acetylene production by the low-pressure portable generator ranges upwards from 30 ft³/hr. The medium-pressure stationary generator can produce up to 6000 ft³/hr.

The acetylene cylinder is a steel container that meets or exceeds the standards established by the ICC for gas-

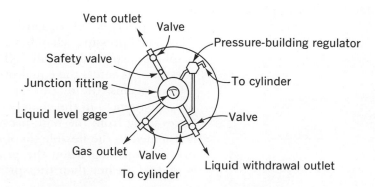

Fig. 2-3 Liquid-oxygen cylinder (top view).

pressure vessels. Safety plugs are located in the top and the bottom of the acetylene cylinder to allow the acetone and the acetylene to escape in case of a fire. These plugs simply melt at 220°F, and the fluids in the cylinder flow out. When the plugs melt, the steel casing of the cylinder will not build up any excessive pressure that would cause the tank to rupture (Figure 2–1).

The acetylene that is stored in these cylinders is considered *dissolved* rather than *generated* acetylene. The most common cylinder sizes for dissolved acetylene are 60-, 100-, and 300-ft^3 containers. These sizes reflect the maximum amount that the containers will hold; however, it is practically impossible to fill a container to its maximum rated capacity. The consumer purchases dissolved acetylene only by the actual amount within the cylinder. The consumer determines the size of acetylene cylinder he needs by the number of cubic feet per hour that will be used. Dissolved acetylene is released from the acetone carrier at the rate of approximately one-seventh of the cubic-foot capacity per hour. For example, a 60-ft^3 cylinder of dissolved acetylene will release approximately $^{60}/_7$ or 8.5 ft^3/hr, and a 300-ft^3 cylinder of dissolved acetylene will release approximately $^{300}/_7$ or 42.8 ft^3/hr. If more than these amounts are drawn from the cylinders, acetone will be siphoned from the cylinder with the dissolved acetylene. This mixture of acetone and acetylene burns erratically and turns the oxyacetylene flame a purplish color. The acetone will mix with the acetylene if the cylinder of dissolved acetylene is laid on its side; therefore, this practice should be avoided whenever possible.

PRESSURE REDUCTION REGULATORS

The basic principle of gas-pressure reduction is founded on the principle of equalizing pressure in a chamber. The equal-pressure chamber is usually regulated by a diaphragm-controlled valve (Figure 2–4a). The high-pressure gas enters the equal-pressure chamber and exerts a force against the diaphragm. As the pressure (pounds per square inch) in the chamber increases, the diaphragm flexes away from the high-pressure inlet. A closing valve is fastened to the flexible diaphragm that closes the high-pressure opening when the pressure inside the chamber is slightly greater than the spring tension supporting the

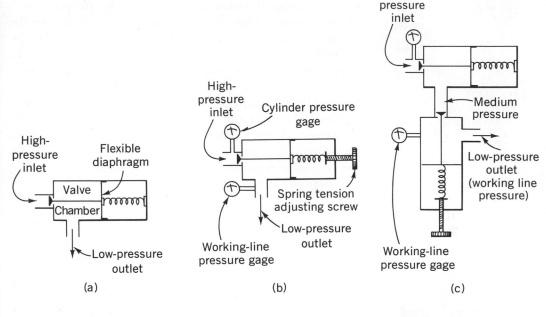

Fig. 2–4 Pressure reduction regulator principle. (a) Basic principle. (b) Single-stage regulator. (c) Two-stage regulator.

diaphragm. The gas that is trapped inside the chamber is allowed to flow out of the low-pressure outlet, which, in turn, relieves the chamber pressure. As the chamber pressure is relieved, the pressure on the diaphragm is relieved, reopening the high-pressure valve.

There are two types of regulators: the single-stage and the two-stage. The single-stage regulator uses the basic principle of pressure reduction with the addition of a spring-tension adjusting screw (Figure 2–4b). This adjusting screw varies the pressure that the spring exerts upon the diaphragm and thus controls the valve chamber or the low-pressure outlet. A pressure-sensitive gage indicates the pressure within the valve chamber. A single-stage regulator is all that is actually needed for both oxygen regulation and acetylene regulation for oxyacetylene welding. For convenience, the cylinder-pressure gage is added to the basic regulator in order to indicate the amount of pressure remaining in the cylinder.

The two-stage regulator (Figure 2–4c) is a combination of the basic principle of pressure reduction and the single-stage regulator. The connection to the high-pressure side of the regulator, the cylinder-regulator connec-

(a)

(b)

(c)

Fig. 2-5 Oxygen regulator. (a) Standard regulator. (b) Master regulator. (c) Line or station regulator. (Courtesy of Victor Equipment Company.)

tion, allows high-pressure gases to flow into an intermediate valve chamber that reduces the pressure to a preset amount. This amount of pressure is built into the regulator by a factory-set spring tension upon the intermediate valve chamber. The chamber furnishes the adjustable valve chamber with an even amount of relatively low pressure gas and minimizes pressure fluctuation to the adjustable second stage. This preset amount of gas enters the second stage of the regulator where the pressure is further reduced by the amount of tension on the spring diaphragm adjusting screw. The last stage of the two-stage regulator is identical in operation to that of the single-stage regulator.

Regulators for different gases are basically the same, except that the pressures that they control differ vastly. For example, regulators are designed to reduce high pressure. Oxygen regulators must withstand 2200 psi at 70°F and sometimes higher pressures when the temperature is above 70°F. Acetylene regulators, on the other hand, need only to control and reduce 350 psi of dissolved acetylene. A safety factor has been built into regulators to control the initial surge of pressure against the valve diaphragm; otherwise this opening surge would rupture the diaphragm and damage the regulator. This safety factor prevents diaphragm damage, but regulators are still not designed to withstand full cylinder pressure upon initial opening of the cylinder valve. The adjusting spring tension screw should always be free, which means that there is no pressure on the diaphragm spring, when high-pressure gas is released into the regulator.

Most manufacturers of regulators provide for three major sizes or regulators: the standard regulator, the master regulator, and the station regulator (Figure 2-5). The standard regulator is the most used because it is portable, and it has the ability to deliver gas pressure demands over a wide range. However, the standard regulator cannot deliver large quantities of gas. The master regulator is designed to deliver large quantities of gases without gas-pressure fluctuation. The construction of the master regulator is the same as the standard except that it is a heavier and larger reducing mechanism. The line or station regulator is one that is usually dependent upon the master regulator for its source of gas. A most common setup that uses station regulators is the manifold system of gas delivery (Figure 2–6).

Manifold systems are used when cubic-foot-per-hour consumption rates are beyond the usage rates of standard regulators. The acetylene manifold is one method to overcome the one-seventh of the capacity discharge rate of dissolved acetylene. For example, if five 300-ft^3 acetylene cylinders were manifolded together, the maximum discharge capacity would be the number of cylinders multiplied by the cubic-foot ratings of the cylinders and then divided by 7, or

$$\begin{array}{r} 300 \\ \times 5 \\ \hline 1500 \end{array} \qquad \frac{1500}{7} = 214.29$$

By manifolding the acetylene cylinders together, the cubic-foot-per-hour rate has been increased to 214.29 before the possibility of injecting acetone into the weld zone occurs. Acetone would, in most cases, contaminate the weld and weaken it.

When manifolding acetylene gas cylinders or installing acetylene generators onto a manifold system, a fire safety device is usually installed between the acetylene supply and work site. This safety device, a flashback arrestor or hydraulic back-pressure valve, is a mechanism allowing gas to flow in only one direction. If a flash or a fire travels

Fig. 2–6 A typical manifolding oxygen-acetylene system. (a) System.

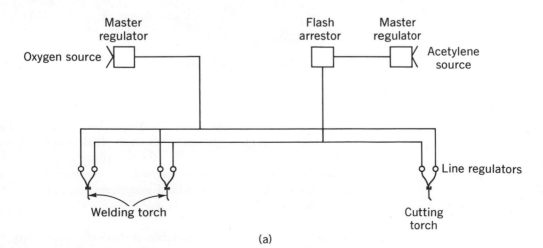

(a)

Fig. 2–6 (cont'd)   (b) Oxygen manifold.   (b)

back up the acetylene line, the shock bubble in front of the fire would close the one-way valve. It is generally good practice to install small in-line flash arrestors either on the connection of the regulator to the manifold line or the connection point of the standard acetylene regulator to the acetylene cylinder.

Hoses   The currently available industrial gases may be piped through steel tubing, stainless steel tubing, brass

Fig. 2–6 (cont'd)   (c) Acetylene
manifold and flash arrestor. (Photographs
courtesy of Victor Equipment Company.)

(c)

tubing, bronze tubing, and cloth-reinforced rubber flexible hose. The only reactional combination and one that should be avoided is copper and acetylene. When acetylene gas comes into contact with copper, copper acetylide forms in low portions of the piping system and turns into a gel. This compound will disassociate violently or explode even when just slightly shocked or tapped.

The most common method of piping both oxygen and acetylene gas is the reinforced rubber hose, which comes

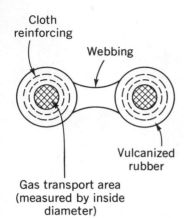

Cloth reinforcing

Webbing

Vulcanized rubber

Gas transport area (measured by inside diameter)

Fig. 2–7 End view of oxygen-fuel hose.

in black, green, and red. The green hose is usually used to transport oxygen; the red hose is usually the fuel-gas hose, and the black hose is usually used to transport other industrially available welding gases. The use of these colors is not standardized, but the colored hoses are many times used in this fashion.

The hose is manufactured in various diameters. The more popular diameters are $3/16$, $1/4$, $5/16$, $3/8$, and $1/2$ in. These industrial hoses are always measured by inside diameter and come in standard lengths of 25, 50, and 75 ft. It is a standard practice to purchase for the oxyacetylene or oxygen-fuel systems hoses that have a webbing to hold the oxygen hose and the fuel hose together (Figure 2–7). Hoses of standard lengths can be purchased with connection fittings attached by the factory or with connections that can be added by the individual. These connections, which are readily available and are simple and easy to install, connect to the hose in the same manner as connections are made on a garden hose.

As a safety precaution, *all* oxygen and compressed-air hoses are right-hand threaded, and all fuel-carrying hoses are left-hand threaded. This standard arrangement prevents oxidants from combining with fuel, which could result in an explosion.

Certain precautions should be taken when using reinforced rubber hoses: only one gas should be used in a hose; the hose should never be patched or repaired; and hot metal should never be placed on the hose.

Torches   Two types of welding torches or welding blowpipes are distinguished by pressures of operation: the injector or low-pressure torch and the medium-pressure or equal-pressure torch. This major difference of pressure is controlled by the chamber in the torch body and often is not readily apparent (Figure 2–8).

The injector-type torches can use acetylene pressures of less than 1 psi because of the venturi effect created by oxygen being expelled from the injector tip. However, to maintain this suction, more oxygen pressure is needed than acetylene pressure. Because of the effect of the oxygen rushing through the injector tip, the low acetylene pressure is stable and the mixture of gases is constant. This feature makes injector-type torches desirable for con-

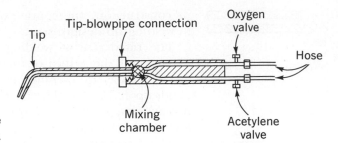

Fig. 2–8 Welding torch or blowpipe and tip.

sumers who have low-pressure acetylene generators. Another feature of the injector torch is its ability to extract more acetylene from a cylinder than an equal-pressure torch; but then, this may not be an advantage because the consumer purchases only the amount of gas needed to recharge the cylinder (Figure 2–9a).

The medium- or equal-pressure welding torch is the more common of the two types of oxyacetylene torches. The mixing chamber in the equal-pressure torch allows both of the gases to flow together in equal amounts (Figure 2–9b). The maximum acetylene pressure used on the equal-pressure torch is the ICC limit of 15 lb.

The exact location of either type of blowpipe mixing chamber is determined by the size of the unit or by manufacturer's whim. However, a rule of thumb is that the smaller torches have the mixing chamber in the removable tip section of the torch and the larger torches have the mixing chamber in the blowpipe or torch body.

Torch welding tips As mentioned previously, welding tips may have the mixing chamber incorporated within the torch bodies or they may not. Other than this difference

Fig. 2–9 Oxygen-fuel mixing chambers. (a) Injector chamber. (b) Equal-pressure chamber.

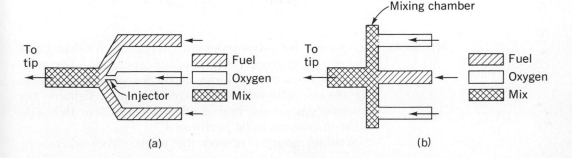

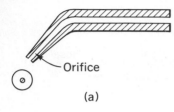

Orifice

(a)

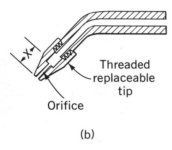

Threaded
replaceable
tip

Orifice

(b)

Fig. 2-10 Welding tips. (a) Solid tip.
(b) Two-piece tip.

in construction, there are two other major differences in welding-tip design: size and type.

The rated size of welding tips is determined by the diameter of the orifice of the tip (Figure 2-10). The diameter of the tip opening or orifice used for welding depends upon the type of metal to be welded, such as whether it is stainless steel, iron, or brass, and the thickness of the metal to be welded.

The selection of type of torch tip is determined by whether or not the operator prefers solid- or multiple-piece welding tips. The advantage of the solid-piece tip is that it is lower in initial cost than the multiple-piece tip. However, if the tip is exposed to more than normal wear, the price of replacing the tip will offset the cheaper original purchase price because the entire tip must be replaced. With the multiple-piece welding tip, instead of replacing the complete tip, it is necessary to replace only a short part at the end of the welding (Figure 2-10b). The overall functioning of both types are identical, and both perform assigned tasks with equal ease.

All the commercial welding tips have basically the same design for the orifice and the welding tip. In essence, there is a passage for the mixture of gases to flow through and an orifice to exit these gases. The diameter of this orifice is determined by the thickness of the metal to be welded (Unit 3). This diameter has been drilled in the tip with a number drill, from 1 to 80, but the tip will not have this diameter stamped upon it. The number stamped on the tip generally indicates the working gas pressures needed for efficient welding. If the torch is the injector type, the number indicates the amount of oxygen pressure to have on the torch body. If the torch is the equal-pressure type, the number stamped on the welding tip indicates the working pressure of the oxygen and the acetylene.

ACCESSORIES   Accessories for oxyacetylene or oxygen-fuel welding fall into two categories, protection and operation. The protection items are such things as goggles, gloves, and welding sleeves. The operation items are spark lighters, tip cleaners, gas savers, and any other equipment that aids in the operations to be performed.

Welding goggles protect the welder from ultraviolet

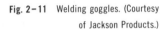

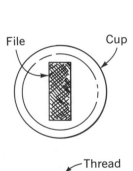

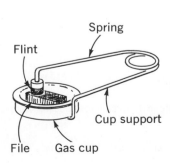

Fig. 2-11 Welding goggles. (Courtesy of Jackson Products.)

Fig. 2-12 Spark lighter.

and infrared rays emitted from the oxyacetylene flame and the molten weld puddle. Welding goggles also protect the wearer from flying sparks and reduce the glare created by the torch flame and the molten weld puddle.

Welding goggles consist of an adjustable headband, a form-fitted spark-proof frame, a green- or brown-tinted filtered lens, and a clear lens on each side of the colored lens to protect the more expensive filter lens (Figure 2-11). Lenses come with shade densities that range from 1 to 14, with no. 1 indicating lightest shade and no. 14 the darkest shade. The average shade intensity for oxyacetylene welding ranges from shade 3 to shade 8, depending upon the light-sensitive characteristics of the welder. The most commonly used shades are no. 4 and no. 5, the types for the school laboratory or medium industry, such as construction. Shades that range in intensity from no. 8 to no. 14 are usually used in electric arc welding.

Welding gloves, sleeves, and jackets may or may not be worn, according to the safety standards of the particular welding facility. However, welding gloves are generally considered a necessary protection from burns. Protective clothing can be made from leather, asbestos, or nonflammable cloth.

The operational equipment that aids in oxyacetylene welding is a necessary part of the total welding system. For example, the spark lighter or striker is the ignitor for the oxygen-fuel flame (Figure 2-12). The striker consists of a cup to catch and hold the gases and a flint and steel to ignite the gases. This same principle was used to ignite the powder in flintlock rifles.

Welding tips exhibit a considerable amount of minor damage because of the heat and the molten metal they are exposed to. The correct welding tip should have a clean,

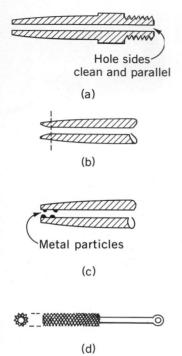

Hole sides
clean and parallel

(a)

(b)

Metal particles

(c)

(d)

Fig. 2-13 Dirty tips. (a) Correct clean tip. (b) Bell-mouthed tip. (c) Metal-splattered tip. (d) Tip cleaner.

smooth, and parallel hole to control the oxygen-fuel flame (Figure 2-13a). If this passage is dirty, cluttered, or mis-shapen, the oxygen-fuel flame will be distorted and difficult to use (Figures 2-13b and c). The welding tip that has become bell mouthed must have its end resurfaced with a file. The welding tip with metal particles within the tip passage must be cleaned (Figure 2-13c). The tip cleaner used for this purpose is a wire with a roughened surface whose diameter corresponds to the diameter of the tip passage. These wires can be purchased singly or in packaged sets. If tip cleaners are not available, a standard number drill can be used, with caution, to clean the welding tip. Special care must be taken not to ream the orifice to a larger size or to ream the orifice out-of-round when using a twist drill.

To use a tip cleaner properly, the oxygen must have a minimum of 5 psi on the working gage, and the oxygen valve on the torch must be open. This will blow the slag and carbon deposits out of the tip when the tip cleaner is extracted from the tip passage. Many times the proper-sized cleaner cannot enter the tip passage and a smaller tip cleaner is used to open the way for the correct-sized cleaner. When the tip is completely clogged, a numbered twist drill may be used to reopen the passage. If the passage cannot be reopened, the tip should then be replaced.

QUESTIONS

1. What are the major components of the oxyacetylene torch?

2. What are the safety devices and how do they work for the acetylene cylinder and the oxygen cylinder?

3. Why do oxygen cylinders have special valves?

4. What are the disadvantages of liquid-oxygen storage?

5. What is the maximum pressure of dissolved acetylene that can be released from a given cylinder?

6. What are the advantages of a two-stage regulator over a single-stage regulator?

7. Why is a master regulator designed to withstand greater loads than a standard regulator?

8. What are the major differences between a low-pressure acetylene generator and a medium-pressure acetylene generator?

9. Why are all fuel lines threaded left-handed?

10. Why should copper lines not be used for acetylene gas?

11. What are the two major types of welding blowpipes? What are their differences?

12. What is the correct procedure for rebuilding a dirty welding tip?

PROBLEMS AND DEMONSTRATIONS

1. After the oxyacetylene torch has been assembled and checked, demonstrate a fusionable weld puddle with the acetylene cylinder in the upright position; then lay the acetylene cylinder on its side carefully and demonstrate a puddle with an acetone mixture. Then compare the two puddles.

2. Select a dirty welding tip, attach it to the blowpipe, adjust the regulator until there is approximately 5 lb of pressure on the working gage, and then demonstrate the correct procedure for cleaning dirty tips.

SUGGESTED FURTHER READINGS

Giachino, J. W., W. Weeks, and E. Brunes: *Welding Skills and Practices,* American Technical Society, Chicago, 1965.

Pender, James A.: "Welding," McGraw-Hill Book Company, New York, 1968.

Rossi, B. E.: "Welding and Its Application," McGraw-Hill Book Company, New York, 1941.

Welding, Cutting and Heating Guide, Victor Equipment Company, Denton, Tex., 1969.

3 OPERATION

SETUP The procedure for setting up the oxyacetylene welding torch must be followed to insure safe operation. Failure to set up the oxyacetylene torch properly may mean a possible chance of explosion and harm to the operator during the setup period. The correct method for the assembly and the dismantling of the oxyacetylene torch includes the following steps:

1. Secure the cylinders to some object so they can not fall over (Figure 3-1a). This is of prime importance, because if the oxygen cylinder were to fall over with the protector cap removed, the valve may be knocked loose and the oxygen cylinder would become a missile. This step, however, does not apply to the acetylene cylinder. The acetylene cylinder must be used only in

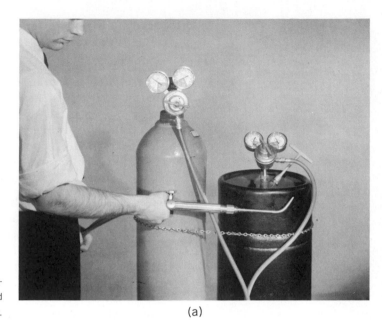

Fig. 3–1 (a) Secured oxygen-
acetylene cylinders. (b) Attached
oxygen regulator.

(a)

the upright position. The reason that it is chained is
so that it cannot fall over. If it were to fall over, the
acetone that carries the dissolved acetylene would mix
with the acetylene gas and contaminate the weld
puddle.

2. After the oxygen and the fuel-gas cylinders have been
secured, unscrew the valve-protection caps to expose
both the oxygen-cylinder valve and the acetylene-
cylinder valve. These valves, especially the valve seats,
must be clean. They can be cleaned with a rag if necessary. Foreign objects may enter either the oxygen regulator or the acetylene regulator and cause damage.
Cloths used to wipe off the oxygen valve and the valve
seat must not contain any grease or oil, not even a
trace. If the rags had even a trace of oil, two of the
three elements needed for combustion would be present, creating extreme fire hazard.

3. Attach the oxygen regulator to the oxygen valve and
the acetylene regulator to the acetylene valve (Figure
3–1b). Remember that the oxygen regulator and all
oxygen connections are right-hand threads and all
gas or acetylene or fuel connections are left-hand

(b)

threads. These connections must be tightened with a wrench but should not be excessively tight.

4. Check the diaphragm adjusting screws to be sure they are in the *out* position. This position is noted by the fact that the screws turn freely. After checking these, open the oxygen cylinder valve all the way. Remember that it is a double valve and must be opened as snugly as it is closed during storage. The acetylene-cylinder valve should then be opened approximately one-quarter to one-half of a turn.

5. Blow out the hoses. Both the oxygen hose, which is green or black, and the acetylene hose, which is red, should be cleaned and checked. To do this, adjust the spring to approximately 5 lb of working pressure on the oxygen cylinder. Do not connect the oxygen hose to the oxygen regulator, however; simply hold it up to the connection and allow the pure oxygen to blow through the hose until all dirt is removed. New hoses are stored with a small amount of talc in the passage to keep the passage free of moisture. The acetylene portion of the hose also has this talc, but acetylene gas should never be used to blow out this hose. Simply hold the acetylene line up to the oxygen line and blow out the new hose in this manner. Once the hoses have been cleaned, connect the green or the black hose to the oxygen regulator and the red hose to the acetylene regulator. Again, remember that all acetylene connections have left-hand threads.

6. Connect the other end of the oxygen hose to the oxygen blowpipe or welding torch that is marked OXY and the red hose to the blowpipe valve that is marked ACET or *acetylene* (Figure 3-1c).

7. Select the proper tip size for the operation to be performed. If the torch is of the injector design, the acetylene pressure should be set at approximately 1 lb and the number stamped on the welding tip will indicate the setting for the oxygen pressure. If the oxyacetylene torch is the equal-pressure design, the number stamped on the welding tip will be the pressure settings for both the oxygen regulator and the acetylene regulator.

Fig. 3–1 (cont'd) (c) Attaching welding
blowpipe.   (c)

8. Test for leaks. After the equipment has been assembled, one should always test for leaks by opening
the oxygen valve all the way and cracking the acetylene
valve a quarter to half a turn; the valves at the torch
body are closed. This operation will not put pressure
on the system until the pressure-adjusting diaphragm
screws have been turned in. It is necessary to put only
about 5 to 10 lb of working pressure on the lines to
check for leaks. To check for leaks, make a soapy solution out of any commercial soap that will make suds,
take a small brush and paint the soap bubbles over
the following connections: (*a*) the oxygen-cylinder
valve, (*b*) the acetylene-cylinder valve, (*c*) the oxygen-
regulator connections, (*d*) the acetylene-cylinder regu-

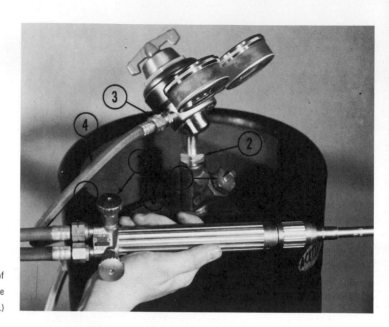

Fig. 3-2 Leak tests. (Courtesy of Union Carbide Corporation, Linde Division.)

lator connections, (*e*) all hose connections, and (*f*) the valves on the torch body (Figure 3-2). Any leakage may cause damage to the equipment, so it would be wise to fix it at this time before lighting the torch. If either the oxygen valve or the acetylene valve leaks, it is mandatory that the oxygen cylinder or the acetylene cylinder be returned to the manufacturer and the manufacturer be notified of the leakage. Remember that old hoses become porous through use, but they are extremely difficult to check with this soap test. If the hoses are in doubt, a test such as one used to find leaks in a tire tube would probably be more effective than the soap test with a brush.

ADJUSTMENT The adjustment of the welding flame depends upon two major factors: the type of material to be joined and the size of the material to be joined.

The type of material, whether it is ferrous (Chart 3-1) or nonferrous (Chart 3-2), determines the flame characteristics for the maximum efficiency of the oxyacetylene flame. For example, the neutral flame does not inject either oxygen or carbon into the weld, and therefore it

CHART 3–1 Data for oxyacetylene welding of ferrous metals

METAL	FLAME ADJUSTMENT
Steel, cast	Neutral
Steel, pipe	Neutral
Steel sheet	Neutral; slightly oxidizing
High-carbon steel	Reducing
Manganese steel	Slightly oxidizing
Cromansil steel	Neutral
Wrought iron	Neutral
Galvanized iron	Neutral; slightly oxidizing
Cast iron, grey	Neutral; slightly oxidizing
Cast pipe	Neutral
Chromium nickel	Slightly oxidizing
Chromium-nickel steel castings (18–8) and (25–12)	Neutral
Chromium steel	Neutral
Chromium iron	Neutral

can be used to weld most metals. The carburizing flame or the slightly carburizing flame is used to weld steel plate, high-carbon steel, and galvanized iron. The excess amount of carbon liberated by the carburizing flame is helpful in the replacement of carbon in the weld zone. The oxidizing flame injects oxygen into the molten weld puddle, thus causing the metal to form an oxide layer that protects the weld zone. However, flux is generally needed to break down the weld zone and clean the surfaces to be joined.

The second major factor in the proper adjustment of the torch flame is the tip size (Chart 3–3). Tip sizes are based on the diameter of the orifice of the welding tip; the thickness of the metal regulates what tip the welder

CHART 3–2 Data for oxyacetylene welding of nonferrous metals

METAL	FLAME ADJUSTMENT
Aluminum	Slightly carburizing
Brass	Slightly oxidizing
Bronze	Neutral; slightly oxidizing
Copper	Neutral; slightly oxidizing
Everdur bronze	Slightly oxidizing
Nickel	Slightly carburizing
Monel	Slightly carburizing
Inconel	Slightly carburizing
Lead	Neutral

CHART 3–3 Welding tip size

TIP SIZE	METAL THICKNESS, in.	OXYGEN	PRESSURE	ACETYLENE	PRESSURE
		psi	ft³/hr	psi	ft³/hr
000	Under $1/64$	1	1	1	1
00	Under $1/64$	1	1	1	1
0	Under $1/32$	1	1.1	1	1
1	Under $1/16$	1	2.2	1	3
2	Under $3/32$	2	5.5	2	5
3	Under $1/8$	3	9.9	3	9
4	Under $3/16$	4	17.6	4	16
5	Under $1/4$	5	27.5	5	25
6	Under $5/16$	6	33	6	30
7	Under $3/8$	7	44	7	40
8	Under $1/2$	8	66	8	60
9	Under $5/8$	9	——	9	——
10	Under $3/4$ up	10	——	10	——

WARNING: ACETYLENE IS EXPLOSIVE. Acetylene is NEVER to be adjusted over 15 psi (15 pounds on regulator).

uses. There are also established amounts of oxygen and acetylene pressures that are necessary if tips are to be used efficiently. For example, if the material or the ferrous metal to be welded were $1/8$ in. thick, the proper tip size would be no. 3. This number would be stamped on the tip. The number 3 indicates that the oxygen pressure and the acetylene pressure are to be set at 3 lb on the working pressure gage. The no. 3 welding tip has a demand of 9 ft³/hr of dissolved acetylene, which means the maximum number of people who could be welding with the no. 3 tip from one 300-ft³ cylinder of dissolved acetylene is four people. If the system is overloaded, the weld zones will be contaminated with acetone.

WELDING TECHNIQUE When welding, an operator can concentrate the heat from the torch either in the weld bead, which is called the backhand technique; or he can concentrate the heat ahead of the weld bead or in the weld puddle, which is called the forehand technique (Figure 3–3a). The forehand welding technique is usually used on relatively thin metals. The torch points in the same direction that the weld is being done so that the heat is not flowing into the metal as much as it could. The tip of the torch is held at approximately a 45° angle, which makes some of the heat deflect

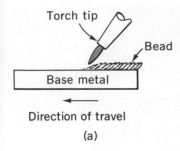

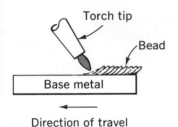

Fig. 3-3   (a) Forehand welding.
        (b) Backhand welding.

away from the metal. Instead of the base metal absorbing all the heat, some of it is reflected off into the atmosphere. In this way, it is possible to weld very thin material. The weld bead appearance is characterized by an evenly flowing, rippled design. The backhand welding technique is one used on heavier or thicker base metals. Basically, in this technique the torch is pointing in the direction opposite to that in which the weld is being done (Figure 3-3b). In this technique, the heat is concentrated into the metal so that thicker materials can be welded successfully. Welds with penetrations of approximately ½ in. can be achieved in a single pass with the backhand technique. The bead is characterized by layers that form a much broader based ripple than that of the forehand technique.

Both of these techniques can be used with or without filler rod. Welding done without filler rod is called puddling. When puddling in the flat position, the torch is usually held somewhere between the angles of 35 and 45° (Figure 3-4a). Even penetration can be determined by observing the amount that the metal sags in the bead's path. The amount of sag should be just enough to be noticeable (Figure 3-4a). Puddling is generally used with metals that are approximately 10 gage to ⅛ in. thick.

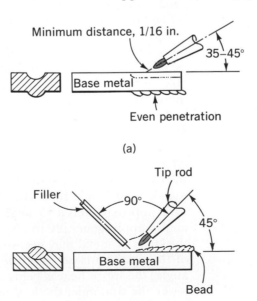

Fig. 3-4   Welding. (a) Without filler rod (puddling). (b) With filler rod.

Fig. 3–4 (cont'd.)   (c) Welding a joint
with filler rod. (Courtesy of Victor
Equipment Company.)

(c)

Metals heavier than this are generally welded with a filler material. When puddling, the tip of the inner cone does not touch the metal. It should be approximately $1/16$ to $1/8$ in. from the metal or the molten puddle at all times. If it is held any closer than this, the torch tip will overheat and pop. Particles of the molten puddle could enter the torch, which means that the torch would have to be shut off and then cleaned. If the tip is too close, it can overheat; in this case, too, the torch would have to be shut off and cooled in water. Puddling is the first operation that should be done by the learner, to help him to develop an acceptable technique of weld manipulation. Whenever the beginner puddles, he should never hold the torch body with both hands. *If the torch body is held with both hands, sooner or later the welder will have to put a piece of filler material or filler rod in either his left or right hand and thus he will have to start all over in developing his technique.* When welding with the filler rod, the welder should hold the filler rod at approximately a 90° angle from the welding tip and the welding tip at approximately a 45° angle from the base metal. These angles apply generally in any welding position; however, thickness of metal may necessitate an increase in the angle of the tip. The closer that the tip is rotated toward a 90° angle, the more heat is being pushed into the metal. The lower the tip angle, the less heat is being pushed into the

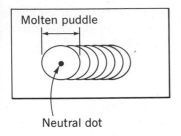

Molten puddle

Neutral dot

Fig. 3-5 Neutral dot.

metal and the more heat is being reflected away from the metal. This principle applies with both the forehand and the backhand welding techniques.

The characteristics of the molten puddle indicate to the welder the penetration of the weld, the torch adjustment, the torch handling, and the torch movements. The puddle then is judged continuously by the welder. The amount of penetration is proportional to the width of the bead. Penetration on thin metal is approximately one-third of the bead width. On heavier metal, with the back-hand technique, the width of the puddle indicates for all practical purposes the depth of penetration of the weld bead. Torch adjustment can be determined by the appearance of the weld bead. If the bead appears scummy, there is too much oxygen. The best way to determine whether the torch adjustment is good is to observe the appearance of the weld puddle. If it has a smooth glossy appearance and if there seems to be a dot floating around the outer edges of the weld puddle, the torch adjustment is good. This so-called neutral dot (Figure 3-5) is a result of the oxides present in the weld, and it floats continuously around the outer edges of the weld puddle. Whenever this dot is absent, there is not a neutral flame. When this dot increases in size, it indicates that there is an over-abundance of carbon present. When this happens, the weld puddle will take on a sooty, dirty, dull appearance result-ing from a carburizing flame. The neutral dot is a good thing to watch for when welding with the neutral flame, which is the most-used flame in welding.

The most difficult skill for the welder to master when he is learning to puddle is to be able to stop and start any place that is desired. To restart the weld puddle or the weld bead, reheat the base metal approximately ½ in. in front of the weld bead and in line with it. Once the metal turns a glossy color and the neutral dot can be seen, the flame can be moved slowly back to the weld zone. Once this weld zone has been reached, directions should be very quickly reversed. Then the weld can be continued in the direction desired (Figure 3-6). When the weld zone has been reached, the normal movements must be increased because of the extra heat the metal has absorbed. If the movements are not increased, the bead width will be extra

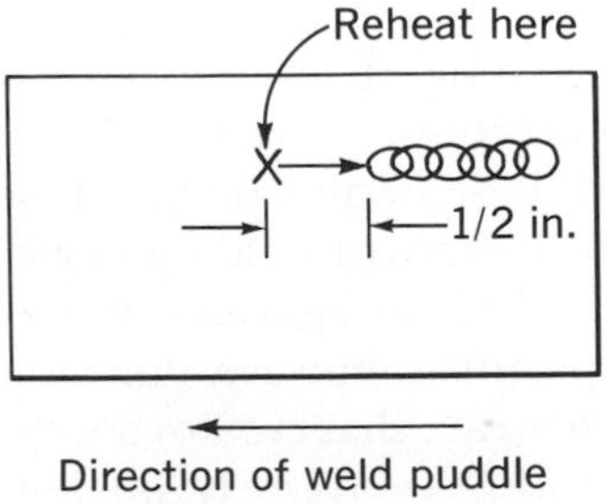

Reheat here

1/2 in.

Direction of weld puddle

Fig. 3-6 Restarting the puddle.

large here. If the flame does not come back far enough, however, there will be a void or a spot in the base metal where penetration will be zero, a situation called an inclusion.

WELD MOVEMENTS

There are as many different weld movements as there are welders. Every person develops his own particular way to move that he finds the easiest after a period of time. Figure 3–7 illustrates six basic ways to move the welding torch. The major point to remember in all of these movements is that the flame tip should not leave the molten puddle and that the welding-torch body should be held with one hand.

The base metal must be preheated and the weld puddle should be established before any movements take place. One method of establishing a weld puddle is with a circling movement to create a spiral of heat and moving the torch constantly until the puddle is established. This continuous motion is many times in the shape of a small circle. The size of this circle will establish the diameter of the weld puddle. After this puddle is molten and the neutral dot is present, many welders use this same small circling motion to form the first welding movement (Figure 3–7). Every time the torch makes one complete circle, the torch is advanced approximately $1/16$ in. The movement is in the direction of the weld bead to be formed. The second welding movement, shown in Figure 3–6, is one that is used often when a person feels he cannot maintain a circling motion. It is simply a matter of moving the torch back and forth in a zigzag manner. Every zigzag advances approximately $1/16$ in. However, extreme care must be taken with this movement in order not to move through the corner sections too quickly. If anything, the welder should pause whenever he changes direction. This pause will insure maximum penetration at these corner points. This advice also applies to the other four types of welding movements presented in Figure 3–7. The straight-line movement seems to be the easiest to perform; however, moving the torch along the base metal at a rate that creates a weld puddle or a weld bead of equal width is extremely difficult. This movement is more suited to the experienced welder or to an automatic welding process. A smooth weld bead results when this one smooth movement is used and the

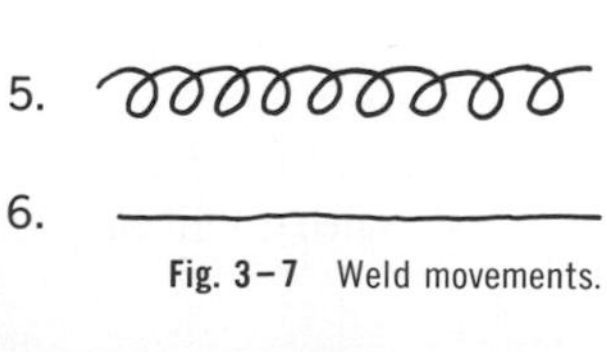

1.

2.

3.

4.

5.

6.

Fig. 3–7 Weld movements.

(a)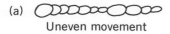
Uneven movement

(b)
Good weld

(c)
Not enough heat

(d)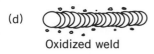
Oxidized weld

(e)
Improper heat control

Fig. 3–8 Weld appearance.

movement can be accomplished after some experience by the welder.

The appearance of the finished weld bead indicates quite a bit of information to the welder. Figure 3–8*a* illustrates the appearance of a weld bead when uneven movements and increased speed of the movement have almost interrupted the weld bead. The operator then noticed this error and automatically overcorrected. In Figure 3–8*c* is illustrated a typical error that most beginners make, caused by either not enough heat being pointed to the base metal, or the base metal not being preheated a sufficient amount, or the movement being so erratic that a molten puddle cannot be maintained. One thing to be noticed on all these welds is that upon solidification of the molten puddle, the neutral dot is no longer visible. It has dissipated into the oxide of the weld zone.

WELD POSITION Oxyacetylene welding or oxygen-fuel welding can be accomplished in any one of four basic positions. These positions include the flat position, indicated by the capital letter *F*; horizontal position, *H*; vertical position, *V*; and overhead position, *O*. Of these four positions, the flat and horizontal positions are the most used. The flat position is the easiest to learn. Many of the movements used in the flat position can be used in the overhead, the vertical, and the horizontal positions. The basic difference is the rate at which the torch is moved. A slow rate in any position other than the flat position will allow the molten puddle to sag from the weld zone which means that there will be a mass of solidified metal that is not fused to the base metal but will be merely overlapping it. Any of these four positions can be welded either by the forehand technique or the backhand technique, with filler rod or without filler rod.

QUESTIONS

1. What is the correct procedure for assembling the oxyacetylene welding torch?

2. Explain in detail the leak test for the oxyacetylene system.

3. What is the proper tip size for metal $1/16$ in. in thickness, $1/8$ in. in thickness, and $1/4$ in. in thickness?

4. What is the consumption rate in cubic feet per hour of a no. 8 tip, a no. 3 tip, a no. 1 tip?

5. What is the forehand welding technique?

6. What is penetration?

7. What is the neutral dot? What does it indicate?

8. What is the correct procedure for starting a weld puddle or restarting the weld puddle?

9. What are the four major welding positions?

10. What is the characteristic appearance of an oxidized weld?

PROBLEMS AND DEMONSTRATIONS

1. Select a piece of ferrous metal about $6 \times 6 \times 1/8$ in. and weld three beads without filler material. Weld one puddle with an extremely uneven movement, one puddle with not enough heat, and the third with the correct amount of heat and correct movement. Contrast the appearance of the three beads.

2. Demonstrate the correct assembly of the oxyacetylene torch system.

3. Demonstrate the correct procedure for checking for leaks of the oxyacetylene torch system.

4. Select a piece of ferrous material approximately $3 \times 3 \times 1/8$ in. and weld a puddle using the forehand technique and a puddle with the backhand technique; then compare the appearances of the beads and the amounts of penetration of the two techniques.

SUGGESTED FURTHER READINGS

Kerwin, Harry: "Arc in Acetylene Welding," McGraw-Hill Book Company, New York, 1944.

Postman, B. F.: *Safety in Welding and Cutting*, American Standard Company, New York, 1958.

Potter, Morgan H.: *Oxyacetylene Welding*, American Technical Society, New York, 1949

Standard Code for Arc and Gas Welding in Building Construction, American Welding Society, New York, 1946.

4 JOINING PROCESSES

The oxyacetylene torch can be applied to three basic types of welding: fusion welding, braze welding, and brazing. One of the major considerations in joint design is the type of welding that will be performed.

Fusion welding occurs when the base-metal molten puddle intermingles with a filler material or with other base metals, combining into a homogeneous structure. This is the strongest type of oxyacetylene weld. Oxyacetylene welding can be accomplished both with filler material and without filler material, as shown in Figure 4-1. If filler material is used, it can be part of the base metal or it can be a filler rod.

In braze welding, the filler material is generally composed of copper and zinc, and sometimes tin is added.

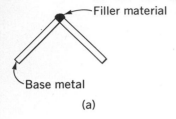

Filler material

Base metal

(a)

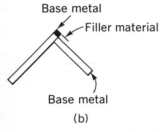

Base metal

Filler material

Base metal

(b)

Fig. 4–1 Fusion welding with filler rod. (a) With filler material. (b) Without filler material.

FUSION WELDING

A requirement of braze welding is that the filler material must have a melting point below that of the base metal but above 800°F. It is generally necessary to use a cleansing flux, and usually a slightly oxidizing oxyacetylene flame is used. The principal limitation of braze welding is that the weld loses its strength as temperature increases; therefore, braze welding can be used only at low temperatures.

Brazing, sometimes referred to as hard soldering, is also done above 800°F. The difference between brazing and braze welding is that braze welding filler material is placed by the welder in the weld joint; whereas in brazing, the filler material is dispersed over the closely fitted surfaces by capillary attraction. The filler material acts as a wetting agent that flows out over the base metal, breaking the surface tension of the base metal, which is not in a molten state as it is in fusion welding. The joint design and the distance between the pieces are extremely important in brazing. The tolerances for the distance between pieces range between 0.002 and 0.010 in. If these tolerances are exceeded, the brazed joint may be weak either because of inadequate filler material or because of an excessive amount of filler material.

Fusion welding with the oxyacetylene torch can be accomplished either with filler rod or without filler rod (Figure 4–1). When filler rod is used, its diameter generally approximates the thickness of the base metal. For example, if the base metal were 1/16 in. in thickness, the filler rod would be 1/16 in. in diameter. However, after reaching 3/16 or 1/4 in., the filler rod maintains its diameter of 3/16 or 1/4 in. even though the thickness of the material increases. When fusion welding a joint without filler rod, the amount of overlap of the base metal approximates the thickness of the base metal (Figure 4–1b).

The four basic joint designs that are used in fusion welding without the aid of filler material are the corner weld, flange weld, double flange weld, and lap weld (Figure 4–2). The lap weld is sometimes called the fillet weld. Generally, the material used in fusion welding without filler material is thin because the edges that are bent up to make the flange can be bent only if the material is thin.

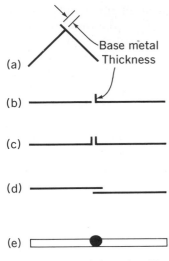

Fig. 4–2 Joint design without filler. (a) Corner. (b) Flange. (c) Double flange. (d) Lap or fillet. (e) Finished flange weld.

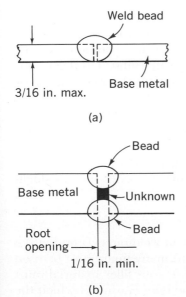

Fig. 4–3 Square butt joint. (a) Thin metal. (b) Thick metal.

If the material were thick, the filler rod would probably be the easier method to use. The finished weld of all of the joint designs in Figure 4–2 yield weldments comparable to welds accomplished with filler material, and the penetration should be the same.

Fusion welding with filler material offers more versatility to the welder than does fusion welding without filler material. However, caution must be taken when welding with filler material because the oxyacetylene weld bead has a maximum penetration of approximately $3/16$ or $1/4$ in. This would mean then that the thickest metal to be welded with a square butt joint would be $3/16$ in. (Figure 4–3a). When welding ferrous material thicker than $3/16$ in., an unknown area between the two beads may cause the total piece to fail (Figure 4–3b). If the square butt joint needs to be improved, an improvement can be made by the root opening (Figure 4–3b). The root opening, which should have a minimum space of $1/16$ in., helps to provide the necessary penetration for the base metal and the weld bead. Whenever the material or the base metal is butted together tightly, the work of the oxyacetylene flame is multiplied, and distortion is harder to control because the two pieces force each other out as they expand when heated. Because of the unknown area between two weld beads when welding thick metal, different joint designs have been found to be of greater predictable penetration. The ideal weld has 100 percent penetration. Because the oxyacetylene flame can penetrate the base metal only approximately $3/16$ or $1/4$ in., the bevel joint was created (Figure 4–4). The bevel joint is a means of removing a portion of the base metal—the portion of the base metal that will be replaced by the filler rod. The angle of the face of the bevel joint generally is between 60 and 90° and the root opening is approximately $1/16$-in. minimum to $1/8$-in. maximum. The root or the square portion of the bevel joint is also $1/16$-in. minimum to $1/8$-in. maximum. Because of the material removed in the bevel design, thicker material can be welded than with a square butt joint. The minimum thickness for the bevel joint design is $3/16$ in., the point at which the square butt weld is no longer possible. The maximum thickness of the material is thicker than approximately $5/8$ in. Materials thicker than this require a different weld joint design.

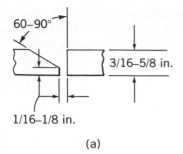

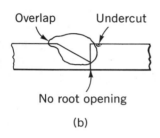

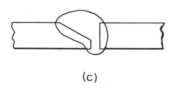

Fig. 4–4 Bevel joint. (a) Design. (b) Poor penetration. (c) Correct penetration.

The major errors in the welding of the bevel joint include the following: not maintaining a root opening, overlapping the weld bead, and undercutting the base metal (Figure 4–4*b*). The root opening can be maintained by always tack welding the base metal before welding begins. These tacks need be just a small puddle every 3, 4, or sometimes 6 in. to hold the pieces to be welded in correct alignment. Overlap occurs when the filler material being added to the puddle has actually overrun the puddle, and fusion has not taken place. Overlap commonly occurs on the opposite side of an undercut. Undercut occurs when the appropriate amount of filler material has not been added to the weld puddle to fill the cavity created by the oxyacetylene flame. Both the overlap and the undercut represent serious errors in welding technique. Errors common to the symptoms include: improper direction, improper direction of the welding tip, or improper speed of advancement. The indication of correct penetration when using the bevel joint is a rippling effect of the weld bead at the top surface of the weld as well as at the bottom surface.

Two other types of joints that can be used instead of the bevel joint are the V groove and the U groove (Figure 4–5*a* and *b*). Either of these two joints is chosen mostly because it is the preference of the designer of the weld joint. However, some authorities maintain that the V groove is the easier joint to machine. However, this depends upon the type of equipment that is surfacing or preparing the weld joint.

For oxyacetylene welding material ½ in. thick or thicker, the double-U groove should be used (Figure 4–6). The major limitation for oxyacetylene welding pieces of material is that the oxyacetylene welding process is relatively slow when compared to other welding processes such as shield arc welding.

BRAZE WELDING   Braze welding is similar to fusion welding, except that the filler material is nonferrous and many times must be used with a fluxing agent. This nonferrous filler material must melt above 800°F but below the temperature at which the base metal turns into a molten state.

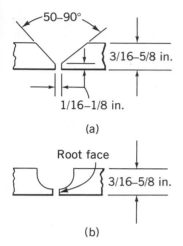

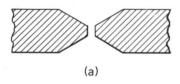

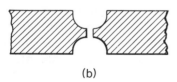

Fig. 4–5 Groove joint designs. (a) V groove. (b) U groove.

Fig. 4–6 Heavy groove joint designs 1/2 in. of thickness or more. (a) Double V. (b) Double-U groove.

The three most important factors when braze welding are the following:

1. The cleanliness of the surfaces to be braze welded

2. The tinning of the base metal with the nonferrous filler material

3. The complete fusion of additional weld material with the previously deposited nonferrous filler material, such as in the case of multiple passes (Figure 4–7*d*)

The basic joint designs that can be used in braze welding are any of the designs that are used in fusion welding (Figure 4–7). The only difference is that the joints cannot withstand the amount of heat that the fusion welds can. The major advantage of braze welding is that dissimilar metals may be joined. For example, high-carbon steel may be brazed to low-carbon steel, or copper may be brazed to steel on one end of the piece and brazed to cast iron on the other end. This opens to the welder a whole world of possibilities that was not available to him with fusion welding techniques. Other advantages of braze welding are that the oxygen-fuel consumption for a given job is less than with fusion welding and that locked-in stresses are not as great because temperatures in braze welding are lower than those in fusion welding. Also braze welding is quite strong up to approximately 500°F; and, because braze welding has a small heat effect of its own, special heat treatments may not be required after braze welding.

The thickness limitation of braze welding is slightly less than that of fusion welding. Butt joints or lap joints may be brazed when they are ³/₁₆ in. or less thick. Material to be braze welded that is thicker than ³/₁₆ in. must be grooved or beveled or must have some mechanical joint prepared.

BRAZING Brazing is differentiated from braze welding by the facts that the nonferrous filler material is attracted to the flux-cleaned areas by capillary attraction and that the joints can be brazed with filler material in thicknesses from 0.002 to 0.010 in. This means that brazing requires less skill,

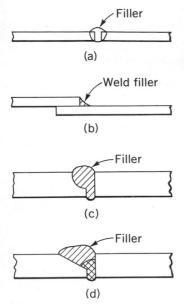

Fig. 4–7 Braze-weld joint designs.
(a) Butt. (b) Lap or fillet. (c) Groove.
(d) Bevel.

virtually no operator manipulation of the filler material; whereas the braze welded joint requires a great deal of manipulative skill because the operator or the welder places the filler material where it is needed. Brazing is differentiated from soldering by the melting temperatures of the filler material. If the filler material melts above 800°F, it is brazing; but if the filler material melts below 800°F, such as in the tin-lead alloys, the process is then known as soldering. The four most common brazing joint designs are the lap, the butt, the butt-lap, and the scarf (Figure 4–8). These joints are common to soldering joint designs as well as brazed joint designs. In fact, all joint designs or preparations that can be used in soldering may also be used in brazing. The determining factor in choosing joint designs is the thickness of the joint itself. When the thickness of the joint is increased, the possibility of voids is increased. Voids within the weld joint are more likely when the amount of filler material is increased.

The major problem of brazing is the alignment of the pieces to be joined. All alignment procedures must be done before heating. Poor laps can occur if proper alignment is not arranged before brazing (Figure 4–9).

EXPANSION OF METALS

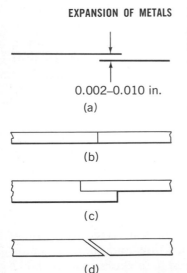

0.002–0.010 in.

Fig. 4–8 Braze joint designs. (a) Lap.
(b) Butt. (c) Butt-lap. (d) Scarf.

Most of the various weld processes inject a certain amount of heat into the metals being joined. The amount of heat injected will indicate how much the metal expands (Chart 4–1). For example, if a 1-in. square of aluminum or a 1-in. cube of aluminum were to be heated to approximately 500°F, the aluminum would no longer be 1 in. size but would have grown in size to approximately 1.006 in.[3]. If this same cube of aluminum were heated further yet to 1000°F, the expansion would double. The aluminum

CHART 4–1 Average expansion of metal (per inch)

METAL	500°F, in.	1000°F, in.	1500°F, in.	Per °F, × 0.00001 in.
Aluminum	0.006	0.012	—	1.360
Brass	0.005	0.010	0.014	1.052
Copper	0.004	0.009	0.014	0.887
Cast iron	0.004	0.007	0.010	0.556
Nickel	0.003	0.007	0.011	0.695
Steel	0.003	0.006	0.010	0.689

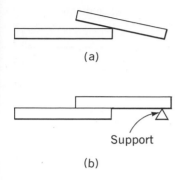

(a)

Support

(b)

Fig. 4-9 Braze. (a) Poor lap.
(b) Correct lap.

would now be 1.012 in. in size instead of 1.000 in.[3]. Aluminum above 1220°F is in a molten state. Each metal has a coefficient of linear expansion, which is generally indicated from a starting point of 20°C or 68°F. For example, the coefficient of linear expansion for aluminum is 0.000013 in. of growth per 1 in. in length per 1°F above the 68°F starting point. Generally when the metal expands above the ambient temperature, the room temperature, it contracts upon cooling to the approximate size that it was before the welding process. This expansion and contraction can lock in stresses in the welding process. This expansion and contraction can lock in stresses in the weld zone, stresses that are strong enough to literally pull the weld or the piece being welded apart. For example, if the break shown in Figure 4-10 were to be welded with fusion weld, the possibility of a break occurring would be very great. The heat injected by the welding would cause the metal to expand. This expansion would be constant while the weld was being performed. Once the weld is completed, however, the metal would start to contract. This contraction would draw in the sides, in the direction shown by the arrows in Figure 4-10, placing stress upon other sections. This could possibly cause the piece to break in a different location. One way to combat these locked-in stresses is to preheat the piece in locations in the same access as the break, such as the locations marked A and A' in Figure 4-10. By preheating before the weld is performed, the broken area or the weld zone and the other preheated locations would contract in the same manner. This then would remove many of the locked-in stresses caused by the heat-affected zone of the weld break.

Expansion and contraction can be categorized into three main groups: longitudinal, angular, and transverse. Expansion and contraction of heat-affected zones result in distortion. Longitudinal distortion is parallel to the weld bead. (Figure 4-11a). Angular distortion elevates the metal parallel to the bead (Figure 4-11b) and transverse distortion is distortion across the bead (Figure 4-11c). Transverse distortion tends to pull the metals being joined together—in fact, pull them together until they overlap.

The control of distortion is simple, requiring only previous thought and a little common sense (Figure 4-12). Distortion may be controlled by clamping the pieces to

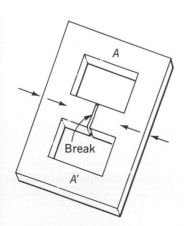

Fig. 4-10 Expansion-contraction.

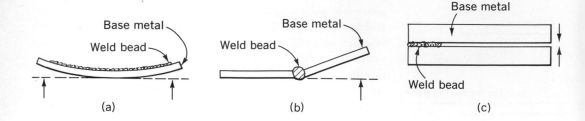

Fig. 4–11   Types of expansion and contraction. (a) Longitudinal. (b) Angular. (c) Transverse.

be welded into a fixture before welding them and allowing them to cool before they are removed from the clamping device. The control of distortion may be accomplished by presetting the materials to be welded. This entails estimating the amount of warpage and is dependent upon the amount of experience of the welder (Figure 4–12). Control of distortion may also be accomplished by placing a strong back or similar type of clamping mechanism on the face of the weld. It can also be accomplished by tack welding or by step welding (Figure 4–12*e*). Step welding means to weld in steps; that is, to weld position 1 first and then to go to the far end of the material; then weld position 2, step back to position 3, and then back to position 4. The arrows shown in Figure 4–12*e* indicate the direction that the weld should be done in order to dissipate the maximum

Fig. 4–12   Control of distortion. (a) Clamping. (b) Presetting. (c) Strong back. (d) Tack weld. (e) Step weld.

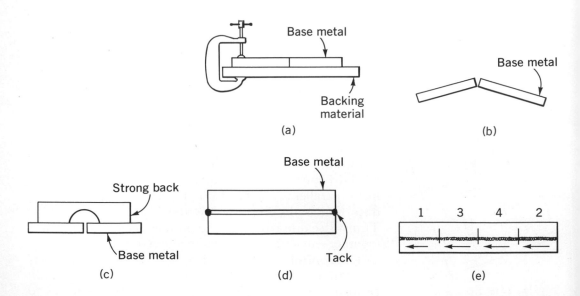

amount of heat. Many times this technique is used in welding items that must be distortion free or in welding such materials as cast iron, for which the heat should be kept to a minimum.

QUESTIONS

1. Differentiate between fusion welding, braze welding, and brazing. What are the major advantages of each?

2. What are the basic joint design differences between designs used with filler material and those used without filler material?

3. When fusion welding, what is the maximum thickness of the square butt joint?

4. What is root opening?

5. What is an indicator of correct penetration when using the square butt joint design?

6. How are V-groove joints used and U-groove joints used?

7. What are the four most common braze weld joint designs?

8. What are the four basic braze welding joint designs?

9. What is the maximum distance possible between pieces of the base metal when brazing?

10. What is distortion?

11. What is coefficient of thermal expansion?

12. How is longitudinal distortion rectified?

13. Name and explain four ways to control distortion in the base metal.

PROBLEMS AND DEMONSTRATIONS

1. Select two pieces of ferrous material approximately $\frac{1}{2}$ in. thick and set them up for a fusion weld. Maintain approximately $\frac{1}{16}$ to $\frac{1}{8}$-in. root face between the two. Weld with the forehand or backhand technique and use filler rod. Do not cool the weld in water. Upon completion of the weld bead, break the weld and note the amount of penetration.

2. Select two more pieces, prepare the joint for a flange weld. Perform the weld and note the appearance of the finished weld bead as compared to a weld bead that uses filler material.

3. Select a piece of ferrous material that is 2 × 6 in. and approximately ⅛ in. thick. Weld a bead with or without filler material. Weld the bead in the middle of the metal for the entire length of the piece. Note the distortion.

4. Select a piece of ferrous material approximately 2 × 6 × ⅛ in. Heat the material, do not weld, just heat the material from one side, approximately halfway up across the 2-in. portion. Allow to cool slowly and note the amount of contraction.

SUGGESTED FURTHER READINGS

Althouse, Andrew D., C. H. Turnquist, and W. A. Bowditch: "Modern Welding," The Goodheart-Willcox Co., Homewood, Ill., 1965.

Pender, James A.: "Welding," McGraw-Hill Book Company, New York, 1968.

Phillips, Arthur L.: *Modern Joining Processes,* American Welding Society, New York, 1966.

Standard Code for Arc and Gas Welding in Building Construction, American Welding Society, New York, 1946.

Stieri, Emanuele: "Basic Welding Principles," Prentice-Hall, Inc., Englewood Cliffs, N.J., 1961.

5 FERROUS WELDING

FUSION WELDING Fusion welding of ferrous materials with the oxyacetylene torch generally employs a neutral flame, which should never be allowed to touch the base metal. The secondary combustion of the flame protects the molten puddle sufficiently. However, in some situations, such as with cast iron and chromium-nickel steel castings, a flux is used to help remove oxides from the molten puddle. Fusion welding can be accomplished in the flat, horizontal, vertical, and overhead welding positions by using either the forehand or the backhand techniques (Figure 5-1). It is applicable to a wide range of ferrous materials (Chart 5-1) and is used more for repairs than as a production welding procedure.

BRAZE WELDING Oxygen-fuel braze welding of ferrous materials is often used to join dissimilar metals or for repair work. All good welding procedures and techniques that apply to fusion welding apply also to braze welding of ferrous materials. Braze welding differs from fusion welding in that the base metal puddle is not in a molten state and the filler material is not ferrous. Common brazing and braze welding alloys used for filler materials include copper and zinc alloys; copper, nickel, and chromium alloys; various silver alloys; copper and gold alloys; magnesium and silicon alloys; and magnesium alloys. More recently, these brazing alloys have included such metals as titanium, beryllium, zirconium, and palladium. These last three exotic metals are being used mainly for joining such metals as those used on supersonic transports, in the space program, and in the missile programs.

Because a dissimilar metal is used as a filler material, a flux is required to prevent the formation of oxides, nitrides, and other undesirable material in the weld zone. At elevated temperatures, metals in the pure form are generally unstable and have a tendency to react with other

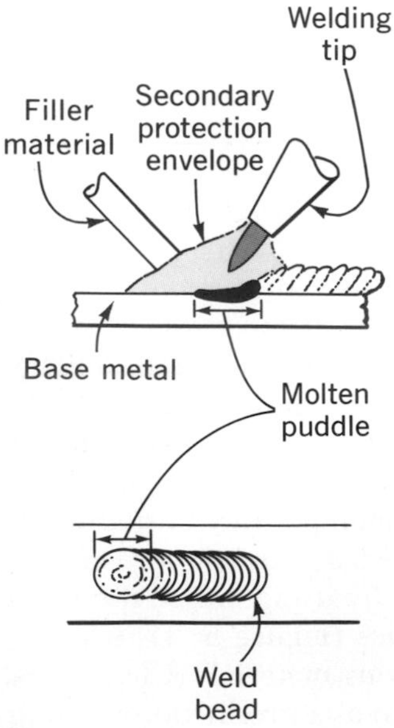

Fig. 5-1 Fusion welding.

CHART 5-1 Fusion welding of ferrous material

METAL	FLAME ADJUSTMENT	FLUX	FILLER MATERIAL
Cast steel	Neutral	——	Steel
Steel	Neutral	——	Steel
High-carbon steel	Carburizing	——	Steel
Manganese	Slightly oxidizing	——	Base metal
Cromansil	Neutral	——	Steel
Cast iron	Neutral	Yes	Cast iron
Chrome-nickel steel castings	Neutral	Yes	Base metal, chrome-nickel, or columbium stainless steel
Chrome-nickel steel	Neutral	Yes	Columbium stainless steel or base metal
Chromium steel	Neutral	Yes	Chrome-nickel or columbium stainless steel
Chromium iron	Neutral	Yes	Chrome-nickel steel or columbium stainless steel or base metal

compounds. The speed of reaction can be controlled by directing the amount of chemical activity through the choice of the fluxing agents and through the control of the temperature fluctuations in the molten puddle. One example of chemical activity is salt air or salt water reacting with iron. If ferrite is exposed to salt for a long time, the metal will corrode. The oxide that results from the corrosion makes it difficult to obtain a uniform metal flow in the weld zone, and the oxide will eventually affect the strength of the weld; therefore, a flux is used to soften and to remove the oxides by floating them to the top of the molten puddle. This procedure gives the filler material a clean bonding zone, and it reduces inclusions that are caused by the oxides.

There are three basic types of fluxes: the highly corrosive, the intermediate, and the noncorrosive. The corrosive fluxes can withstand high temperatures without burning and charring while the other types cannot. Because gas welding has a high operating temperature, it is restricted to the use of the highly corrosive flux. Highly corrosive fluxes consist of inorganic acids or salts that react to the oxide, sometimes at room temperature. Some start reacting when heat is applied to them. The more common fluxes are made of sodium, potassium, lithium, and borax. Fused borax, fluoborates, boric acids, and alkalis are also used; but these fluxes usually require higher flame temperatures as in gas welding. However, they are much easier to remove from the finished weld areas. The most common braze welding flux is borax or boric acid, which is borax and water. Flux can be applied as a paste, a powder, a liquid, a solid coating, or a gas. Paste and powder are the most commonly used, and even these two forms can be applied in various ways. As a paste, the flux can be brushed on the metal to be welded. Powder can be either sprayed on or applied by dipping the preheated filler rod into the powder. Liquid can be sprayed on the surface or fed to the metal through some gas source, or the gas can even be supplied directly into the acetylene or the inlet and allowed to mix with fuel in the torch body before being ejected directly from the torch welding tip.

The most common ferrous materials that are braze welded are steel and its alloys, galvanized iron, cast iron, and chromium-nickel cast iron (Chart 5–2).

CHART 5-2 Braze welding of ferrous material

METAL	FLAME ADJUSTMENT	FLUX	FILLER MATERIAL
Steel	Slightly oxidizing	Yes	Bronze
Galvanized iron	Slightly oxidizing	Yes	Bronze
Cast iron	Slightly oxidizing	Yes	Bronze
Chrome-nickel cast iron	Slightly oxidizing	Yes	Bronze

Braze-welding procedure All ferrous materials are braze welded by following a procedure that is much like that used for fusion welding of cast iron except that tinning qualities must be present (Figure 5-2). Tinning is the application of a metal coating between the base metal and the filler material. In most braze welding, tinning is accomplished automatically by bringing the base metal up to what is known as a tinning temperature, a temperature above 800°F but below the melting point of the specific ferrous metal being welded. The tinning temperature is generally indicated by a dull-red color. When the brazing filler material is touched to this dull-red base metal, a very thin coating flows out from the brazing rod. This flowing out is called tinning or wetting. The tip size used for braze welding corresponds to that which would be used if the base metal were steel and the welding procedure were fusion. The tip size should be made with the torch set at a neutral or a slightly oxidizing flame. The major welding should be accomplished with a secondary combustion envelope, however, rather than with the inner cone. The distance between the inner cone and the base metal should be maintained at approximately ¼ in. or double the distance maintained during fusion welding. Before the base metal is heated for braze welding, the brazing filler rod is heated and dipped approximately 2 to 3 in. into the brazing flux. This procedure will cause the flux to adhere to the filler material. The heat of the welding tip is concentrated on both pieces of the base metal (or both sides of the crack if it is a repair weld) until the ferrous base metal reaches a dull-red point. The preflux brazing material that will form the braze bead is then added. The welding technique may be either forehand or backhand. Generally, the forehand method is preferred for braze welding because too much heat results from the

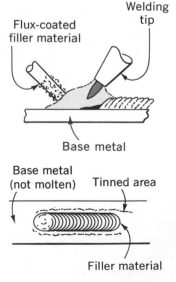

Welding tip

Flux-coated filler material

Base metal

Base metal (not molten)   Tinned area

Filler material

Fig. 5-2 Braze welding of ferrous material.

backhand technique, and too much heat will cause the bronze to run out of the edge of the puddle, leaving an oxidized area in its center. Conversely, too little heat will cause the bronze to ball up in small round balls running over the surface of the metal. A properly deposited bronze weld will be one in which the ripples are smooth and even. It will have a small excess of flux and tinning present at the edges of the bead. Sometimes this excess extends out as much as $\frac{1}{8}$ or $\frac{1}{4}$ in. from the bead. The bead will be slightly convex in shape and will be smoother than the weld bead in the fusion welding process.

OXYACETYLENE FUSION WELDING OF GREY CAST IRON

Cast iron is a complex alloy of six or more elements in the following approximate proportions: 96 percent iron, 3 percent carbon, 1 percent silicon, 0.2 percent sulfur, 0.75 percent phosphorous, and 1 percent manganese.

Castings of cast iron are used for automobile blocks, automobile heads, water pump bodies, pipe fittings, machinery frames, housings, and many other machined parts. There are four broad classifications of cast iron: pig iron, white cast iron, grey iron, and malleable iron.

Pig iron is the product of the blast furnace. It is usually remelted into one of the other three types of cast iron.

White cast iron is hard, brittle, and magnetic. It is readily broken by sharp blows with the hammer; and it is called white cast iron because when the metal is broken, it appears silvery and white. White cast iron is produced by cooling the molten iron so rapidly that the iron carbide compound does not separate from the carbon. Many times a softer white cast iron is made by allowing the molten iron to cool at a slower rate. White cast iron can be so hard that it is nonmachinable. It is used for plowshares, ball-mill agitators, stamping shoes, wear plates, and sometimes car wheels for heavy cars.

Grey cast iron, which is the most common form of cast iron, is used extensively in machinery castings. When it is broken, the break appears a dark grey color, and the bright skin by which malleable cast iron can be identified is absent. Grey cast iron is made by allowing the molten iron to cool slowly, which separates the iron and carbon. The carbon separates out as tiny flakes of graphite that are scattered homogeneously throughout the grey cast iron. This carbon is sometimes called a free or graphitic carbon. Since graphite is an excellent lubricator, grey cast iron is used in such things as auto cylinder blocks that require lubrication. Because of the presence of these scattered flakes of graphite throughout the metal, it is more easily machined than other types of cast irons. Because of the tiny flaky particles of graphite, cast iron needs little lubrication when being machined. Grey cast iron has four basic subdivisions: pearlitic cast irons, austenitic cast irons, alloy cast irons, and Meehanite cast irons.

Pearlitic cast irons have the combined carbon in a pearlite matrix. Generally, pearlitic cast irons contain from 0.60 to 0.90 percent combined carbon. Austenitic cast irons are those in which the grains, or the flakes of graphite, are interspersed within austenitic structures. Cast iron of this type generally includes such alloying elements as aluminum, nickel, chromium, vanadium, titanium, or zirconium. Meehanite cast iron is a trade name applied to a group of grey cast irons. Cast iron of this type has been made by or produced by adding calcium silicide to the molten metal, which produces a cast iron somewhere between a grey cast iron and a white cast iron. Some very high strength irons have been produced by adding calcium silicide.

Malleable iron withstands great strains and rough usage. It will deform before breaking. When it is broken, however, its color has a white steely skin extending from the surface toward the center, which is usually dark and dull. Malleable iron is made by heat treating white cast iron. The heat treatment is a reheating of the iron under closely controlled conditions to about 1400°F for a certain length of time. The heat acts upon the graphitic carbon crystals or flakes so that they do not separate out as flakes but as round particles of carbon, causing a malleable cast iron that withstands rough usage and will deform before it breaks. It is tougher than other irons, a higher ability to resist shock loads.

The oxyacetylene welding of cast iron can be either fusion welding if the welding rod is cast iron or bronze welding if a bronze welding filler material is used. Both processes require a fluxing material. Fusion welding should always be used in cast iron when the item to be welded will be subjected to temperatures greater than 650°F.

Since the bronze filler material used in braze welding of cast iron loses its strength above 650°F, it is not recommended for use on items that are in service at or above this temperature.

WELDING OF CAST IRON

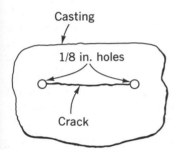

Fig. 5–3 Hole placement.

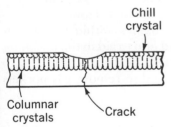

Fig. 5–4 Chill crystals (small equiaxed crystals).

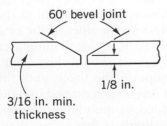

Fig. 5–5 Bevel joint.

When a welder is called upon to weld cast iron, it is usually for the purpose of fixing a broken or cracked casting. Before a cracked casting can be welded, the area around the break must be cleaned thoroughly with kerosene. After a few minutes, the kerosene should be wiped off with a dry cloth. By rubbing a piece of white chalk over the cracked area, fine hairline cracks may be made more visible. This method of crack identification by chalk is called a wet-line type of method, for it makes a wet appearance on the crack. To prevent the cracks from extending during the welding operation, holes 1/8 in. in diameter are drilled a short distance from the end of the cracks (Figure 5–3). A narrow strip along each end of the joint then must be ground to remove the surface layer from the casting. This surface layer is known as a casting skin, a chill crystal zone, or a small equiaxed crystal zone (Figure 5–4). This crystalline zone ranges in thickness from 0.001 to 0.050 in. and must be removed before welding can be done even on very thin pieces. If the metal is less than 3/16 in. thick, the material can be butt welded. However, if the material is more than 3/16 in. thick, the crack area should be beveled (Figure 5–5). This beveled area should extend to within 1/8 in. of the bottom of the crack. This 1/8 in. should be left to prevent the oxyacetylene weld puddle from melting through the bottom of the metal. This unground area will make it easier for the operator to control the weld puddle and also make it easier to build up a sound base for the weld. Many times a crack extends completely through the material, necessitating the use of carbon blocks or similar types of heat sinks that will absorb the excess heat. These carbon blocks would be placed below any openings to prevent the molten weld puddle from escaping through the casting (Figure 5–6).

Occasionally a joint stronger than the beveled joint is desired. Different methods are used to obtain this stronger joint. A hole can be drilled in the beveled edges of the casting, then threads tapped in the holes so that

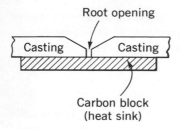

Root opening

Casting   Casting

Carbon block
(heat sink)

Fig. 5-6 Carbon block.

Studs

(a)

(b)

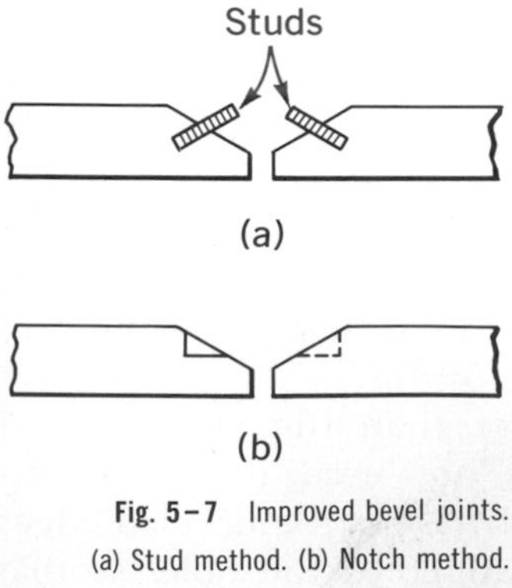

Fig. 5-7 Improved bevel joints.
(a) Stud method. (b) Notch method.

studs can be screwed into the holes. Then the joint is welded (Figure 5-7a). Another method is to bevel the edges and then cut notches into the beveled edges. Then the joint can be welded (Figure 5-7b).

In the oxyacetylene fuse welding of cast-iron objects, the objects must be preheated before welding in order to prevent new cracking. Preheating also assures uniform expansion of the metal to be welded. Annealing of the weld is also more easily accomplished if the casting is preheated. Preheating may be done with an oxyacetylene torch on a small casting or in a furnace if a large casting is involved. Grey iron castings should be preheated to a temperature of about 1500°F, or to a dull red; care should be taken that the temperature does not exceed this point. All welding must be done while the casting has this dull-red appearance. If the casting is allowed to cool at an uneven rate, stresses that will be locked in will come out after the welding has been accomplished and will be destructive to the piece. The tip used for grey cast iron should be about the same size used for mild steel of the same thickness. The weld should be made with the torch set at a neutral or a slightly carburizing flame. The filler rod for fusion welding of cast iron should be of the same composition as the base metal being welded. However, a flux is needed to break down the oxides within the parent metal and the filler material and also to aid in the puddling of the cast iron. Another function of the flux is to add to the fluid qualities of the metal and to remove gas pockets and oxides.

Welding procedure   The procedure for welding cast iron is to direct the flame at the bottom of the V or the bevel joint until the base metal starts to melt. The welding torch is held at about the same angle as in welding mild steel, with the tip of the inner cone about 1/8 to 1/4 in. from the metal. When a puddle is established at the bottom of the beveled joint, the flame is moved up the sides of the beveled joint to melt the sides. After the entire welding area has become molten, the welding rod should then be introduced into the outer envelope of the flame until the rod is hot. The hot end of the rod is then dipped into the cast iron flux which, when added to the molten puddle, helps keep the molten puddle fluid. If flux is not used, infusible slag mixes with the iron oxide that forms on the puddle, causing inclusions and blow holes in the weld. Once the rod has been dipped into the flux, it should be introduced into the molten puddle, which should be maintained at all times by the oxyacetylene torch. The rod gradually melts from the heat of the molten puddle and the filler material fuses with the base metal. If white spots or gas bubbles appear in the puddle, more flux should be added and the flame should be played around the specks until the impurities float to the surface. The impurities should then be skimmed off the puddle with the welding rod. If the impurities adhere to the welding rod, it should be tapped against the welding bench so that the impurities will fall off. When the weld is completed, the casting should be reheated until it is at the dull-red stage again, approximately 1500°F. Then the entire casting should be allowed to cool very slowly until it reaches room temperature. A casting that has been preheated, welded, reheated again, and allowed to cool slowly is readily machinable both in the weld zone and in the area around the weld.

OXYACETYLENE WELDING OF STAINLESS STEEL

The American Iron and Steel Institute recognizes more than 40 grades of stainless steel, which fall into three broad divisions based on their predominate crystalline structures —ferritic, martensitic, and austenitic (Chart 5–3). These three categories of stainless steel exhibit different properties, which determine how these three types of stainless steel are used. Ferritic stainless steel has the properties of being nonhardenable and of being magnetic. Martensitic stainless steel has the properties of being hardenable by heat treating and of being magnetic. Austenitic stainless steel has the properties of being extremely tough and of being ductile in the welded state. Strictly speaking, stainless

CHART 5–3 Composition range of stainless steel

CLASS	CARBON, %	CHROMIUM, %	NICKEL, %
Ferritic	0.35 max	16–30	——
Martensitic	0.05–0.50	4–18	——
Austenitic	0.30–0.25	14.30	6–36

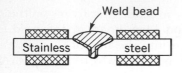

Copper heat sinks

Fig. 5–8 Welding jig.

steel should contain a minimum of 11 percent chromium, which forms a thin, passive, slightly adherent surface layer of chromium oxide that gives stainless steel its corrosion resistance.

All methods of welding can be used when joining stainless steel, although the oxyacetylene method is the most effective. However, some of the newer inert-gas processes have replaced oxyacetylene welding of stainless steel in the past few years. No matter what method is used, the problem of reducing the effects of heat still exists. One of the methods used for reducing heat effects in stainless steel welds is the use of chill plates that are made of copper pieces; these chill plates help conduct the heat away. They are many times incorporated into ridged jigs and fixtures that also aid in the elimination of warping or distortion (Figure 5–8).

If jigs or fixtures cannot be used, special procedures such as skip or stepback welding should be incorporated.

Before starting to weld stainless steel, it is a good idea to choose a strong joint design that will be satisfactory for the job. The flange joint design is probably the best for thin stainless steel. Stainless steel sheets from up to approximately 1/8 in. in thickness may be butt welded. With a plate thicker than 1/8 in., the edges should be beveled or grooved in some manner so as to provide maximum fusion to the bottom of the weld or complete penetration of the base stainless metal.

When stainless steel is welded with oxyacetylene, it is generally necessary to use a torch tip that is one or two sizes smaller than the tip that would normally be used to weld mild steel. Since the heat has a tendency to remain in the weld zone of stainless steel, a smaller flame is mandatory. If a smaller tip is not used, the large flame of the oxyacetylene torch might possibly destroy many of the properties of stainless steel, especially the chromium elements within the metal. The neutral flame is essential for welding stainless steel because even a slightly oxidizing flame would oxidize the chromium in the steel, thus reducing its corrosion-resistant qualities. An excess carburizing flame is undesirable because it increases the actual carbon content in the weld zone, causing the weld to be weak, brittle, and sometimes porous. A good rule to remember is that since the neutral flame is subject to regulator fluctuation, it is often better to use a slight excess of acetylene, thereby insuring in all cases that the oxidizing flame will not be present. If the oxidizing flame does touch the chromium steels or stainless steels, a chromium oxide will be present; 4000 to 5000°F of heat will then be necessary to penetrate this layer of oxide. Therefore, a flux is needed to break up this insulating barrier between the oxyacetylene flame and the work. This flux also insures a better control of the molten weld puddle and a sound, clean weld. Fluxes are prepared by mixing a powdered flux material, a chemical, with water to the consistency of a thin paste. Then the flux is applied by brushing it on the seam to be welded on the rod. Care should be taken to get it on both the seam and the rod. It is a good policy to coat the underside of the seam with flux in order to prevent oxidation on the bottom half of the base metal, which allows a more perfect union to be made along the bottom of the seam or the weld joint. A popular flux contains 1/2 lb zinc, 1/2 oz (ounce) potassium bichromite, approximately 1/3 oz hydrochloric acid, and 16 oz water. The filler metal used for the weld joint is of extreme importance. Whenever possible a special columbium 18-A or titanium welding rod is essential for achieving satisfactory welding properties when welding stainless steel. It is also a wise precaution to use a rod that contains 16.5 percent more chromium content than the base metal being welded. A rod of this type will allow for the slight oxidation losses that may occur in the welding operation. If it is impossible to obtain either of these two types of rods, the next best substitute is to cut strips from the base metal and use these as filler rods. However, this brings about the chance of other materials being injected into the weld zone. More specifically, the weld will sometimes undergo intergranular cracking because of the chromium carbide present in the weld as the result of not using a special filler rod.

A rule of thumb for welding stainless steel is to use the forehand technique on light sheets of metal and either the backhand or forehand technique on the heavier plate. If a smaller torch tip is not available, a welding tip that is the same size used for mild steel can be used, except that the torch should be adjusted to a soft flame to compensate for the increased tip size. This substitution will reduce the amount of chromium reduction in the stainless steel; it is

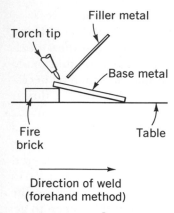

Direction of weld
(forehand method)

Fig. 5–9 Welding stainless steel. Flux
both sides of the base metal and filler
metal.

not, however, a preferred method because the soft welding
flame may fluctuate. The torch should be held at an angle
of about 45° to the base metal and the tip of the inner cone
of the flame should be kept to within 1/16 in. of the molten
puddle. This technique, which is standard practice in all
good welding, is vital in the welding of stainless steel because it helps greatly to prevent oxidation. It is also helpful
to elevate the starting end of the joint so that the line of
the weld inclines slightly downward toward the direction
of the finishing end or in the direction of the weld bead.
This practice permits the flux that fuses at a lower temperature than the stainless steel to flow forward and downward providing protection for the metal as it fuses. The
filler rod should be held close to the end of the cone of
the oxyacetylene flame to preheat its end. When withdrawn
from the puddle, the rod should be moved entirely from
the oxyacetylene flame until it is almost ready to dip back
into the puddle. Columbium or titanium rods flow freely
and care must be taken not to direct too much heat upon
them. If stainless steel strips are used, they should be
maintained within the secondary combustion zone of the
oxyacetylene flame (Figure 5–9).

Welding should be done from one side only and the
seam should be completely filled in the first pass. The
welder should try to refrain as much as possible from going
back over a weld because success in welding stainless steel
depends upon keeping the heat at a minimum. Going back
over a hot weld produces excessive heat and thus increases
the loss of corrosion-resistant elements in the stainless
steel.

QUESTIONS

1. What are the major differences between braze welding and brazing?

2. What are the advantages of brazing over braze welding?

3. What are the advantages of brazing or braze welding over fusion welding?

4. Why is a flux needed in either braze welding or brazing?

5. What should be the basic flame setting or adjustment for braze welding?

6. What is the maximum thickness for square butt joints in braze welding?

7. What are the design limitations for brazed joints?

8. What are the design limitations for brazed welded joints?

9. What are the major alloys used for brazing and braze welding?

10. What are chill crystals?

11. What types of fusion weld need a fluxing agent?

12. What are the major classifications of cast iron?

13. What are the major classifications of stainless steel?

14. What is a heat sink?

15. Why can either braze welding or brazing join dissimilar metals? And why cannot fusion welding join dissimilar metals?

PROBLEMS AND DEMONSTRATIONS

1. To demonstrate the chemical activity of flux, place a small amount of sodium or potassium in a small dish of water and note the reactions. Care must be taken not to stand too close to the water because of the violent reaction of these chemicals.

2. Braze weld a piece of ferrous material without a fluxing agent. Then braze weld a piece of ferrous material with the flux and compare the results.

3. Break a piece of grey iron, a piece of white cast iron, and a piece of malleable iron to show the different grain structures and colors. Identify these structures.

SUGGESTED FURTHER READINGS

Blanc, G. M. A., J. Colbus, and C. G. Keel: Notes on the Assessment of Filler Metals and Fluxes, *Welding Journal*, vol. 40, pp. 210–211, May 1961.

Graham, D. Frank: *Welder's Guide*, Theo Audel Company, New York, 1946.

Houck, L. H.: *Selfwelding*, The Lincoln Electric Company, Cleveland, 1958.

Jefferson, T. B., and Gorham Woods: *Metals and How to Weld Them*, The James F. Lincoln Arc Welding Foundation, Cleveland, 1954.

Reed, H. T.: Ground Rules for Stainless Welds, *American Machinists-Metal Working Manufacturing*, vol. 40, pp. 84–85, 1962.

Rossi, E. Boniface: "Welding Engineering," McGraw-Hill Book Company, New York, 1954.

6 NONFERROUS METALS

Common nonferrous metals are aluminum, copper, nickel, magnesium, and lead. Nonferrous metals can be fusion welded, braze welded, or brazed. Common fuel gases that are used for the welding of nonferrous metals are acetylene, hydrogen, natural gas, and propane. The oxyacetylene flame is probably the most used because of its availability for welding of other metals. The equipment for oxygen-fuel welding includes standard torches, hoses, and regulators. The basic difference between welding ferrous and nonferrous metals is in the size of the tip used in welding the nonferrous metals (Chart 6-1).

ALUMINUM   Aluminum, chemical symbol Al, is a slightly magnetic, ductile, malleable metal that resembles tin. Aluminum is

CHART 6-1   Welding of nonferrous metal

METAL	TYPE OF WELD	FLAME ADJUSTMENT	FLUX	FILLER MATERIAL
Aluminum	Fusion	Slightly reducing	Yes	1100 series 4000 series
Brass	Fusion	Oxidizing	Yes	Bronze
	Braze weld	Slightly oxidizing	Yes	Bronze
Bronze	Fusion	Neutral	Yes	Bronze
	Braze weld	Slightly oxidizing	Yes	Bronze
Copper, deoxidized	Fusion	Neutral	No	Base metal
	Braze weld	Slightly oxidizing	Yes	Bronze
Copper, electrolytic	Fusion	Neutral	No	Base metal
	Braze weld	Slightly oxidizing	Yes	Bronze
Nickel	Fusion	Slightly reducing	No	Base metal
Inconel	Fusion	Slightly reducing	Yes	Base metal
Monel	Fusion	Slightly reducing	Yes	Base metal
Magnesium	Fusion	Neutral	Yes	Base metal

not found in a metallic state but must be processed through electrolysis from bauxite ore. Aluminum has approximately 60 percent of the electrical conductivity of copper. It is a strongly electropositive metal that corrodes rapidly. In fact, aluminum corrodes so rapidly that an oxide is formed on its surface when it is simply exposed to the oxygen in the atmosphere. This oxide on the aluminum is one of the major deterrents in its successful welding. The oxide melts at approximately 3600°F and the aluminum itself melts at approximately 1210°F, which means that if the oxide is heated until it is molten, the base metal will almost boil since aluminum boils a little over 4000°F. Aluminum is used in the production of steel as a deoxidizer to remove oxygen in the steel. Aluminum is also used to control the ring growth in steels by forming nucleation points from which the crystals of steels may grow.

Aluminum may be classified into three main groups: commercially pure aluminum, wrought aluminum, and casting aluminum. The Aluminum Association has devised a four-digit numbering system for the identification of aluminum (Chart 6-2). This system assigns index numbers that identify aluminum alloys, including such alloys as 99 percent pure aluminum, copper, manganese, and silicon magnesium. The system also has openings for further alloying elements. The first digit of the number always indicates the major alloying element of the alumi-

CHART 6−2 Aluminum Association index system

ALUMINUM ALLOY	NUMBER
Aluminum, 99.00% pure	1xxx
Copper	2xxx
Manganese	3xxx
Silicon	4xxx
Magnesium	5xxx
Magnesium and silicon	6xxx
Zinc	7xxx
Other element	8xxx
Unused index	9xxx

num. The second digit ranges from 0 through 9. The zero indicates that no special control has been maintained in the alloying constituents. The numbers 1 through 9 indicate some type of special control, which depends on the manufacturer of the aluminum. The last two digits in the series indicate the minimum amount of aluminum not under special control. For example, 1075 aluminum indicates aluminum that has 99.75 percent controlled aluminum, or pure aluminum, with 0.25 percent without any control. Also, 1180 aluminum indicates an alloy that is 99.80 percent pure aluminum; the remaining 0.20 percent has had some kind of special control over it. The 1 as the second digit signifies what kind of control, but what the 1 in this case means depends on what manufacturer is producing the aluminum.

In the 2xxx series through the 9xxx series, the last two digits do not have any special meaning except the meanings assigned by the manufacturer. The last three digits in these cases serve to identify the various alloying ingredients added by the manufacturer. For example, 3003 aluminum or an aluminum-manganese alloy contains about 1.2 percent manganese and a minimum of 90 percent aluminum. Another example is 6151 aluminum, an aluminum-silicon-magnesium-chromium alloy; in this number, 6 is the designation for magnesium and silicon, and the numbers 151 indicate the special alloying ingredient and the percentage of the alloy. If the 1 as the second digit signifies chromium, then this code number indicates that 0.49 percent of the uncontrolled impurities is chromium, with 99.51 percent being aluminum, magnesium, and silicon.

Aluminum may further be categorized by whether or not it is heat treatable. Aluminum that is not heat treatable includes pure aluminum, or the 1000 series; manganese, or the 3000 series; and magnesium, or the 5000 series. In heat-treatable aluminum alloys the principal alloying ingredient is either copper, magnesium, silicon, or zinc. The 4000 series is the silicon series of aluminum alloys or those that are used mainly for welding purposes and for filler materials in weld joints. The temper designation of a heat-treatable alloy is sometimes indicated by a symbol after the four-digit number for the alloy. Some of these symbols include F, which means as fabricated; O, annealed; H, strain hardened, or cold-worked hardening. The symbol H can be further subdivided into H1x, strain hardened only; H2x, strain hardened and partially annealed; and H3x, strain hardened and stabilized. These H designations correspond with and have replaced the older designations of half-hard, three-quarters hard, and full-hard. The H designations go up to the number 8, indicating full-hard temper, which allows a wider range of hardness values than the older designations.

Other temper designations include the symbols W, which means solution heat treated, and T, which means heat-treat temper. The symbol T is always followed by one or more digits, which denote basic heat treatments. The symbol T3 means solution heat treated, cold worked; T4, solution heat treated and naturally aged; T5, artificially aged; T6, solution heat treated, then artificially aged; T8, solution heat treated, cold worked, then artificially aged; T9, solution heat treated, artificially aged, and then cold worked; and T10, artificially aged, and then cold worked.

Casting alloys of aluminum have as major alloying ingredients copper, silicon, magnesium, zinc, nickel, manganese, tin, chromium, and beryllium. A different numbering system is used for casting alloys. A typical number would be a sand casting aluminum alloy number 43, which has 5 percent silicon in it and 95 percent aluminum. Many times, a letter precedes the two- or three-digit number, indicating that a standard casting alloy has been further modified in composition. Therefore, it is possible to have a casting alloy numbered 140 and another numbered A140.

Aluminum alloys that are not heat treatable are easiest to weld. These alloys are usually welded with a no. 1100,

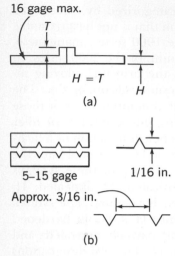

16 gage max.

$H = T$

(a)

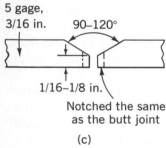

5–15 gage   1/16 in.

Approx. 3/16 in.

(b)

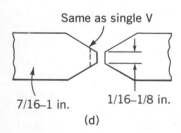

5 gage,
3/16 in.   90–120°

1/16–1/8 in.

Notched the same
as the butt joint

(c)

Same as single V

7/16–1 in.   1/16–1/8 in.

(d)

Fig. 6-1 Aluminum/magnesium joint designs. (a) Flange. (b) Notched butt. (c) Single-V notch. (d) Double-V notch.

no. 4043, or no. 4047 welding rod, all of which require flux. Heat-treatable aluminum alloys, especially those alloys that contain copper and zinc, such as alloys no. 2024 and no. 7075, generally are not fusion welded, because of the difficulty in tracking encountered during the solidification of the weld zone.

Joint design   The three basic joints used in the oxyacetylene welding of aluminum are the lap joints, the flange joints, and the butt joints. The lap joint is not recommended unless no other type of joint can be used because the lapped area will retain the flux plus the oxides. This area will corrode rapidly unless it can be cleaned, and the large contact area of the lap joint generally prohibits this necessary cleaning. One way to use a lap joint is to make all lap joints 100 percent weld enclosed.

The flange joint is used mainly on material that is 16 gage or thinner (Figure 6–1a). The height H of the flange is usually equal to the thickness I of the base material. If the base material were 16 gage, the flange height would be approximately 0.016 in. Many times the height of the flange will be slightly in excess of the thickness of the metal. The higher the flange, the more metal will be available to the welder. The flange many times is on only one piece of metal, and the other piece of metal is simply butted up to the flanged portion. The flange is formed by the use of a wooden mallet or many times a bending break, such as a sheet-metal break.

There are basically three types of butt joints: the notched butt joint (Figure 6–1b), the single-V notched butt joint (Figure 6–1c), and the double V (Figure 6–1d). Butt joints are designed for metal from 15 gage up to approximately 1 in. thick. The notched square butt joint is used generally with metal that ranges from 15 or 16 gage minimum to approximately 5 gage maximum. Instead of being brought up to touch as in the welding of ferrous material, the pieces to be jointed are notched 1/16-in.-deep with a cold chisel approximately every 3/16 or 1/4 in. These notches provide a mechanical means for holding the weld zone intact and act as a means of bonding for the weld bead. The single V is a standard V with a bevel ranging between 90 and 100°. It is used for aluminum between 5 gage and approximately 3/16 or 1/4 in. thick. The double

V is used only for very heavy material that ranges from approximately 7/16 in. up to 1 in. in thickness. The reason the double V is used rather than a very deep single V is that the amount of deposited materials will be less with a double V. The cross-sectional area of the joint design is much smaller in the double V than in the large single V.

Welding procedure Before the beginning welder can proceed to the fusion welding of aluminum, he must know three basic points about this particular metal. First, aluminum does not give any indication of change in temperature as ferrous material does. Its color does not change until all at once it suddenly reaches the melting point and collapses. Second, aluminum has extreme hot-shortness. Hot-shortness means that aluminum is weak at elevated temperatures—when it is hot, its strength is short. Aluminum also is highly thermal conductive, which means that heat travels through it quickly and spreads rapidly through the mass of the aluminum. Third, there is a tenacious oxide on the surface of all aluminum that is exposed to the atmosphere. Aluminum oxide has a melting point of slightly over 4000°F. This oxide, because of its higher melting temperature, will be unaffected far beyond the 1200°F melting point of the base metal; therefore, a flux, an agent to break down the oxides and to change the oxides into a fusible slag, must be used when fusion welding any aluminum. The oxide, once it is turned into fusible slag, is lighter than the base metal and floats to the surface of the weld puddle. It can be ladled off with a puddling stick or some such device with which the welder can manipulate the weld puddle.

The lack of an intermediate or mushy stage when aluminum goes from a solid to a liquid can be controlled by the use of a welding jig (Figure 6–2), which acts as a heat damp to absorb some of the heat of the base metal. It also prevents the metal from buckling and from possible distortion during welding. When using a welding jig, remember that the aluminum facing must be protected with asbestos or some other material and that the base metal must be allowed to cool before the piece is removed from the jig.

There is no color change in aluminum that has been heated to a welding temperature; however, the welder can use other methods to determine the temperature of alumi-

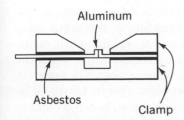

Aluminum

Asbestos

Clamp

Fig. 6–2 Aluminum welding jig.

num. Six methods of determining the proper preheating or the proper welding temperature for fusion welding of aluminum include the following:

1. By rubbing a pine stick on the base metal until it leaves a char.

2. By applying a carpenter's blue chalk to the base metal and preheating the base metal until the chalk mark turns white, which signals the proper temperature for welding.

3. By striking the cold aluminum with a metal hammer. Aluminum gives out a metallic ring which decreases as the temperature of the metal increases. When the metal is at the welding temperature, it will no longer give out a metallic ring when struck by a hammer.

4. By measuring with commercially available temperature measuring devices, such as Tempil sticks.

5. By tapping the oxyacetylene-heated area with the end of the welding rod until a certain mushiness can be felt at the end of the welding rod in the heated zone. This technique requires a great deal of experience.

6. By coating the base metal with the black carbon residue of the oxyacetylene flame before oxygen is applied. When the carbon residue, the black carbon, burns off, the aluminum is ready to be welded.

Preheating is an important step in good aluminum welding. Since aluminum dissipates heat approximately twice as fast as iron-based metals, preheating is a necessary step. Preheating makes it easier to maintain a good flowing puddle in aluminum when uniform heat is applied to the base metal, and it reduces overall costs. Another advantage in preheating is that a smaller difference between the temperatures of the base metal and the weld area eliminates to a great extent distortion and buckling in the metal, thus also eliminating internal stresses. Since the parts to be welded come up to a welding temperature faster when the area is preheated, basic initial heating costs are lowered.

There are several ways to preheat. The metal might be put into a preheating furnace and brought up to 600 to 800°F; or, it might be preheated with a torch, such as the oxyacetylene torch. Remember that the metal should be kept between 600 and 800°F during the full oxyacetylene welding process. After the welding is completed, the entire piece is allowed to cool at a uniform rate, which unlocks many of the stresses inherent in the base metal because of temperature differences.

The primary problem in welding aluminum is that of the refractory oxide skins, which prevent the base metal and the filler metal from fusing properly. This oxide skin is removed by the use of fluxes consisting of sodium or potassium chloride. Alkali fluorides are added to increase the rate of oxide removal, and lithium chloride is added to lower the melting point. Fluxes work mainly by penetrating the oxide in metal and separating the oxide film from the base metal. Oxides are partly dissolved into usable slag and partly dispersed in the oxide layer. Fluxes for aluminum are available primarily in powdered form which is mixed with water to form a thick paste. The paste should be kept in earthenware, glass, or aluminum containers to avoid contamination of the mixture. Welding rods should be uniformly fluxed by dipping the rod or by painting the flux on the rod. The joints also should be coated with flux to protect the base metal from contamination and to aid in the removal of oxide from the base metal.

In the welding of aluminum, the torch flame is adjusted to a slightly reducing neutral flame, which insures a sound weld with the greatest speed and economy, especially for the beginning welder. The oxidizing flame would cause the formation of an excess of aluminum oxide and would result in a defective weld. A neutral flame could be used, but the fluctuation of the pressure within the torch body or the regulators could cause the neutral flame to become oxidizing. For these reasons, a $1\frac{1}{2} \times$ reducing flame is generally used for the oxyacetylene welding of aluminum.

The welding technique considered best for aluminum is the forehand welding technique with the torch held at approximately a 30° angle. No movement other than the forward movement of the torch is required on thin aluminum. On thicker aluminum, the torch is given a uniform lateral movement to distribute the weld metal over the width of the weld joint. A slight back-and-forth motion along the direction of the bead assists in oxide removal. The filler rod should be dipped periodically in the weld

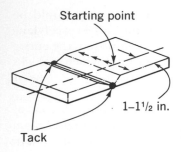

Starting point

1–1¹/₂ in.

Tack

Fig. 6–3 Welding procedure for plate aluminum.

puddle and drawn from the puddle with a forward motion. Oxyacetylene welding of aluminum does not differ to any great extent from the welding of ferrous material except for the use of the flux. After some practice, the inexperienced operator can manipulate the molten aluminum weld puddle satisfactorily.

After the weld is completed, all residual flux must be removed. The flux, having absorbed moisture, now becomes corrosive. If it is left on the weld, the corrosive action will attack the weld zone. Flux may be removed by washing the aluminum with hot water and scrubbing it with a stiff brush or by dipping the aluminum in an acid solution, such as nitrate hydrofluoric acid, sulfuric acid, nitric acid, nitric dichromate acid, phosphoric acid, or chromium trioxide. The first three are used mainly for general-purpose flux removal. Nitric dichromate acid and phosphoric acid are especially good for cleaning such thin metal parts as fins and tubes on heat exchangers. All six types of flux removers work well, but nitric dichromate and phosphoric acid are not used on any piece of aluminum that will come into contact with food. The addition of dichromate to the cleaning acid is effective as a means for preparing surfaces to inhibit corrosion. It is used primarily in such applications as aircraft fuel tanks. After the acid dips, the acid is generally washed off by water.

Porosity in the weld is another main problem in the welding of aluminum. Porosity in fusion welding is caused mainly by the rejection of hydrogen in the weld zone, and it can be controlled in varying degrees. The joint design and welding position affect the degree of porosity, which appears most frequently or most severely at the start of weld runs or weld beads. One way to combat porosity at the beginning and end of weld zones is by starting the weld approximately 1¹/₂ in. from the edge of the base metal. After that area is welded, the weld is restarted by remelting the beginning of the weld and proceeding in the other direction (Figure 6–3).

COPPER BASE   Copper is one of the oldest metals known to man. Records indicate that copper was successfully worked as early as approximately 4500 B.C. Today, copper is one of the most-used commercially available metals. It is used in wiring, sheets, rods, tubes, and castings. When it is mixed with silver, it is the most electrically conductive metal known. Silver is rated 100 on the electrical conductivity scale; copper has a rating of 93. Copper melts at 1981°F. Its coefficient of thermal expansion is 9.2, approximately 1½ times that of steel. This high coefficient of expansion means that copper expands more during heating than steel. As a result, during the cooling process, the metal locks in stresses and tends to crack if precautions are not taken before, during, and after the welding process. To overcome this great expansion of copper, the welder can use three methods to control the locked-in stresses, or the chilling effects of the base metal. The base metal can be insulated with material such as asbestos, or it can be pre-heated. Another method is to use multiple blowpipes or oxyacetylene welding torches to weld the base metal. Of these three methods used to cope with the problem of expansion, the most used is preheating the metal with multiple torches.

In terms of weldability, copper could be practically classified into two main groups: electrolytic copper and deoxidized copper. Electrolytic copper is 99 percent pure but contains 0.01 to 0.08 percent oxygen. This oxygen in the electrolytic copper is scattered homogeneously throughout the base material and makes a good weld almost impossible. When the base metal is heated to a melting point, oxygen is dispersed around the grain boundaries and develops into cuprous oxide, Cu_2O. This cuprous oxide coats each grain, making it more difficult for the heat to penetrate, acting as a chilling effect. This oxide also breaks down the bonding between the grains and causes a reduction in the strength of the fusion-welded zone. This reduction in strength may be as high as 60 percent.

The oxygen in the atmosphere does not add much to the formation of cuprous oxide. The oxygen entrapped in the base metal does the damage. Therefore, electrolytic copper should not be welded when high strength of the weld zone is required.

Oxidized copper has had the oxygen extracted from it. It represents no problem for fusion welding of the base metal. This oxygen-free copper is commercially available as rods, tubes, sheets, and plates. Oxide copper is recommended for all fusion welding requirements and can support 100 percent strength fusion weld when used with proper welding techniques.

The joint designs used for copper and copper alloys are the same as those used in the welding of steel or ferrous-based materials. The spacing between the various joints is identical to that used in the welding of steel. The one major difference between the different joints in the welding of steel and copper is that a backup is used many times with copper because of the fluidity of the molten copper. Asbestos makes an ideal backup material for the welding of copper. When using a backup material, remember to groove the backup material directly underneath the joint to allow 100 percent penetration of the weld zone. Failure to do this will many times result in a weak joint.

Copper and the many copper alloys can be successfully fusion welded, braze welded, and brazed. The copper alloys provide a means for the brazing of many other materials besides the copper-based metals. Chart 6-3 identifies the major copper-based filler materials that are used in the welding of both ferrous and nonferrous metals.

Welding procedure   The welding technique used for fusion welding or braze welding copper is the same as that used for the welding of steel; the only difference is the high fluidity of the molten copper. The tip size for the oxyacetylene torch should be one or two sizes larger than the ones used for the same thickness of steel. A neutral flame should

CHART 6-3 Oxyacetylene copper-based filler material

MATERIAL	USE
Electrolytic copper	Commercial wiring and furnace brazing
Deoxidized copper	Castings, tubing, sheets, welding rods for copper
High-zinc alloys	Braze welding of cast iron, copper, etc. (welding rods)
Phosphor-tin alloys	Wear-resistant surface applications
Copper-silicon alloys	Base-metal rods
Nickel alloys	Base-metal rods
Aluminum alloys	Base-metal rods, wear-resistant surfaces
Beryllium alloys	Base-metal rods, wear-resistant surfaces

be used for the fusion welding and a slightly oxidizing flame for the braze welding of copper and copper alloys. A neutral flame is used for the fusion welding of copper alloys because of the transition of the excess oxygen into a cuprous oxide which deteriorates the grain boundaries in the base metal. The slightly oxidizing flame used in the braze welding of copper and the copper alloys does not affect the base metal because borax-based fluxes are used in the braze welding procedure, as mentioned earlier in this section. Whenever possible, either in braze welding or in fusion welding of copper and its alloys, the weld should be accomplished with one pass. If the material is too thick to allow for this, subsequent passes should be started approximately 4 in. from one edge of the metal. The metal should be welded in one direction until the weld is completed; then the pass is restarted at this 4-in. beginning point and welded in the other direction to the other edge of the metal. In general, braze welding needs no aftertreatment of the weld zone; but after fusion welding either electrolytic copper or deoxidized pure copper, it is preferable to peen or hammer the weld zone in order to reduce grain size, to break up the cuprous oxide grain boundaries, and to reduce locked-in stresses. This peening or hammering may be done with a ball-peen hammer, and it should be done when the weld zone is cool. Peening refines the grain and as a result raises the tensile strength of the copper, thereby also decreasing its ductility. Therefore, many times the copper must be annealed to restore it to its original condition. Copper is annealed by heating it until it is red or black, then quenching it immediately in water. If the copper is thicker than approximately 14 gage, the peening should be accomplished while the metal is hot. When peening, remember that the material on the other side of the area to be peened should always be supported. The area to be peened should extend approximately ½ in. on both sides of the weld. This extension will take in the area on each side of the weld bead suffering possible hydrogen embrittlement.

NICKEL   Nickel is an important commercial metal that compares with stainless steel in its corrosion-resistant properties and that, when compared to ordinary steels, can exceed the mechanical properties of steel. Nickel is available commercially both in wrought shapes and in casting ingots. Wrought nickel is 99.5 percent pure with the remainder being composed of copper, silicon, carbon, iron, or manganese. Cast nickel is composed of approximately 96 percent nickel, with alloying elements the same as those in the wrought metal. While nickel and its alloys can be successfully joined by either braze welding, fusion welding, or brazing, braze welding and brazing are not frequently done. The brazing filler material will not exhibit corrosion-resistant properties. Nickel and its alloys are highly susceptible to impurities on the metal surface, which lead to embrittlement, porosity, or center-bead cracking. Common contaminants are grease, oil, paint, ink, temperature-indicating crayons, or other types of material found around the work area. These undesirable elements must be cleaned from the surface of the metal before it is welded. An easy way to clean the surface of nickel and its alloys is with a wire brush and a spray of hot water since water is the most effective means of removing chemicals from the surface of the metal.

Joint design   The joint designs for nickel and its alloys may be the same designs as those used in the joining of steel. However, when designing joints for nickel plates, the V joints have been found to work better if they have a 75° angle as a minimum (Figure 6–4).

Welding procedure Equipment used for oxyacetylene welding of nickel differs from that used for other nonferrous or ferrous materials in that generator acetylene should not be used as a fuel gas, because it contains approximately 0.5 percent sulfur, which causes embrittlement to nickel-based material. Acetylene gas that has been commercially generated and stored in cylinders has been scrubbed free of everything other than pure acetylene gas and is stored in acetone. The oxyacetylene flame adjustment for the welding of nickel and all nickel-based alloys is the slightly excess feather acetylene because this adjustment prevents any equipment malfunction that would result in an oxidizing flame, which would burn out the oxidizing elements in the base metals and cause an embrit-

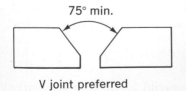

75° min.

V joint preferred

Fig. 6–4   Nickel joint design.

tlement in the weld zone. This embrittlement, when it occurs, is characterized by a center-line splitting of the weld bead. To further decrease the possibility of an oxidizing flame, a soft flame rather than a hard one should be used for welding nickel. Because of this soft flame, many welders use the next tip size larger and allow just enough gas for this soft flame to flow through the torch tip. The low-pressure oxygen in acetylene necessitates the one-size-larger tip for an adequate amount of heat to be expelled by the flame.

The weld manipulation procedure should be approximately the same as that used for mild steel except that puddling of the metal should be avoided whenever possible. Also, the welder should add filler material to the puddle and never allow it to drop from the filler rod to the base metal. The blowpipe should be held closer to the base metal than in welding steel. The inner cone of the welding tip should slightly touch or just touch the surface of the nickel, and the welding filler material should always be held within the protective secondary combustion envelope, which will prevent the filler material from oxidizing before its insertion into the weld puddle. Fusion welding of pure nickel requires no flux; however, alloys of nickel, such as inconel and monel, require a flux to further cleanse the base metal and to break up the oxides that are formed as a result of the alloying agents.

MAGNESIUM Seawater is the principal source of magnesium, a light, hard metal not found in nature. Magnesium is extracted from seawater in the form of magnesium chloride. This chloride is subjected to electrolysis, and pure magnesium collects on the cathode. Magnesium does not have sufficient mechanical properties to be used in the pure state; however, when it is alloyed with zinc, silicon, aluminum, manganese, or sometimes tin, it exhibits qualities comparable to aluminum while weighing only approximately 75 percent as much. Since magnesium is subject to hot-shortness just below its melting point, it is restricted to low-temperature usage. Magnesium has a relatively high coefficient of expansion: 15.0, compared with 6.5 for iron and 13.1 for aluminum. Therefore, magnesium weldments

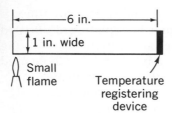

Fig. 6-5 Thermal conductivity check.

must be designed to avoid high stresses or distortion, and the weld should be accomplished in a fixture whenever possible (Figure 6-5).

Joint design The joint designs for magnesium follow closely those used for aluminum (Figure 6-1). The major difference is that the entrapped slag or flux must be free for removal, so that, as in the welding of aluminum, lap joints and fillet joints should not be used. The major point to remember in the joint design for magnesium is that there should not be any areas for entrapped slag or areas that would be impossible to clean of flux.

Welding procedure Before welding magnesium, any dirt, grease, or oil in the immediate weld zone should be removed. Magnesium has a tenacious oxide film that must also be removed either by wire brushing or filing. Failure to remove this oxide would result in a porous weld. After the weld area has been mechanically cleaned, flux must be applied to all edges to be welded and to the welding rod. Preheating of the base metal is to be avoided because of the high coefficient of thermal expansion. Preheating will cause excess buckling and distortion in the base metal. Distortion can be controlled by frequent tacking. It is common practice to tack weld magnesium parts together every ½ to 3 in. Many times, after the parts have been tack welded together, they are realigned with a wooden mallet while the tack welds are still hot. The neutral flame is recommended for the welding of magnesium; however, a slightly excess neutral flame is used many times to control the possibility of any oxidizing flames that would inject oxygen into the base metal. The welding technique for magnesium alloys is generally forehand with the torch held approximately at a 30° angle to the base metal for this material and up to a 45° angle for thicker materials.

The tip of the inner cone should always be just touching the surface of the magnesium alloy which will help the secondary combustion envelope to protect the base metal and to keep out oxidants. Whenever possible, the weld should be completely accomplished in one pass. The flux used for the welding of magnesium is highly corrosive; therefore, the flux must be completely removed upon finishing the weld bead. Failure to remove it will cause the flux to attack the weld very quickly and destroy the inherent strength of the magnesium alloy. Flux can be easily removed by scrubbing with a wire brush and hot water. When it is impossible to use a wire brush and hot water, these areas must be soaked in boiling water that contains about 5 percent sodium dichromate for approximately 2 hr.

QUESTIONS

1. What are the common nonferrous metals?

2. What are the major alloys of heat-treatable aluminum?

3. Why is the lap joint not recommended for aluminum?

4. Describe the ways to tell the welding temperature of aluminum.

5. Why is flux used when welding aluminum?

6. What are the two main types of copper?

7. Why is deoxidized copper preferred for welding?

8. How is copper annealed?

9. What major factor leads to the embrittlement of nickel fusion welds?

10. Why is a soft flame used when welding nickel?

11. What are the restrictions in welding magnesium?

12. Why should the tip of the inner cone always touch the magnesium being welded?

PROBLEMS AND DEMONSTRATIONS

1. Demonstrate the welding temperature for aluminum by completing the following:

   (a) Rubbing a pine stick on the base metal until it chars

   (b) Applying a carpenter's blue chalk to the base metal and watching it turn white

   (c) Continuously striking the aluminum being heated until it thuds instead of rings

   (d) Measuring with commercially available temperature-measuring devices, like Tempil sticks

   (e) Tapping the heated area with a welding rod until mushiness is felt

(f)   Coating the base metal with the black carbon residue of the acetylene flame and watching it disappear when the correct welding temperature is reached

2. Test the thermal conductivity of several nonferrous metals by attaching one end of a thermocouple to an identically sized piece of copper, aluminum, or other nonferrous metal, and one end of the metal strip to a low-temperature heat source.

3. Purposely contaminate a coupon of nickel and perform a fusion weld on it. Compare the resultant weld with one that has been welded on a coupon free from impurities.

4. Demonstrate how magnesium will ignite when the secondary combustion zone of the oxyacetylene flame does not protect the base metal.

SUGGESTED FURTHER READINGS

Lancaster, J. F.: "The Metallurgy of Welding, Brazing, and Soldering," George Allen and Unwin, Ltd., London, 1959.

Oberg, Erick, and F. D. Jones: "Machinery's Handbook," The Industrial Press, New York, 1964.

Welding Alcoa Aluminum, Aluminum Company of America, Pittsburgh, 1954.

Welding Kaiser Aluminum, Kaiser Aluminum and Chemical Sales, Inc., Oakland, 1967.

7 OXYGEN–FUEL CUTTING

Thomas Fletcher is credited with the discovery in 1887 of using oxygen for the cutting of ferrous metals. Fletcher heated ordinary wrought iron until it turned to a bright cherry red, which happens in the temperature range between 1600 and 1800°F. He then directed a stream of oxygen toward the heated portion causing a rapid oxidation process to take place. This rapid oxidation is now known as oxyacetylene cutting. Around the turn of the century, the first combination preheat tip and pure-oxygen jet-type apparatus was designed. Although it was handier than any of the previous tools, machine-guided equipment was needed to improve the accuracy and speed of oxyacetylene cutting.

An interesting sidelight in the history of oxyacetylene cutting is that the potential of some of the first cutting torches released was recognized by safecrackers; as a consequence, there was much talk at the time of legislation that would restrict the ownership of cutting equipment. This restriction, however, was never imposed.

Before the introduction of the cutting torch, the steel and iron industries were seriously hindered by the limited thicknesses of ferrous material that they could cut. Before the oxyacetylene cutting torch, the only way to separate plate steel was by shearing it, a method limited to cutting low-carbon steels up to a maximum of $1\frac{1}{2}$ in. thick and even thinner for alloy steels. Also, the mobility of large steel plates created a problem. The oxygen cutting torch eliminated all these problems for the steel industries. The cutting torch was the lightest and most mobile shearing apparatus that had been conceived up to that date. Also, the thickness limit for oxygen cutting is approximately 48 in., which means that very heavy, thick materials can be cut with the process. In the early days of oxyacetylene cutting, some opposition was voiced, based on the possible deterioration of the metal adjacent to the cut. Extensive research proved that this deterioration did not occur; in fact, research showed that in the proximity of the cut there was little or no impairment of the base metal. However, certain metallurgical phenomena related to the heating and cooling of the cut metal that were not completely understood at that time have become more comprehensible through developments in the heat-treating processes.

CHEMISTRY OF OXYGEN CUTTING All metals oxidize when exposed to the oxygen in the atmosphere. In this metallic oxidation, particles of base metal and the oxygen combine into a sometimes protective coating for the metal. In most ferrous metals, the oxide is a very loose, or porous, coating. Since this looseness allows more of the base metal to be exposed to the atmosphere, more and more of the oxide is formed—the process continuing in ferrous material until eventually the whole structure turns into ferrous oxide. An increase in the amount of pure oxygen directed against the metal increases the rapidity of the reaction. Heat also increases the reaction time, especially in ferrous metals. Many of the metals that have a close oxidation structure, or a tight oxide structure on the outside, such as nickel steel, are generally considered corrosion-resistant.

When heated to approximately 1600°F or above, ferrous material becomes a cherry-red color. At this temperature, it oxidizes very rapidly when it is brought into contact with a high-pressure stream of pure oxygen. This intense reaction has an extrathermic effect but does not provide enough heat to sustain the reaction. Therefore, preheating is necessary in order to sustain the cutting. The principle of the cutting process includes the reaction of the metal to the extreme heat which causes oxides to form and a small amount of metal to liquefy. Both the oxide and the small amount of metal are blown away, exposing more metal to the action of the oxygen which in turn releases a small amount of heat, liquefies a small amount of metal, and also oxidizes a small amount of ferrous metal, which causes the reaction to maintain itself, provided oxygen and a preheat source are present. This process, then, is called cutting of iron or steel. The chemical reaction at approximately 1600°F is

$$3Fe + 2O_2 = Fe_3O_4 + 26{,}691 \text{ cal (calories)}$$

These residual calories, however, are not enough to keep the intense oxidation reaction continuing because there is a chilling effect from the surrounding metal that has not been heated. The heat is also dissipated in the atmosphere, and the cool high-pressure oxygen strikes the metal creating an additional cooling effect. As a result of the oxidation of the ferrous material, ferrous residue is present on the surface of the metal being cut. In this kerf area, the thickness of the reaction layer is approximately 0.001 in. (Figure 7–1). The residue material contains Fe, FeO, Fe_2O_3, Fe_3O_4, or other derivatives of iron and oxygen. Oxygen cutting, with oxyacetylene as a fuel source, requires approximately 70 to 100 ft³ of pure oxygen an hour plus the acetylene for the neutral flame. This fuel will generate an inner cone temperature of 5480°F preheat flame, approximately 5300°F 2 in. below the end of the tip, and approximately 4800°F 4 in. below the tip, sufficient heat for cutting for most fields. The formula of three molecules of iron to combine with two molecules of oxygen

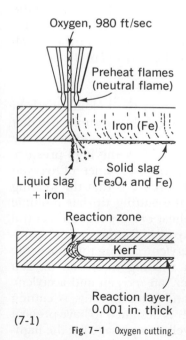

Oxygen, 980 ft/sec

Preheat flames (neutral flame)

Iron (Fe)

Solid slag (Fe₃O₄ and Fe)

Liquid slag + iron

Reaction zone

Kerf

Reaction layer, 0.001 in. thick

(7-1)

Fig. 7–1 Oxygen cutting.

to form one molecule of iron oxide plus a heat residue, yields the result that approximately 4.6 ft³ of oxygen is needed to react with iron in order to dispose of 1 lb of iron cut from the kerf. However, when the equipment is in proper order, the oxygen is pure, and the cutting operation is done correctly, less oxygen is required. In fact, 30 to 40 percent of the iron will not be oxidized but will be washed away by the jet flow of the oxygen (Figure 7–1).

Under average conditions of oxygen cutting, the depth of hardening caused by heating and cooling is not too great. These average conditions are not met when steel with a carbon content of more than 0.30 percent is used. A chemical reaction that takes place in oxygen-fuel cutting requires that approximately 2900 Btu (British thermal units) be dissipated per pound of steel oxidized. Thus, the total heat of the cutting edge is roughly this amount plus the heat of the preheating tip. Some of this heat will radiate into the atmosphere and some will go into the slag stream. But the amount of heat energy left is still enough to raise the temperature of the steel to above the critical temperature. This increase above the critical temperature will mean a microstructure change upon the cooling of the metal. This structural change will depend upon the severeness of the quench, the amount of carbon in the metal, and what temperature is attained as a result of the heating.

EQUIPMENT The cutting torch is different from the welding torch in two ways. The cutting torch has a high-pressure valve that is activated by a lever, trigger, or button. In addition to the oxygen and acetylene adjustment valves, this high-pressure oxygen valve, when pressed, releases a jet of oxygen through the tip. The other difference is that the welding tip has one orifice and that the cutting tip has multiple orifices. Cutting tips have preheat orifices and one central orifice that directs the flow of oxygen for the cutting operation (Figure 7–2). The oxyacetylene cutting torch uses the same basic components as a welding torch: an oxygen cylinder, an acetylene cylinder, an oxygen and acetylene regulator, adequate hosing, and a torch body. A cutting head attached to the torch body is designed to have preheat orifices in the cutting tip as well as a passage for the high-pressure, pure oxygen (Figure 7–3). The cutting ap-

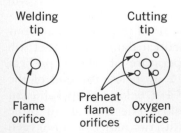

Fig. 7–2 Oxyacetylene torch tips.

paratus may be in a torch designed specifically for cutting, or the proper cutting apparatus may be attached to a standard torch body (Figure 7–4).

Cutting torches may be of the premixed or head-mixed type. In the premixed type, the oxygen and fuel are mixed in the body of the torch. In the head-mixed type, the most widely used type, the oxygen and fuel are mixed just before entering the tip. Torches are further classified as medium-pressure torches, in which acetylene is used in the range of 1 to 15 psi, and low-pressure, in which the acetylene is 1 psi or less. These two pressure classifications are the same as those used for welding torches.

FUEL GASES   Several fuel gases can be used with oxygen in oxygen cutting. Selection of these gases depends mostly on their cost and their availability. The common gases used in oxygen-fuel cutting are acetylene, hydrogen, propane, natural gas, and city gas.

The chemical formula for acetylene is C_2H_2, which means that there is a hydrocarbon chain of two carbons held together by a triple bond. Each carbon atom is bonded to one hydrogen atom. From this balanced equation, $2\frac{1}{2}$ molecules of oxygen combine with 1 acetylene molecule to produce combustion. The formula for natural gas or methane is CH_4. In this reaction, one carbon atom is bonded to four hydrogen atoms. Two molecules of oxygen are required to combine with the natural gas molecule, which is a decrease of one-half an oxygen molecule requirement in comparison to acetylene for complete combustion.

Propane is different because of the C_3H_8 molecular composition. In this reaction, three carbon atoms are bonded together and eight hydrogen atoms are bonded to the carbon atoms. The oxygen requirement is five molecules for the complete combustion of propane.

The flame temperature for acetylene is 5880°F; for natural gas, 5500°F; and for propane, 5650°F. Low-carbon steel ignites at a temperature between 1600 and 1800°F, which means that a cut can be started faster with acetylene. However, in natural-gas experiments conducted in a laboratory, it was found that a hand torch can start a cut in a piece of cold mild steel in as few as 3 sec when the

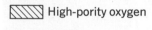 High-pority oxygen

Mixed gases

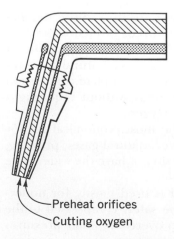

Preheat orifices
Cutting oxygen

Fig. 7–3 Cutting head.

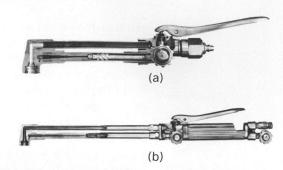

(a)

(b)

(c)

Fig. 7–4 Cutting torch types. (a) Combination cutting head. (b) Standard cutting torch. (c) Standard cutting torch in operation. (Courtesy of Victor Equipment Company.)

proper tip size and gas pressure are used. Since the majority of production cuts are hundreds of inches long, the piercing time of the gases becomes relatively insignificant in comparison to the total cutting time. Natural gas, propane, and acetylene are capable of complete combustion and can be used in confined areas without requiring a supplemental supply of air or oxygen.

Natural gas is at present the most economical gas for oxygen cutting of steel. However, natural gases, propane, and fuels other than acetylene do not have the wide range of applications that acetylene does.

Hydrogen is preferred and is used mostly for underwater cutting because it can be safely compressed to the pressures that are necessary to overcome water pressures at depths at which salvage operations are undertaken.

Hydrogen is also good for heavy cutting because of its characteristic long flame. Its disadvantage, however, is its relatively low Btu value per cubic foot. It is also difficult to adjust, and it has a high oxygen consumption for cutting or for the preheating flames.

Oxygen cutting can be accomplished manually or by machine. Manual cutting is done forehand with a traditional oxygen cutting torch. Machine cutting uses a mechanical or electrical device to control the cutting torch head. Machine cutting often uses multiple heads (Figure 7-5). Oxygen cutting by machines operates similarly to manual cutting in regard to the cutting process itself. Oxygen cutting machines are further divided into two classes, portable and stationary. Portable machines are used mostly for straight-line cutting. On a portable machine (Figure 7-5a) the carriage supports the torch and the adjustable mounting. It is usually run by an electric motor on a straight track. The speed of the motor is adjustable to the size of the metal being cut. These machines are usually used for the preparation of plate for welding; and, generally, the portable torches can be used for bevel cuts.

The stationary types of cutting machines are designed on two different mechanical principles for controlling the cutting torches. One is the pantograph design and the other uses a cross-carriage mechanism (Figures 7-5b and c). Both designs can cut almost any shape of any size. These machines use many cutting heads which can cut the same number of identical shapes at the same time, fulfilling demands for cutting on a production basis. Stationary machines are driven by motors and are designed to follow outlines, or templates, of the parts to be cut. Pantographic cutting machines can increase or decrease from the template size the part to be cut (Figure 7-5d).

In addition to cutting single thicknesses of metal, oxygen cutting machines can cut through several sheets of metal at the same time. This operation is known as stack cutting. When plates are cut in a stack, they should be clean and flat with the edges in alignment where the cut is to be started. It is also important that the plates be in good contact. In other words, there should be no insulating material between the sheets. If the plates are not in good contact, the cutting operation will not be efficient and the cuts will

Fig. 7–5 Machine cutting. (a) Portable single-cutting installation. (b) Stationary multiple-cutting installation. (c) Numerically controlled flame cutting machine. (d) Stationary pantographic multiple-cutting installation. (Courtesy of Air Reduction Company.)

(a)

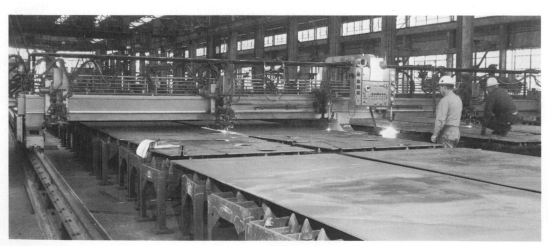

(b)

(c)

(d)

not extend completely through the stack of material. Good contact can be assured by using clamping devices, and sometimes good cuts are assured by placing a waster plate on top of the stack to be cut. This waster plate is simply one that will be discarded after the cut is made. In the past, heavy cutting, defined as the cutting of metal having thicknesses from greater than 12 in. sometimes ranging up to 4 ft or more, was successfully accomplished, but results were usually uncertain. It was considered an art to be practiced only by those who had a great deal of experience in the field of cutting and welding. Now, heavy cutting can be done by a person with little experience as long as he knows the basic fundamentals and principles of oxygen cutting. The principles and procedures for heavy cutting are not much different from those of regular thickness cutting. When cutting thicker materials, the tip should have either larger holes or larger preheat orifices. The tip may have six to eight preheat holes for adequate preheating, which is one of the secrets for successful heavy cutting. Also, greater attention must be given to those signs visibly indicating correct or faulty cutting conditions when heavy materials are being cut. One important point to remember, contrary to popular belief, is that heavy cutting does not require high gas pressures. Oxygen flow is the controlling factor in heavy cutting. It usually takes from 80 to 125 ft³/hr per inch of thickness.

CUTTING TIP SIZE Cutting tips are classified by the size of either the orifice diameter of the preheat holes or the orifice diameter of the oxygen cutting orifice, or by the number of preheat holes in the tip, or by any combination of these characteristics, depending on the manufacturer. The preheat hole size may range, by number, from 000 through 16 for a total of 20 tip sizes. The oxygen cutting orifice has approximately 18 different orifice diameters, ranging from a no. 72 diameter hole up to a $^9/_{32}$-in. diameter hole. The number of preheat holes in a cutting tip is determined by the thickness of the material to be cut and normally ranges from 4 up to 12. The cutting-tip size selected depends on whether speed or economy is the major consideration. The highest rate of cutting and the lowest gas consumption will be achieved by experienced operators on good clean metal on long cuts. Chart 7–1 gives the relationships of the aver-

CHART 7-1 Hand cutting tip size

METAL THICKNESS, IN.	TIP SIZE	EQUAL PRESSURE			INJECTOR TYPE		
		0*	A†	SPEED‡	0*	A†	SPEED‡
1/8–3/8	0	20–30	3	14–18	15–25	1	14–18
3/8–3/4	1	30–40	5	12–15	25–35	1	12–15
3/4–1	2	40–45	5	10–12	35–40	1	10–12
1 1/2–2	3	45–50	5	9–10	45–50	1	9–10

*0 = oxygen pressure required.
†A = acetylene pressure required.
‡Speed = in./min.

age tip sizes to the metal thicknesses for both low-pressure torches and injector-type torches, with the estimated speed range for experienced operators. Pressure settings for both the equal-pressure and the injector-type torches are for 25 ft of hose. Longer hoses require higher pressures.

The characteristics of a correctly made oxygen cut are drag lines that are vertical and edges that are square. Also, the drag lines should not be too pronounced and should be evenly spaced (Figure 7-6a). Preheat flames that are too small for a cut cause bad gouging at the bottom of the cut (Figure 7-6b). If the preheat flames take too long to melt the top edge of the cut, an excess of slag results (Figure 7-6c). When the oxygen pressure is too low, the top edge of the cut melts over, which rounds the top edge of the cut (Figure 7-6d). If the oxygen pressure is too high and the tip size too small, the kerf will be irregular and the cut can lose its penetration and fail to go completely through the stock (Figure 7-6e). The kerf lines indicate if the cutting speed is too slow and if a zigzag motion rather than a straight-line motion is used (Figure 7-6f). If the cutting speed is too high, a fast break in the drag lines and an irregular cut will result (Figure 7-6g). Another indication that the cutting speed is too fast is that a large amount of slag will collect on one end and the drag or lag lines will change at rapid angles (Figure 7-6g). When the travel is erratic, zigzag, or wavy, a gouging effect on one end will be created (Figure 7-6h). Bad gouging indicates that the cut was lost and had to be restarted (Figure 7-6i).

Whenever a cutting torch removes metal, it leaves an area called a kerf (Figure 7-6j). The markings on the side of the kerf are called the drag (Figure 7-6a). These lines indicate the cutting action of the oxygen and are expressed

(a)

(b)

(c)

(d)

(e)

(f)

(g)

(h)

(i)

(j)

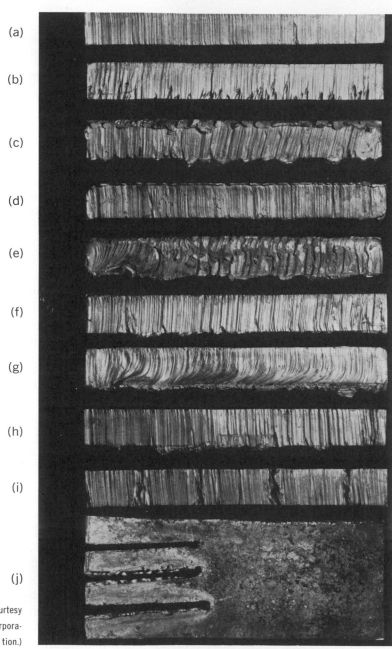

Fig. 7-6 Cutting techniques. (Courtesy Linde Division, Union Carbide Corporation.)

as a percentage of the thickness of the metal being cut. Large percentages of drag or lag are an indication of poor cutting technique (Figure 7-7).

OPERATION The oxygen cutting torch can be used with excellent results on any kind of steel or wrought iron. Some welders can get fair results on cast iron; however, this cutting process cannot be used on nonferrous metals and many of the corrosion-resistant steels. Steel having a carbon content greater than 0.25 percent or any alloy that is hardenable should always be preheated when using this process; otherwise it will crack.

During cutting, the operator should carefully watch the volume of oxygen fed to the cut and the speed of the torch motion across the metal. If too much oxygen is fed to the metal being cut, the cut will widen as the oxygen jet stream penetrates the thickness of the metal, leaving a bell mouth on the side of the metal away from the torch. If the torch is moved too rapidly across the metal, it will be difficult to preheat the metal on the farther side, which will not burn away, causing a turbulent action of the torch gases and resulting in a very rough cut. If the torch is moved too slowly across the work, the torch will not preheat the metal sufficiently as it proceeds, causing the oxidation action to cease. The oxygen jet will have to be interrupted until the preheating flames can bring the work piece back up to the proper cutting temperature, which is 1600 to 1800°F and is indicated when the metal is approximately a bright cherry-red color. The operator will find it is necessary to clean the surface dirt and oxide crust before starting the cutting operation as these will also slow the cutting speed or make the kerf of the cut rough and irregular.

The same procedures are followed to light the oxyacetylene torch as when lighting the oxyacetylene welding torch, with one exception: after the neutral flame has been achieved, the high-pressure oxygen lever must be depressed and the neutral flame achieved once again because the oxygen pressure differs slightly when the high-pressure oxygen valve is depressed and this valve will be depressed when cutting. After the steel has been marked sufficiently to indicate where the cut is to be made, the

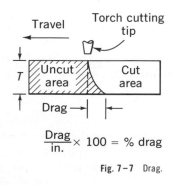

$$\frac{Drag}{in.} \times 100 = \% \text{ drag}$$

Fig. 7-7 Drag.

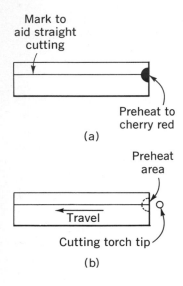

Mark to aid straight cutting

Preheat to cherry red

(a)

Preheat area

Travel

Cutting torch tip

(b)

Fig. 7-8 Cutting steel. (a) Preheat. (b) Cut.

torch should be moved so that the preheat flame heats about ¹⁄₁₆ in. of the base metal at the edge of the metal where the cut will be started. This area should be preheated until the zone is a cherry-red color (Figure 7-8). Then, very quickly, the torch tip should be moved so that the high-pressure oxygen will not cut, the high-pressure oxygen lever is then depressed and the cut is quickly begun. By using this method, the edges of the kerf will not be rounded but will be square to the edges cut. When the cut is started, the torch should be moved in the direction of the cut. Moving too fast loses the cut; moving too slowly will cause the slag to fuse the steel back together. The thickness of the metal being cut determines the cutting angle of the torch tip (Figure 7-9). The thinner the material being cut, the lower the cutting tip is held. When cutting extremely thin material, the cutting tip is held at an angle between 15 and 20°. When cutting material ³⁄₈ in. or more thick, the torch is held perpendicular to the metal being cut. The thickness of the material also determines the speed of the cut. The thicker the material, the more slowly the torch should progress across the metal.

PIERCING HOLES

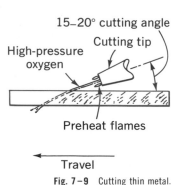

15–20° cutting angle

Cutting tip

High-pressure oxygen

Preheat flames

Travel

Fig. 7-9 Cutting thin metal.

With the oxyacetylene torch, piercing holes in steel is a more difficult operation than straight cutting. A hole is pierced by preheating the metal to be pierced with the torch in a perpendicular position (Figure 7-10). After the material has been preheated to a cherry red, the tip is tilted slightly, which allows the sparks in the jet stream to bounce away from the torch tip. Failure to do this will probably result in the clogging of the tip with the slap residue rebounding from the area being pierced. After the jet stream completely penetrates the metal, the torch is moved back to the vertical position and the cut is finished. If the metal will not pierce all the way through, not enough oxygen pressure is being used. After the piece has been pierced, however, the oxygen pressure can be reset if need be.

CUTTING CAST IRON

Very successful cutting has been performed on cast iron in salvage shops and foundries. The most important thing to do when cutting cast iron is to preheat the whole casting

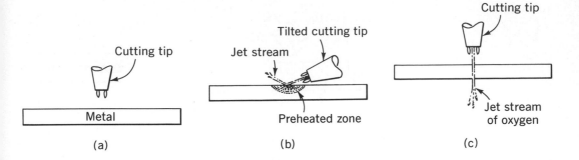

Fig. 7-10 Piercing holes. (a) Preheat position. (b) Piercing position. (c) Cutting position.

before the cut is started. The preheating flame of the torch should be adjusted to a carburizing flame to prevent any oxidation from forming on the surface before the cutting starts. Once this cutting is started, the kerf for cast iron should always be wider than a like cut in steel because of the oxidation difficulties. Then, after the cutting has been completed, it is equally important to allow the cast iron to cool slowly. The cast iron is protected while cutting by setting the preheating flames of the oxyacetylene cutting torch to an excess of acetylene, which prevents much of the oxidation. After the cast iron is preheated, the torch should be held at an angle of approximately 40 to 50° when starting the cut. After the cut has been started, the torch is slowly rotated until it is at an angle of approximately 60 to 75° (Figure 7-11b). During the complete cutting process of cast iron, the cutting stream of the torch, or the torch itself, must be oscillated constantly to help remove the oxides formed when cast iron is cut. As experience is gained, the width of these oscillations can be decreased until the oscillations are quite small. Wide oscillations are the mark of a beginner in cutting cast iron.

OXYGEN LANCE The oxygen lance is a method of cutting that is used extensively on thick metal. This apparatus allows metal up to 5 to 6 ft thick to be cut. The apparatus consists of an oxygen cylinder, a rubber hose, an oxygen-pressure regulator, and a length of $\frac{1}{8}$- or $\frac{1}{4}$-in. pipe. When the lance is used for large-sized cuttings, several oxygen cylinders are manifolded to provide sufficient oxygen. The principle of the operation of this lance is identical to that of the cutting torch, with some modifications. First, the preheating flame

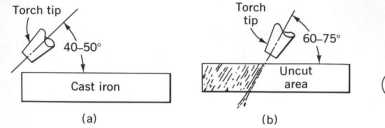

(a)

(b)

(c)

Fig. 7–11 Cutting cast iron. (a) Starting the cut. (b) Maintaining the cut. (c) Oscillation of torch cutting tip.

is not included in the lance but is separate. Oxygen-acetylene or other gas torches are used. The lance is itself just a means of delivering a high-pressure jet stream of pure oxygen to one zone or spot. Pipe that directs this flow of oxygen is also subjected to the excessive temperatures in the presence of the oxygen at the end of the lance and is burned away along with the metal being removed. The procedure for using the lance is to preheat the metal with a standard oxyacetylene torch. After the metal has been preheated, the lance is then turned on and the jet stream of oxygen is directed to the point at which the cut is to be made. Preheating becomes more difficult as the depth of the cut increases. A cutting lance has been used to pierce the center of a freight car axle shaft throughout its entire length without burning through the side of the axle—however, success in this kind of operation requires a great deal of skill.

POWDER CUTTING Powder cutting is a process that was created for the corrosion-resistant metals and steels, such as nickel steels and stainless steels. This process has been used also for the cutting of cast iron. The basic working principle of powder cutting is the injection of an iron powder into the oxygen stream before its ejection from the cutting tip. This iron powder is preheated as it passes through the oxyacetylene preheat flame, and it bursts into flame in the stream of pure oxygen. The heat created by the burning of the powder makes it possible to start cutting without preheating. The unburned iron powder in the jet stream acts to reinforce the cutting action. The burning of the iron powder also acts as a flux, and this fluxing action enables

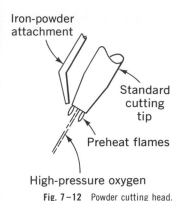

Iron-powder attachment

Standard cutting tip

Preheat flames

High-pressure oxygen

Fig. 7–12 Powder cutting head.

the cutting oxygen stream to oxidize the base metal continuously without the help of outside oxidizing agents. The apparatus for powder cutting generally includes a supply of oxygen-acetylene, a cutting blowpipe, and a means of supplying the powder into the oxygen stream. There are two types of blowpipes. One is a converted standard blowpipe (Figure 7–12) and the other is a special blowpipe that has been designed especially for powder cutting. Powder cutting can be done either manually or by machine. However, it is a process that has currently been replaced by other means.

QUESTIONS

1. What is rapid oxidation?

2. What is the maximum thickness that an oxyacetylene torch will cut?

3. Explain the term KERF.

4. Explain the differences between an oxyacetylene welding tip and a cutting tip.

5. What are the two major types of torch bodies?

6. What are the fuel gases used in cutting? Their advantages?

7. Why doesn't compressed air work as well as oxygen for the oxygen lance?

8. What are the relationships between tip sizes and metal thicknesses?

9. Why is the cutting head of the torch tipped at an angle when cutting thin ferrous stock?

10. Why is cutting cast iron more difficult than cutting mild steel?

11. What is powder cutting?

12. What is the function of the oxygen lance?

PROBLEMS AND DEMONSTRATIONS

1. Demonstrate the oxidation principle by filling a metal canlike container with pure oxygen and then placing a preheated, red-hot wire in the pure oxygen atmosphere.

2. Vary the cutting-head angle as a cut is made on a piece of ferrous metal. Notice the

different results of the various angles. Look at such things as kerf width, results of varying speeds of travel, and slag removal.

3. Pierce a hole in ferrous material by tilting the torch head for the initial cut. Compare this method of hole piercing with the method of simply increasing the distance from the tip to the work.

SUGGESTED FURTHER READINGS

Althouse, Andrew D., et al.: "Modern Welding," Goodheart-Wilcox Co., Homewood, Ill., 1970.

Morris, J. L.: "Welding Processes and Procedures," Prentice-Hall, Inc., Englewood Cliffs, N. J., 1958.

Rossi, Boniface E.: "Welding Engineering," McGraw-Hill Book Company, New York, 1954.

Worthington, J. C.: Natural Gas-Oxygen Cutting, Theory and Application, *Rep. Am. Welding Soc. Natl. Fall Meet.*, Detroit, Mich., September and October 1959.

Ⅱ SHIELD ARC WELDING

8 INTRODUCTION

Arc welding had its beginnings with the electric dynamo, which was invented in 1877; and since then, the field of arc welding has grown to one of great importance. Electric arc welding was first used to connect the various parts of storage battery plates, a process established in 1881 by Auguste de Meritens, a French inventor. In 1886, resistance welding was first used. It was developed by Elihu H. Thompson and was called butt welding because it consisted of placing the metal pieces to be welded into a large clamp which brought the edges of the metal together while at the same time a heavy current was passed through the two metals. At the joint of the metals, the resistance of the metal to the electric current produced high heat which fused the pieces together.

The use of arc welding depended naturally upon the

development of electricity, so that the improvements in dynamos or generators in the 1880s permitted the advancement of arc welding. The first actual arc welding, defined as the melting and fusing of metal by the means of electrodes, was developed by N. V. Bernado, who created a mechanism using carbon electrodes. This method produced an arc between the carbon electrode and the metal edges to be welded, thus heating the metal to a fusible state. The development of better carbon electrodes, making them more consistent electrically, improved carbon arc welding tremendously; however, newer welding processes have replaced this method almost completely.

A metallic electrode was tried by N. G. Slavianoff in 1895, but he experienced little success with this innovation because he used bare metal electrodes. With the invention of the flux-coated electrode in 1905, the success of the metallic electrode was assured. Since 1905, there have been few new developments in shield arc welding. The major development has been the production of portable machines and automatic welding machines.

Present-day shield arc welding is accomplished by producing an electric arc between the work to be welded and the tip of the electrode. This type of welding has the advantages of less heat loss and less oxidation than that of the oxyacetylene flame. For these reasons and because of rapid development in the equipment which makes shield arc welding suitable for a variety of purposes, the practice of making welds electrically is increasing rapidly. Most shield arc welding is done with metallic electrodes.

ARC COLUMN THEORY   The arc column is generated between an anode, which is the positive pole of a dc (direct current) power supply, and the cathode, the negative pole. The electrical theory of the arc column is that the ions pass from the positive pole to the negative pole because metal ions are positively charged so they are attracted to the negative pole.

The cathode and anode are touched together, and an air gap is established after they are drawn apart. Ions pass through this air gap because of their attraction to the negative cathode. They are seeking a negative charge by which to balance the atomic structure disrupted by electricity. As they travel through this air gap, the ions collide with gas

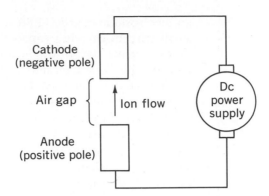

Fig. 8–1 Ion theory of the arc column.

molecules in the atmosphere which produces a thermal ionization layer. This ionized gas column acts as a high-resistance conductor that enables more ions to flow from the anode to the cathode. Heat is generated as the ions strike the cathode (Figure 8–1).

However, this ion theory does not completely explain the arc column. Perhaps the electron theory of the arc column explains what happens more fully (Figure 8–2). Basically, electrons have a small mass, 9.1×10^{-28} gram. Electrons are easily disassociated from the metal at the negative pole, or the cathode. This small amount of negatively charged mass is accelerated away from the cathode to the positive pole, or the anode, striking it at a highly accelerated velocity. This path of the negatively charged mass is generally in the interior of the arc column which is the hotter portion of the arc column. The electrons carry

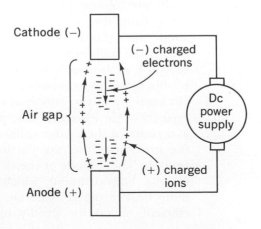

Fig. 8–2 Electron theory of the arc column.

an electrostatic charge, 4.80×10^{-10}. This electrostatic or small current-carrying capacity is multiplied thousands of times, causing part of the heat of the arc column. Another part of the heat liberated by the arc column is directly related to kinetic energy. The formula for kinetic energy is one-half the mass times the square of the velocity ($\frac{1}{2}MV^2$). In this case, M equals the mass of each electron. The electrons accelerate to a high velocity. Their masses are extremely small but there still is a very high amount of energy stored within them. As they strike the anode, heat energy is released. Intermingling with the negatively charged electrons, ions that are positively charged are returning from the anode to the cathode and producing the ionized gas layer which further protects the electrons and the electrostatic unit within the electron. The electrostatic unit is induced into the anode causing in the anode an electromotive force (emf), which is directly transferred into heat energy. Approximately 1 kWh (kilowatt hour) of electricity will create 3413 Btu. One Btu is the amount of heat required to raise the temperature of 1 lb of water 1°F. The energy or kilowatt hour rating of the arc column is easy to predict because there is a direct relationship between the energy supplied to the arc column and the heat that will be liberated by the arc column. This rating is determined by computing current, in amperage, times voltage, times time in hours. Time can also be broken down into seconds. When it is time in seconds, the heat is expressed in units of joules.

Because of the characteristics of the electrons and ions in the arc column, three areas of heat are liberated in the arc stream: the cathode area, the plasma area, and the anode area (Figure 8-3). Of the three areas, the anode area is the high-heat area where approximately 10,000 to 11,000°F of heat is liberated. The liberation of the heat results from the combination of the impingement of the electrons upon the anode anvil and the current-carrying capacity of the electrons. The plasma area is heated mainly as a result of the atomic collision of the few electrons and the many ions that are passing through the ionized gas column. The cathode or the negative pole is subject mainly to ionic bombardment, which produces the state of medium heat in the arc column. The arc is one of the most efficient means for producing heat that is available to

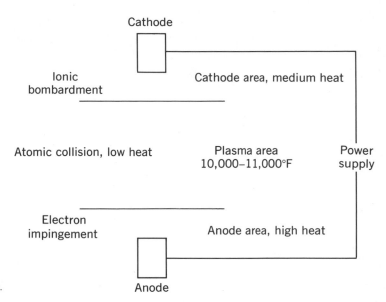

Cathode

Ionic
bombardment

Cathode area, medium heat

Atomic collision, low heat

Plasma area
10,000–11,000°F

Power
supply

Electron
impingement

Anode area, high heat

Fig. 8–3   Heat liberation.

Anode

modern technology. Approximately 50 percent of the electrical energy put into the arc system comes out in the form of heat energy. The major point to remember in the selection of the heat zone is that approximately two-thirds of the energy released in the arc column system is always at the anode or the positive pole. This is true in all dc systems. Another type of arc power source used in shield arc welding is alternating current (ac). When an ac power supply is used, the heat in the arc column is generally equalized between the anode and the cathode areas, so that the area of medium heat is then in the plasma area (Figure 8–3).

Reviewing the nature of atomic structure might help in understanding the electron theory of the arc column. All atoms are believed to have three components: neutrons, protons, and electrons (Figure 8–4). The neutron has a neutral charge, the proton a positive charge, and the electron a negative charge. Protons and neutrons compose the nucleus of an atom, which is always positively charged. This nucleus floats in space and is surrounded by electrons. The electrons orbit around the nucleus in paths called shells. The negative charge of the electrons balances the positive charge of the nucleus. Protons have a larger electromagnetic charge than do electrons; consequently, in

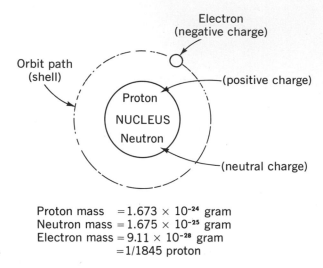

Proton mass $= 1.673 \times 10^{-24}$ gram
Neutron mass $= 1.675 \times 10^{-25}$ gram
Electron mass $= 9.11 \times 10^{-28}$ gram
                $= 1/1845$ proton

Fig. 8–4 The neutral atom.

most cases, more than one electron is required to balance the electrical charge of the atom.

In welding, electrons are forced from the end of an electrode onto a base metal by an electrical charge. Theoretically, the electrical charge disrupts the flow of the electron in its orbit causing it to fly off at a tangent from its orbital path to seek another positive nucleus to which it will reassociate itself automatically. Metals characteristically release electrons readily, especially when the temperature of metal is increased. This characteristic is called thermionic emission. When an atom loses its characteristic electrical balance by the addition or subtraction of electrons from their orbits, it is considered an ion. An atom in an ionized state will seek to rebalance itself to the neutral state by associating with an electron. These reactions result in the flow of electrons and ions in the arc column during welding.

The welding circuit consists of a power source; two cables, the electrode cable and the ground cable; the ground clamp; the electrode holder; the stinger; and the electrodes or rods (Figure 8–5). The two basic types of power supplies for arc welding are the dc generator rectifier and the ac transformer. Each of these two power supplies has distinct advantages. In dc welding, the electron flow is in one direction; in ac welding, the electron flow is in both directions. In dc welding, the direction can

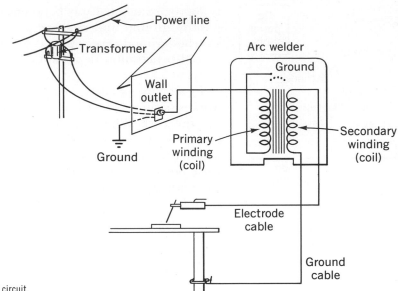

Fig. 8–5   Typical ac welding circuit.

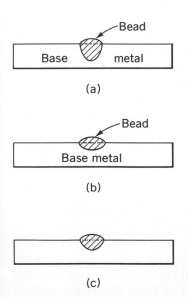

Fig. 8–6   Penetration. (a) Dc (positive ground). (b) Dc (negative ground). (c) Ac.

be changed by simply reversing the cables at the terminals located on the generator. The different settings on the terminals indicate that the electron flow will be either from the electrode to the work, which is the positive ground, or from the work to the electrode, which is the negative ground.

Two-thirds of the heat is developed near the positive pole while the remaining one-third is developed near the negative pole. As a result, an electrode that is connected to the positive pole will burn away approximately 50 percent faster than one that is connected to the negative pole. This is helpful in obtaining the desired penetration of the base metal (Figure 8–6). If the positive ground is used, the penetration will be greater because of the amount of heat energy supplied by the electrode force to the work. At the same time, the electrode will burn away slowly. If the poles are reversed and there is a negative ground, two-thirds of the heat will remain in the tip of the electrode. For this reason, the penetration of the heat zone in the base metal will be shallow when compared to the penetration depth of the positive ground arc column. Alternating current combines the characteristics of both the deep penetration and the shallow penetration, yielding a penetration depth that

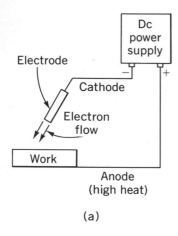

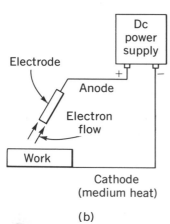

Fig. 8−7 Dc polarity. (a) Straight polarity. (b) Reverse polarity.

is approximately halfway between that achieved by the dc positive ground and negative ground. The electron flow switches grounds every time the ac cycle changes, yielding a bead penetration depth approximately between the two dc types.

In straight polarity, the electrode is negative and the work is positive. The electron flow goes from the electrode into the work. When the electrode is positive and the work is negative, the electron flow is from the work to the anode, which is called reverse polarity (Figure 8−7). When reverse polarity is used, the work remains cooler than when straight polarity is used.

In both the ac and dc power sources, the arc serves the same purpose. It produces heat to melt metal. If two pieces of metal that are to be joined are placed so that they touch or almost touch one another and the arc from the electrode is directed at this junction, the heat generated by the arc causes a small section of the edges of both pieces of metal to melt. These molten portions along with the molten portions of the electrode flow together. As the arc column is moved, the molten puddle solidifies, joining the two pieces of metal with a combination of electrode metal and base metal.

In the past, most electrodes were bare wire. With these bare electrodes, the arc was difficult to control and cooled too quickly; as a result, the weld bead absorbed from the atmosphere oxygen and nitrogen, which turned into oxides and nitrides, producing a brittle, weak weld. Most modern electrodes are coated. The function of the coating is to form a gaseous shield around the weld to protect the molten metal from these contaminants in the atmosphere. Oxygen and nitrogen are the main gases which must be kept away from the weld zone. If these gases are kept out of the weld zone, the chances are high that the deposited metal will contain properties that are comparable to those of the base metal. The coatings on the electrodes burn as the electrode wire melts from the intense heat of the arc. As the electrode wire melts, the electrode covering, or the flux, provides a gaseous shield around the arc, preventing contamination (Figure 8−8). The force of the arc column striking the work piece digs in the base metal a crater, which fills with molten metal. As the flux melts, part of it mixes with the impurities in the molten pool causing them

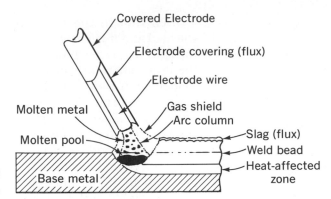

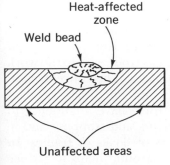 **Fig. 8-8** Covered electrode metal deposit.

to float to the top of the weld. When this mixture of impurities and flux cools, it forms slag, which also does its part to improve the weld. It protects the bead from the atmosphere and causes the bead to cool more uniformly. The slag also helps to design the contour of the weld bead itself by acting as an insulator. By insulating the heat-affected zone, located in the parent metal or the base metal and completely surrounding the weld bead, the slag allows an even rather than erratic heat loss from this heat-affected zone, thus helping to control the crystal or grain size of the metal. The arc column reaches temperatures generally of 5000 to 7000°F, which have a harsh effect on the parent metal. The molten pool, which must maintain a temperature of approximately 2800°F, radiates heat outward and changes the crystals surrounding the weld bead. Many times, after welding, the part must be heat treated to change the size of the grains in the weld bead and the surrounding area (Figure 8-9). The heating of the base metal by the arc stream and the resultant molten weld puddle or crater generally extend deep into the base metal. The extent of the heat-effected zone can be observed by studying the crystalline structure of the base metal in this zone. It is generally represented by a large grain. The grains in the unaffected areas of the metal are smaller. Because of the protection of the flux, the weld bead itself has medium-sized grains that extend to large grains at deeper penetration. It is not necessary to heat treat mild steel; but, with many metals, this heating will result in locked-in stresses that must be relieved either through peening or further heat treatment of the entire piece of metal.

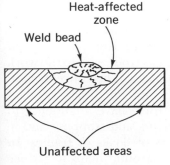

Fig. 8-9 Heat-affected zone.

Even though the major portion of the arc column is shielded and flux covers the solidifying portion of the weld, the light that is given off from the arc column is harmful to the eyes. Therefore head shields are used to protect the welder from radiation. Lenses for eye protection are extremely important because large quantities of infrared and ultraviolet radiation are given off by the arc. This radiation can burn the retina and even cause third-degree burns on the skin. All good welding lenses will stop 99.5 percent of the infrared and ultraviolet rays. Although the visible light given off by the arc does not contain harmful rays, it is uncomfortable to the eye and will cause blurred vision.

MAGNETIC ARC BLOW

Magnetic arc blow is experienced generally when direct current is employed for welding; however, this peculiarity is sometimes experienced when using alternating current for the power supply. When using alternating current, there is only approximately 1 percent chance that magnetic arc blow will be encountered. When current flows through a conductor, it produces a magnetic flux that circles around the conductor in perpendicular planes. The centers of the flux circles are located at the center of the conductor (Figure 8–10). The magnetic flux is produced in the steel and across the arc gap. The arc column is mainly influenced by the lines of force crossing the gap. As the weld joins the pieces together, there is less and less chance that the magnetic field will concentrate in the arc gap. As the weld is filling the gap of the joint, it pushes the magnetic flux ahead of the arc. As long as the flux can travel, no serious arc blow will interrupt the weld. When the flux ceases to move, however, it piles up, and a magnetic field of considerable strength develops. The buildup of the flux causes a deflection of the arc column as it pulls away from this heavy concentration of magnetic force. Ionized gases that carry the arc from the end of the electrode wire to the work piece are acting as a flexible conductor. This concentration of flux that pulls the arc from its intended path is called arc blow. Areas where lines of force have a tendency to concentrate are at points of starting and stopping and in such places as the inside corners of boxes or frames. In some instances, relocation of the ground helps to reduce the strength of the magnetic fields. The effect of the

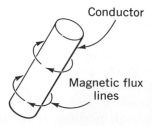

Conductor

Magnetic flux lines

Fig. 8–10   Magnetic flux motion.

ground placement is much more apparent in narrow material than in wide material. The wider the base metal, the less important ground placement is. The types of electrode used also affect the magnetic field. Electrodes that have iron-powder coatings or other heavily coated electrodes will produce a large slag formation that is troublesome when arc blow occurs. The slag formation runs under the arc when the arc blow occurs, causing incomplete fusion and excessive weld splatter.

Also when the slag runs under the arc, continuity of the weld bead is broken. The arc column is not self-starting, so if it is extinguished it must be reestablished. There are two essential factors involved in proper continuity. The arc column must be maintained, and the electrode must be fed into the weld puddle at a continuous rate in order to maintain the proper spacing between the electrode and the work. As the arc column becomes longer, more arc voltage is required to maintain the arc. The type of coating on the electrode also changes the amount of emf or arc voltage that is required to maintain an arc column within a range of optimum fusion. If there is an insufficient amount of arc voltage or an oversupply of arc voltage, there will be insufficient fusion of the weld bead. Oxidation and porosity can be kept to a minimum also by keeping the arc column within an optimum fusion range, which helps to control the problem of arc blow.

The base metal, which is also a conductor, has a flux field around it as current passes through it. These lines of force are perpendicular to the current passing through the work. Magnetic lines of force circle around the electrode, around the arc column, and around the work piece. The "right-hand rule" used for finding the direction of the flux is that when the thumb of the right hand points in the direction of the current flow, the forefinger points in the direction of the flux lines. There are three areas of magnetic field travel. The first field is created by the current passing through the electrode. The second field is created in the base metal by the ground. The third is created by the electrode arc column that comes in contact with the base metal. The current is passed through the arc column into the metal. Of these three types of flux fields or magnetic fields, the second type created in the base metal by the ground is desirable because this field causes a slight

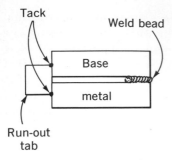

Tack

Weld bead

Base

metal

Run-out
tab

Fig. 8–11   Run-out tabs. Note: Extend
the weld bead well onto the run-out tab.

forward pull in the arc column. The first type and the last type should be controlled.

Magnetic fields created by the flux can never be removed but they can be controlled by various methods. One method is to set up a magnetic field of sufficient strength to neutralize the force caused by the flux. Other methods of controlling arc blow include welding away from the earth ground connection; changing the position of the earth connection on the work; wrapping the welding electrode cable a few turns around the work; using run-out tabs (Figure 8–11); reducing the welding current or the electrode size; welding toward a heavy tack or portion of the weld already completed; reducing the rate of travel of the electrode; shortening the arc column length; or changing the power supply to alternating current.

QUESTIONS

1.   What is the electron theory of the arc column?

2.   What is the ion theory of the arc column?

3.   What are the components of an atom?

4.   What are the major heat areas in the arc column?

5.   What effects do electrons have upon the penetration of the arc into the base metal?

6.   What is straight polarity? Reverse polarity?

7.   What are the effects of magnetic arc blow?

8.   What are run-out tabs?

9.   What is ionized gas?

10.   What is the function of the flux coating on shield arc electrodes?

11.   How does the atmosphere support the arc column?

PROBLEMS AND DEMONSTRATIONS

1.   Select a piece of ferrous metal about $4 \times 6 \times 1/4$ in. and weld a bead with a $1/8$-in. electrode using dc straight polarity, dc reverse polarity, and alternating current. Compare each bead as to estimated penetration, bead size, bead height, and ease of welding.

2. Select two pieces of ferrous metal and tack them together with no space at one end of the joint and a ³⁄₈-in. space at the other end. Use a dc power supply and start the arc at different joint openings. Notice how arc blow hinders the arc column.

SUGGESTED FURTHER READINGS

Austin, John B.: *Electric Arc Welding,* American Technical Society, Chicago, 1952.

Kuger, Harold: *Arc Welding Lessons for School and Shop,* The James F. Lincoln Arc Welding Foundation, Cleveland, Ohio, 1957.

Linnert, George E.: *Welding Metallurgy,* vol. I, American Welding Society, New York, 1967.

Patton, W. J.: "The Science and Practice of Welding," Prentice-Hall, Inc., Englewood Cliffs, N. J., 1967.

Sachs, R. J.: "Theory and Practice of Arc Welding," Van Nostrand Reinhold Company, New York, 1943.

Stiere, Emanuele: "Basic Welding Principles," Prentice-Hall, Inc., Englewood Cliffs, N.J., 1952.

⑨ ELECTRODES

Arc welding is the predominant welding process used in the United States, and shield arc welding is used more than all the other different types of arc welding processes. Almost all the shield arc welding done is with covered electrodes because they are easily manipulated and leave contamination-free weld deposits. For a long time, bare steel rod was used for electrodes; but, in 1912, experimentation with coated electrodes was begun. These coated electrodes were first made commercially available in 1929. Development of coated electrodes fostered extensive use of arc welding because they made welding easier and increased the strength of the welds. However, it was 1937 before the American Welding Society (AWS) and the American Society for Testing Materials (ASTM) began to establish a manufacturer's code for the standardization of welding electrode wires and coatings. The National Elec-

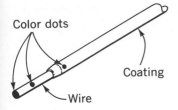

Color dots

Coating

Wire

Fig. 9-1 NEMA coding of electrodes.

trical Manufacturer's Association (NEMA) also attempted to standardize electrodes by a color coding. Dots of color were applied to the electrode coating to indicate various electrode characteristics (Figure 9-1). This color coding, however, has been superseded by the AWS classification. There are now two major classifications of coatings on the electrodes: lightly dusted electrodes and semicoated electrodes—the classifications based on the thickness of the flux covering on the electrodes. The coatings on electrodes can be made of many different materials and can serve many purposes. Several materials commonly used are titanium dioxide, ferromanganese, silica flour, asbestos clay, calcium carbonate, and cellulose, with sodium silicate often used to hold ingredients together.

Electrode coatings do much to increase the quality of the weld. Part of the coating burns in the intense heat of the arc, providing a gaseous shield around the arc that prevents oxygen, nitrogen, and other impurities in the air from combining with the molten metal to cause a poor-quality weld. Another portion of the coating mixes with the impurities in the metal and floats them to the top of the weld where they cool in the form of slag, which also does its part to improve the weld. It protects the weld, or bead as it is called, from the air as it cools and causes the bead to cool more uniformly. This slag also helps to control the basic shape of the weld bead. The type of electrode used depends upon the type of metal to be welded, the position in which the weld is to be done, whether the power supply is ac or dc, and what dc polarity the welding machine is using.

In the process of metal arc welding, the primary problems are controlled by three variables. The first of these variables is the speed of travel in producing the bead. The other two variables deal with the amperage and voltage characteristics. In manual welding, the welder controls the speed of travel and the arc voltage. The metal to be welded is melted by the intense heat of the arc at the same time the electrode tip is melted and is transferred across the arc to the molten pool on the base metal (Figure 9-2). The current for manual operation usually ranges from 15 to 500 amps (amperes). The voltage ranges from 14 to 24 volts with bare or lightly covered electrodes and from 20 to 40 volts with the semicoated electrodes.

Flux coating   Weld
Arc   bead
Wire   column
Gas
Metal   shield
transfer   Slag
area

Molten pool   Base metal

Heat-affected zone

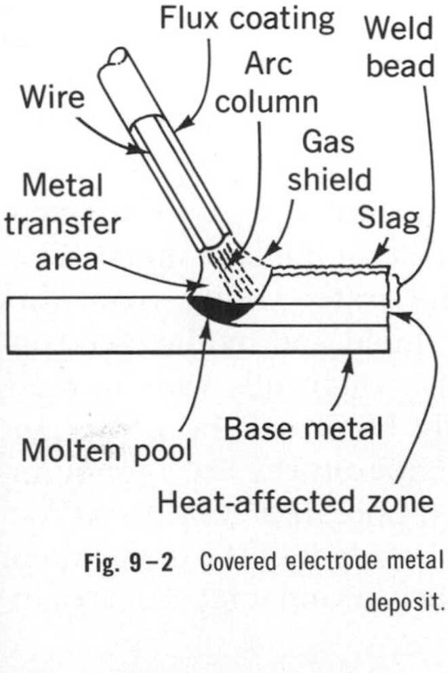

Fig. 9-2 Covered electrode metal deposit.

MECHANISM OF TRANSFER

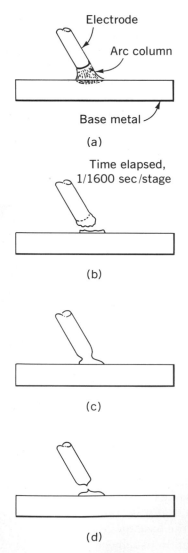

Electrode

Arc column

Base metal

(a)

Time elapsed,
1/1600 sec/stage

(b)

(c)

(d)

Fig. 9-3 Transfer of electrode metal. (a) Heat stage. (b) Deform stage. (c) Contact stage. (d) Pinch-off stage.

The transfer of metal from the tip of the electrode to the base metal is actually a method of short circuiting. The electrode makes contact with the base metal every $1/400$ sec. This creates a problem in the welding machine where the short circuit plays havoc with the amperage control within the machine. This short-circuiting effect or mechanism of metal transfer results, basically, from various distinct forces. One is the electromagnetic force, which is responsible for the pinch-off effect. Another is the emf (electromotive force). Another force, resulting from the ion-electron travel, directs the metal from the electrode to the base metal. The force of gravity, especially in the flat welding position, the force resulting from the rapid expansion of gases, and the force of capillary attraction resulting from the lessening of the surface tension of the base metal also aid in this transfer of metal.

The short circuiting of the electrode has four basic stages: the heat stage, the deform stage, the contact stage, and the pinch-off stage (Figure 9-3). After the arc column has been created between the electrode and the base metal, the electrode tip and the target area heat, causing the metal to become plastic. The forces listed above aid in the deformation of the heated tip of the electrode and the heated spot in the base metal until the electrode metal comes into contact with the base metal. The arc column is extinguished at this time; however, current flow is maintained. The current flow creates a pile-up of electromagnetic lines of force which pinch off the electrode. The arc column is restarted in order to begin another short-circuit cycle. This four-stage cycle occurs in as little time as $1/400$ sec.

The short-circuit transfer mechanism, coupled with the magnetic arc pull, causes the weld metal to flow in the direction of the ground (Figure 9-4). An arc crater is formed by the force of the arc column striking the base metal. The digging action that produces the crater results from the expanding gases of the fluxing shield and by the electron stream striking the metal. This crater fills with molten metal. The electrons continually bombard the center of the molten pool. Therefore, temperatures are higher in the center of the crater or molten pool than they are at the edges. The edges solidify first. The edges of the crater then control the width of the weld. The varying temperatures in

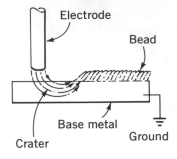

Fig. 9-4 Electrode metal flow.

different parts of the crater determine the amount of waves or ripples on the weld surface. The amount of metal that accumulates in the molten pool can be controlled by the amount of electrical energy supplied to the arc column. The melting rate of metal in the arc column is directly related to the amount of electrical energy supplied to the arc column. The melting rate of the electrode, or the burn-off rate, is directly affected by the amount of current applied. The arc voltage has little effect on the melting rate of the electrode; therefore, the current setting is of prime importance in the selection of the burn-off rate for an electrode.

Metal transfer is also accomplished through the spray arc. The short-circuiting method is based on low arc voltage, which ranges from 14 to 26 volts. The spray arc method of metal transfer is also based on arc voltage, but of a higher range, usually from 20 to 40 arc volts. The increase of potential voltage means an increase in the emf, which explodes particles of metal off the electrode tip. The higher the emf, the finer the particle size; in fact, the particles leaving the electrode tip in the form of a spray have much the same appearance as the paint ejected from a paint spray gun. The electrons leaving the surface of the electrode have a higher velocity than those leaving the electrode in the short arc method of metal transfer. This high velocity is thought to create small air pockets under the surface of the tip of the electrode, making it easier for the particles to fly off from the surface of the electrode.

ELECTRODE CLASSIFICATION Electrodes are classified and given an identifying number by the American Welding Society. This number gives the strength of the weld, the weld position in which the electrode will give the best results, and the current requirements of the electrode.

There are four basic positions in which arc welding can be done: flat, horizontal, vertical, and overhead. Of the four, the flat position is the easiest, most economical, and generally results in the strongest weld joints. There are times, however, when it is not practical to weld in a flat position. For example, a building cannot be turned on its side to accommodate a welder. Other positions must be used. If the horizontal welding position is required, a

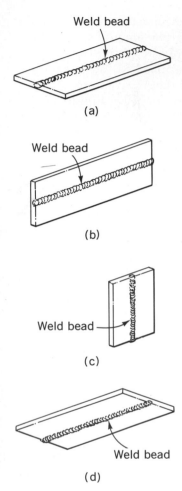

Weld bead

(a)

Weld bead

(b)

Weld bead

(c)

Weld bead

(d)

Fig. 9-5 Welding positions. (a) Flat
(F). (b) Horizontal (H). (c) Vertical (V).
(d) Overhead (O).

shorter arc column should be used, which is true for the flat position too. A shorter arc column helps prevent the molten puddle of metal from sagging. If the welding is to be done in the vertical position, the welder can choose whether to deposit the bead in uphill or downhill direction. Welding downhill is preferred for a thin metal because the penetration characteristics are not so great and it is much faster than uphill welding. Uphill welding is generally thought to result in a stronger weld. If the metal to be welded is ¼ in. or more thick, the welding should be done in the uphill direction. Welding in the overhead position necessitates caution to assure that the welder is not burned by drops of falling metal. Although this is the most hazardous welding position, it is not the most difficult. Generally, welders think of the vertical uphill position as being the most difficult (Figure 9-5).

The chemicals used for the coatings on electrode are varied and depend on three conditions that the welder has to control. The electrode works as an electrical, a physical, and a metallurgical control in the welding operation.

Electrically, the resistance of an arc is dependent upon the state of the ionized gases in the arc column. This arc is unstable because of its negative electron characteristics. As resistance crosses the arc column, the arc decreases while the current or amperage increases. These rapid current changes can be avoided by adding ionization agents to the flux coating. These agents act as resistance elements and stabilize the arc. Chemicals commonly used for flux coatings are sodium and potassium.

Physically, the electrode coating evolves gases that actually hold the molten pool in position. Thus the viscosity of the molten metal is controlled, permitting welding in positions other than in the flat position. For example, a low-viscosity slag must be used in order to keep the weld pool from dripping through the slag when welding overhead.

Metallurgically, the coating produces a slag that protects the molten pool from the atmosphere and yields reducing agents and alloys that are transmitted into the molten pool, improving the mechanical properties of the weld bead. For example, manganese, which will oxidize, is added to the molten pool in preference to iron. During the heat reaction, the manganese becomes manganese oxide;

and, in the form of a slag, it will float to the surface. The parent metal will remain oxide free. Other ingredients in the flux coating that cause a fluxing or cleansing action on the metal are the ferromanganese agents. Fluxing agents clean the metal by disposing of the oxides, by restricting further oxidation or by reducing the present oxides.

Because the welder is able to control the electrical, physical, and metallurgical conditions existing in nature by the use of the coated electrodes, the possibilities of shield arc welding are practically without limit. Correct knowledge of the electrode and its uses and functions are extremely important if the welder is to maintain effective welding practices.

The AWS electrode classification system is the most efficient that has yet been devised (Chart 9-1). An electrode numbered E6010 reflects by the letter E that it is an electrode. It has a tensile strength of 60,000 psi, indicated by the first two digits of the four-digit numbering system. In a five-digit numbering system, the first three digits times 1000 psi give the tensile strength of the electrode. The third digit indicates that the electrode can be used in the flat, horizontal, vertical, and overhead welding positions. If the number 2 were used, the electrode could be used only in the flat and the horizontal positions. If the digit 3 were used, the electrode could then be used only in the flat position. The last digit in the AWS classification system identifies the type of flux used to coat the wire electrode. The type of flux action determines to a great extent the penetration characteristics as well as the arc digging action, which is the result of electrons impinging upon the base metal. The last digit also determines the power supply that should be used with the electrode. In the case of the example E6010, the last digit is 0; because the next to last digit is 1, the power supply would be dc$^+$, or DCRP (direct current reverse polarity). The arc action would be strong and would create a deep crater, yielding in turn a deep penetration of the metal. The penetration is indicated by the type of flux which is represented by the last two digits also. Chart 9-1 applies to electrodes for all low-alloy ferrous metals. The E6010 electrode is often referred to as universal or as an all-purpose electrode; it is used a major portion of the time because of its adaptability and the strength of its weld bead.

CHART 9-1 AWS electrode classification system (Exxxxx or Exxxx)*

LAST DIGIT	LAST-DIGIT CHARACTERISTICS†			
	POWER SUPPLY	ARC ACTION	TYPE OF FLUX	PENETRATION CHARACTERISTICS
0	10, dc+; 20, ac; or dc	Digging	10, organic; 20, mineral	10, deep; 20, medium
1	ac or dc+	Digging	Organic	Deep
2	ac or dc	Medium	Ductile	Medium
3	ac or dc	Soft	Ductile	Light
4	ac or dc	Soft	Ductile	Light
5	dc+	Medium	Low hydrogen	Medium
6	ac or dc+	Medium	Low hydrogen	Medium
7	ac or dc	Soft	Mineral	Medium
8	ac or dc+	Medium	Low hydrogen	Medium

*E = electrode. First set of digits (xxx or xx) = tensile strength × 1000 psi. Next-to-last digit: 1 = all positions; 2 = flat and horizontal positions; and 3 = flat position only.
†T. B. Jefferson, "The Welding Encyclopedia," p. E-4, Monticello Books, Morton Grove, Ill., 1968.

The E6011 electrode duplicates many of the operating characteristics of the E6010 electrode, but it uses alternating current as a power supply. Because of the advantages of alternating current, the weld deposit of the E6011 electrode usually yields a weld bead with slightly higher ductility, higher tensile strength, and a higher weld-bead strength.

The E6012 electrodes are often referred to as the poor fit-up electrodes because they provide smooth welds over broad root gaps. They are characterized as yielding a smooth even ripple on the weld face and as producing a heavy slag with a medium penetration pattern and a very quiet arc. The E6012 electrode is usually suitable for high-speed, single-pass, horizontal, fillet welds.

The E6013 electrode is similar to the E6012 electrode, except it has a quieter arc, a smoother ripple on the bead, and more easily removed slag. The E6013 electrodes are replacing the E6012 electrodes because lighter metal can be welded more easily with them.

Other commonly used low-alloy-steel electrodes are the E6014, the E6024, and the E6027. These electrodes have a large amount of iron powder in the flux coat covering.

Much of the iron powder within the flux becomes part of the weld upon solidification of the weld zone. This feature provides welding speed approximately twice that of conventional electrodes. Slag removal when using iron powder rods is exceptionally easy. The welder simply taps the bead and the slag falls off. Iron-powder rods are known as contact rods. For welding, the rod or electrode is simply drawn across the surface and the flux covering allowed to touch the base metal (Figure 9–6). Drag or contact welding is generally accomplished with electrodes that have thick flux coatings, like the semicoated electrodes. If the electrode is dust coated, the standoff method is used to achieve correct arc length. Electrodes of the low-alloy series are available in E45xx, E60xx, E70xx, E80xx, E90xx, E110xx, and E120xx series. The E120xx series produces a bead that has the tensile strength of 120,000 psi.

CORROSION-RESISTANT ELECTRODES

The numbering system used for corrosion-resistant welding electrodes is similar to that used for mild-steel electrodes. The prefix E at the beginning of the numbers indicates that the electrode is a gun and arc welding electrode; for example,

E308–16

where   E = electrode
       308 = 18 percent chromium and 8 percent nickel
         1 = all position
         6 = ac or dc

The first three digits indicate the chemical composition, which is determined by chemical analysis of the base metal. The number 308 in the example refers to a stainless steel that is 18 percent chromium and 8 percent nickel. The 1 indicates that this electrode may be used in all positions, and the 6 indicates that the power supply may be either alternating current or DCRP.

Whenever chromium is subjected to high temperatures such as are found in an arc column, an affinity for carbon is created, meaning that it is attracted to carbon, which causes impurities within the weld. Therefore, carbon derivatives are not used as fluxing agents. Manganese and silicon replace the carbon, providing the necessary de-

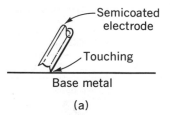

(a)

(b)

Fig. 9–6 Arc length. (a) Drag or contact. (b) Standoff.

oxidizers. Titanium oxides are used to promote arc stability and slag manipulation.

The classification of corrosion-resistant electrodes is determined by the chemical analysis of the metal to be welded and not by the filler metal. Therefore not all base metals have corresponding electrodes. The most common electrodes are presented in Chart 9-2. Generally, corrosion-resistant electrodes deposit a bead that contains more than 4 percent chromium and less than 50 percent nickel. As an example, the 330-15 series electrode has a normal composition of 35 percent nickel and 15 percent chromium. It is designed for welding base metals of a similar composition. The electrode deposit will be extremely resistant to heat and scaling, and the resultant bead will withstand temperatures above 1800°F. The remaining digits are 1 and 5, indicating that the current would be DCRP and that this electrode could be used in all positions. Chart 9-3 indicates the remaining classification digits that are used for corrosion-resistant steel arc welding electrodes.

NONFERROUS ELECTRODES Nonferrous electrodes do not have the same classifications as ferrous electrodes. Nonferrous electrodes are dependent upon the chemical composition of the base metal

CHART 9-2 Corrosion-resistant electrodes

SERIES	COMPOSITION	BASE METAL	APPLICATION
308	18% chromium, 8% nickel	Similar	Corrosion resistance
309	25% chromium, 12% nickel	Dissimilar	Corrosion resistance
310	25% chromium, 20% nickel	Similar	Surface hardening
316	18% chromium, 12% nickel, 2% molybdenium	Similar and high alloy	High temperature above 1100°F
317	18% chromium, 12% nickel, 5% molybdenium	Similar and high alloy	High temperature above 1100°F
330	15% chromium, 35% nickel	Similar	High temperature above 1800°F
347	15-17% chromium	Similar	Corrosion resistance
410	12% chromium	Similar	Abrasion
430	19% chromium, 9% nickel	Similar	Abrasion
502	4-6% chromium, 0.50% molybdenium	Similar	Abrasion

CHART 9–3   Amperage requirements for corrosion-resistant electrodes

CLASSIFICATION	CURRENT	POSITION
E330–15	DCRP	All
E330–16	Ac or DCRP	All
E330–25	DCRP	Flat and horizontal
E330–26	Ac or DCRP	Flat and horizontal

to be welded. The major nonferrous electrodes are copper, aluminum, bronze, and nickel (Chart 9–4). The composition of the copper electrode is either phosphorus bronze A, for low-strength electrolytic or oxidized copper, or phosphorus bronze C for high-strength electrolytic or oxidized copper. The most commonly used aluminum arc welding electrode contains 95 percent aluminum and 5 percent silicon and can weld successfully all alloys of aluminum, both heat treatable and not heat treatable. The power supply for the aluminum electrodes is generally dc, either DCRP or DCSP. Aluminum bronze electrodes have three classifications based on the hardness of the weld deposit. Aluminum bronze electrodes that have 8 percent aluminum, 1 percent iron, with copper making up the balance, are the most ductile of the aluminum bronze electrodes and are particularly suited for the joining of dissimilar metals or for applying hard surfaces to the base metal. Even harder are the aluminum bronze electrodes with 9 to 10 percent aluminum and 3 to 5 percent iron; these are designed for welding where the base metal is similar to the composition of the electrode. They are useful for such jobs as the welding of bearings. The aluminum bronze electrodes that have from 12 to 13 percent aluminum and from 3 to 5 percent iron with copper making up the bal-

CHART 9–4   Nonferrous electrodes

ELECTRODE	COMPOSITION	BASE METAL	POWER SUPPLY
Copper	Phosphorus bronze A	Electrolytic or deoxidized copper	Ac or dc⁻
Copper	Phosphorus bronze C	Electrolytic or deoxidized copper	Ac or dc⁺
Aluminum	95% Al, 5% Si	All alloys	Dc⁺
Aluminum bronze	–Cu, 8% Al, 1% Fe	Dissimilar	Ac or dc⁺
	–Cu, 9–10% Al, 3–5% Fe	Similar	Ac or dc⁺
	–Cu, 12–13% Al, 3–5% Fe	Dissimilar	Ac or dc⁺
Nickel	Greater than 50% nickel	Similar and dissimilar	Ac or dc⁺

ance are designed for welding metals that are dissimilar. The weld bead is highly resistant to corrosion and is extremely hard; these are desirable qualities for such welding as repairing dies.

Nickel electrodes generally contain a minimum of 50 percent nickel coupled with other alloy agents such as copper. They can be used to weld nickel and its alloys or can be used to weld an overlay on dissimilar metals. The nickel-based electrodes are capable of welding monel, inconel, nickel-chromium alloys, and hastelloy.

QUESTIONS

1. What are the differences between lightly dusted electrodes and semicoated electrodes?

2. What is the main function of sodium silicate?

3. What is NEMA?

4. How does metal transfer from the electrode to the base metal?

5. What is burn-off rate?

6. Explain the AWS electrode classification.

7. Why are there temperature differences in different areas of the welding crater?

8. How does the flux covering on an electrode determine the characteristics of the electrode?

9. What is meant by a digging arc?

10. How does the contact or drag method of welding affect the arc length of the electrode?

11. What is meant by standoff arc welding?

12. Why are corrosion-resistant electrodes used?

13. What are the major nonferrous electrodes?

PROBLEMS AND DEMONSTRATIONS

1. Select a piece of ferrous metal and weld one bead each with the E6010 electrode, the E6011 electrode, the E6012 electrode, and the E6013 electrode. Use a DCRP power supply. Compare the weld beads, the ease of welding, the slag removal, and the amount of splatter of each of the weld beads.

2. Secure a lightly dusted electrode and a semicoated electrode. Weld with each of the electrodes on a piece of ferrous scrap metal. Use both the standoff and the contact welding method. Compare the ease of welding with the semicoated electrode to welding with the lightly dusted electrode.

SUGGESTED FURTHER READINGS

Henry, O. H., and G. E. Claussen: *Welding Metallurgy, Iron and Steel,* American Welding Society, New York, 1949.

Mikulak, J.: *Welding Handbook,* American Welding Society, New York, 1964.

Nicholes, Herbert, Jr.: "Heavy Equipment Repair," North Castle Books, Greenwich, Conn., 1964.

Séférian, D.: "The Metallurgy of Welding," John Wiley & Sons, Inc., New York, 1962.

10 EQUIPMENT

POWER SUPPLIES   Power is needed to supply the current that supports the arc column for fusion welding. There are three types of welding power supplies: dc motor generators, ac transformers, and ac transformers with dc rectifiers.

The motor generator produces dc power like the generator in a modern day automobile furnishes current for ignition and lights. The welding generator is driven by an electric, gasoline, or diesel motor. The gasoline- and diesel-driven generators are useful when there is no electrical power. If internal combustion engines are used to drive welding generators, speed governors are mandatory if the equipment is to operate satisfactorily. Generators are designed to operate at speeds of 1500, 1800, or 3600 rpm (revolutions per minute) in order to give optimum current values.

The dc generator is designed to furnish a wide range of current settings, to provide nearly constant voltage and amperage characteristics, and to be able to change voltage rapidly to meet the changing requirements of the arc column. Direct current power supplies are available in portable, stationary, and multiple-station units (Figure 10-1).

The size and type of electrodes that are used and the penetration and welding speed that are desired determine the current supply requirements. Generator capacity is usually designated as 150, 200, or 400 amps. The anticipated maximum current requirements and the length of time the welder will be used govern the generator size. The duty cycle, based on the length of time the welder will be used, is given in percentages based on 10-min periods. For example, a 10 percent duty cycle would mean that the welder would be operated 10 percent of 10 min or 1 min. A 30 percent duty cycle refers to a welder that would be operated 3 min out of every 100. The maximum amperage and voltage usually accompany the percentage of the duty cycle. These figures then give the maximum operating current and voltage values that can be obtained during the duty cycle. For example, a machine may have a maximum of 40 volts and a maximum output of 250 amps for a duty cycle of 60 percent. To operate that machine at higher settings for greater lengths of time would deteriorate the insulation within the machine and cause its early failure.

The dc generator is designed so that it will compensate for any change in the arc column voltage, thus insuring a stabilized arc. When globules of metal flow across the arc, the arc is short circuited. This short circuiting causes the arc voltage to drop to 0 during the welding operation; and it can have adverse effects, because a large increase in arc current causes an excessive amount of metal splatter, and a decrease in the arc current makes the arc flutter and go out. Therefore, a dc welding machine must be able to remedy the fluctuating current resulting from the mechanism of metal transfer and maintain an even arc column.

There are three major current voltage characteristics commonly used in today's arc welding dc machines to help control these fluctuating currents. They are the drooping-arc voltage (DAV), the constant-arc voltage (CAV), and the rising-arc voltage (RAV) (Figure 10-2). The machine that

Fig. 10–1 Types of dc power supplies. (a) Portable gas engine motor generator. (Courtesy of Hobart Brothers Company.) (b) Stationary electric-motor-driven motor generator. (Courtesy of Hobart Brothers Company.) (c) Multiple-station power supply. (Courtesy of Miller Electric Manufacturing Company.) (d) A multiple-station power supply on the job site. (Courtesy of Miller Electric Manufacturing Company.)

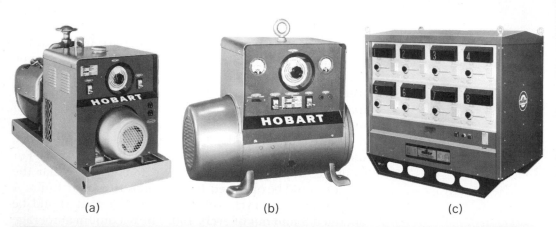

(a)   (b)   (c)

(d)

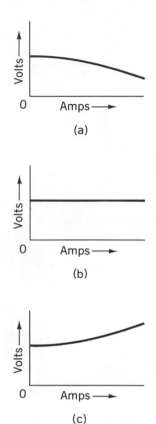

Fig. 10–2 Dc voltage/amperage characteristics. (a) Drooping-arc voltage (DAV). (b) Constant-arc voltage (CP or CAV). (c) Rising-arc voltage (RAV; modified CAV).

is designed with the DAV characteristics provides the highest potential voltage when the welding current circuit is open and no current is flowing. As the arc column is started, the voltage drops to a minimum, which allows the amperage to rise rapidly. With DAV, when the length of the arc column is increased, the voltage rises and the amperage decreases. The DAV is the type of voltage-amperage relationship preferred for standard shield arc welding that is manually done. The CAV and a modification of it called the RAV are characteristics preferred for semiautomatic or automatic welding processes, because they maintain a preset voltage regardless of the amount of current being drawn from the machine. These types of voltage-amperage characteristics are sensitive to the short-circuiting effect of the shield arc mechanism of metal transfer. Therefore, the spray arc is used rather than the short-circuit arc method of metal transfer. The spray arc is much like a spray gun spraying paint, working on a principle different from that of the short-circuit pinch-off effect of welding with a stick electrode. An advantage of the RAV over the CAV characteristic is that as the amperage requirement is increased, the voltage is automatically increased, thus helping to maintain a constant-arc gap even if short circuiting occurs. This RAV is adaptable to the fully automatic processes.

Controls for dc welding machines may be of several types, ranging from a simple rheostat in the exciter circuit to a combination of exciter regulators and a series of field taps. A third means for controlling current is a mechanical provision for shifting the generator brushes on a self-excited machine to vary the current output. On certain types of generators, leakage of the magnetic field may be imposed artificially to give the desired current output. The rheostat, however, has a maximum number of fine adjustments and is beginning to be the preferred method of amperage control (Figure 10–3).

Some available arc welders are equipped with remote-control current units that permit the machine to be located in a remote part of the job site or plant. The machine operator may adjust the voltage amperage to the desired level without leaving his work station. These units are useful when the operator must climb up, down, in, or out

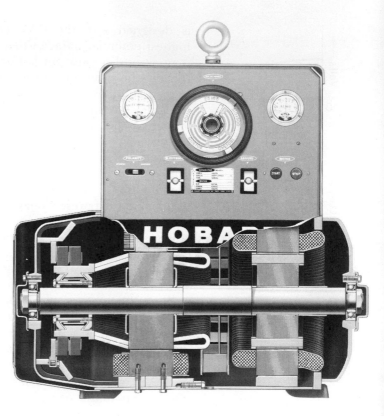

Fig. 10-3 Dc generator cutaway.
(Courtesy of Hobart Brothers Company.)

of a work station for the adjustment or readjustment of the current (Figure 10-4).

Voltage and amperage meters are sometimes available on welding machines. On some machines, they indicate the polarity in addition to the efficiency of the current values and the potential values. Some types of machines have individual meters for the voltage and the amperage. Volt meters on dc welders range from 0 to +100 to −100. The amperage meters register both positive and negative current values with the scale going above the rated capacity of the machine (Figure 10-5).

The polarity indicates the direction of the current in the dc circuit. Since some welding operations demand that the flow of current be changed, a polarity switch or indicator must be on a dc welding machine. When the electrode cable is fastened to the negative pole of the generator and the piece to be welded is the positive pole, the polarity is negative; or, more commonly, is referred to as straight

Fig. 10-4 Remote current controls. (Courtesy of Hobart Brothers Company and Miller Electric Manufacturing Company.)

polarity. On early dc welders, the change of polarity involved reversing the cables on the output-input terminals. Modern machines equipped with polarity switches eliminate the need of disconnecting the cable, for the switching is achieved internally.

Alternating current power supplies The transformer welder is usually limited to ac power; however, rotation generators that are engine driven can be designed to produce alter-

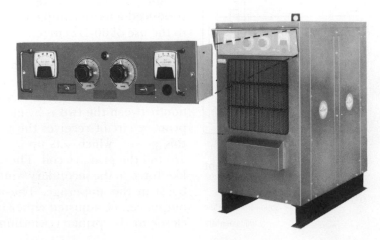

Fig. 10-5 Meters. (Courtesy of Miller Electric Manufacturing Company.)

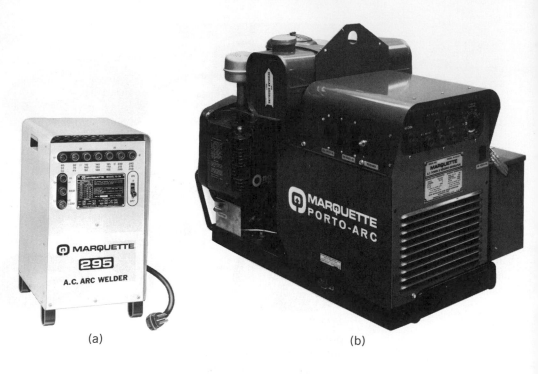

(a)

(b)

Fig. 10-6 Ac transformer welder.
(a) Stationary—ac power supply.
(b) Portable—gas-engine-driven ac
power supply. (Courtesy of the Marquette
Corporation.)

Line power, 60 Hz

Electrode

Workpiece

Secondary coil adjustment
(by taps, rheostat, or
adjusted core coil)

Fig. 10-7 Solid-core ac welder
fundamentals.

nating current (Figure 10-6). Alternating current transformers are usually used in the shop.

Early efforts with ac welding indicated a need for frequencies higher than the common 60-Hz (Hertz, cycles per second) current. A higher frequency can be obtained by several methods, but all involve further expenses and equipment. The development of electrode coatings that produced a more complete ionization in the arc stream led to the use of 60-Hz current as a satisfactory means of direct transmission of electrical energy.

The ac transformer welder has a primary and a secondary circuit (Figure 10-7). The primary circuit is independent of the secondary circuit and the only connection between the two is by electromagnetic induction. The primary circuit receives the 60-Hz line power and boosts this power, which sets up a perpendicular magnetic field around the primary coil. These lines of force then induce a like force in the secondary windings, thus causing a further boost in the amperage. The amount of boost or current output can be adjusted either by moving the secondary coil closer to the primary winding, consolidating the lines of

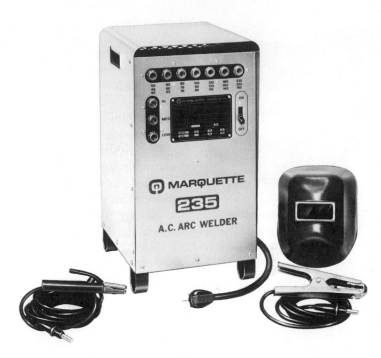

Fig. 10−8 Tapped-reactor welder.
(Courtesy of the Marquette Corporation.)

force, or by inserting a reactor coil into the circuit. If the reactor coil is inserted into this circuit, it generally fits on the electrode side of the secondary coil. If a reactor is used for adjustment, it can be a tapped reactor (Figure 10−8), a reactor with an adjustable coil, or a reactor with a rheostat that controls the amperage output.

Rectified power supplies A rectified power supply is an ac transformer that has a one-way current valve installed on the electrode side of the secondary coil. This valve allows current to flow through it in only one direction (Figure 10−9). The typical rectified arc welder has no moving parts, giving it high longevity. The only moving part in a rectified welder is a fan that cools the transformer. The fan is not a basic part of the electrical system. The rectified transformer welder is capable of supplying ac straight polarity or DCRP with equal ease, making this power supply the most adaptable for shield arc welding (Figure 10−10). The rectified transformer welder generally plugs into single-phase or three-phase line voltage as do standard ac transformer welders. Single-phase power yields a power cycle to the welder every half-cycle. Three-phase power

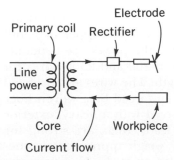

Fig. 10−9 Ac transformer rectified to dc.

yields 3 times the power output because three single power
phases are entering the power cycle at three different times
(Figure 10–11).

There are two kinds of rectifiers. One uses a silicon
diode that is more efficient and is hermetically sealed, thus
impervious to time. Silicon diodes are almost ageless in that
they maintain rectifying characteristics indefinitely. The
larger, less-efficient rectifier has selenium plates, which are
larger than the silicon diode and take up more space. Since
these plates are larger, they are damaged more easily and
can be attacked by moisture or fumes, causing them to age
rapidly. The output characteristics of the selenium rectifier change when it is but a few years old. The silicon
diode is the preferred type of rectifying unit for welding
machines.

Of the three basic types of welders (the dc welder, the
ac welder, and the ac rectified welding unit), each has its
own distinct advantages. The dc welder permits portable
operation, is better suited for welding thin metal, and
efficiently uses a larger variety of stick electrodes. The ac
welder has a lower initial cost, as well as low operation and
maintenance cost. It has only one moving part, the cooling
fan; it can use common single-phase, 60-Hz current; and
it operates quietly. The ac transformer that is rectified has
a simple design and is compact. It also has the quiet operation and high efficiency of the dc welder, coupled with the
efficiency of the ac transformer equipment.

EQUIPMENT ACCESSORIES   The equipment accessories for the arc welding machine
are the cables or leads; the electrode holder, or stinger;
appropriate lead connectors; and the ground attachments.

The leads that carry the welding current to the work are
very flexible and are generally made of copper or aluminum wire. The wires that carry the power are generally
very fine, but consist of between 800 and 2500 wires for
maximum flexibility and strength. The wires are insulated
by a rubber covering that is reinforced by a woven fiber
covering that is in turn reinforced with a heavy, exterior
rubber coating. The welding leads, then, are strong for
their weight. Aluminum cables weigh approximately one-
third as much as copper cables or leads. However, the
current-carrying capacity for aluminum cables is ap-

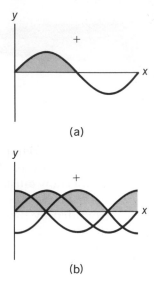

Fig. 10-11 Rectified ac power. (a) Single-phase (full wave—360° power every $\frac{1}{120}$ cycle). (b) Three-phase (full wave—360° power every $\frac{3}{120}$ or $\frac{1}{40}$ cycle).

proximately 60 percent that of copper cables. The proper aluminum cable size can be determined by simply going to the next higher step for similar copper welding lead, and the current capacity will be approximately the same (Chart 10-1). The shielding of the metal leads is sufficient because the potential voltage carried by the leads is not excessive. The voltage carried by the leads varies between 14 and 80 volts. Connectors for the electrode lead should be designed so that the current-carrying capacity of the lead will be allowed for. Welding leads are connected by mechanical connectors, soldering, welding, or brazing. The mechanical connectors are the connection leads that are probably most used because they can be more easily assembled and disassembled (Figure 10-12).

Electrode holders or stingers come in a variety of sizes. They are generally matched to the size of the lead, which is in turn matched to the amperage output of the arc welder. Electrode holders come in sizes that range from 150 to 500 amps. They are a means of securing the electrode to a mechanical holder that is easy to use. Most electrode holders have grooves cut into the jaws which enable the electrode to be held at various angles for ease in manipulation. Larger electrode holders, when high amperage is used, provide a heat shield in order to protect the operator's hand from the excessive amount of heat and the radiation that is liberated during the arc welding process (Figure 10-13).

The ground clamp that completes the circuit between the electrode and the welding machine is generally

CHART 10-1   Copper welding lead characteristics

SIZE NO.	CAPACITY	
	0-50-FT LEAD,* AMPS	100-250-FT LEAD,* AMPS
$\frac{4}{0}$	600	400
$\frac{3}{0}$	500	300
$\frac{2}{0}$	400	300
$\frac{1}{0}$	300	200
1	250	175
2	200	150
3	150	100
4	125	75

*Combined length of both the stinger and the ground lead.

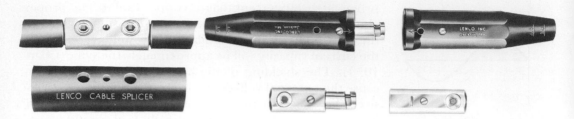

Fig. 10-12 Welding cable connectors. (Courtesy of Lenco Inc.)

fastened to the metal being welded either with a clamp, a bolt, or some other means, depending on the size of the metal (Figure 10-14). Generally, the connection must be clean and easy to install. Many times, if the arc welding is done at a welding table, the table is grounded by having the ground clamp bolted directly to the cable. Then the material to be welded is simply laid on the table to complete the circuit.

OPERATOR ACCESSORIES

Operator accessories fall into two groups: they are either weld bead cleaning accessories or safety accessories. The main implements for cleaning the weld bead are the chipping hammer and the wire brush (Figure 10-15). The chipping hammer is chisel-shaped and is pointed on one end to aid in the removal of slag. The wire brush, which removes small particles of slag, is generally made of stiff steel wire embedded in wood. Of course, power wire wheels, when available, may be used in place of wire brushes; however, many times these are not as available to the welder who is working in the field as to the welder working in a shop.

Provisions for the safety of the operator are most important to insure workable situations, especially in arc welding, since infrared and ultraviolet rays may be harmful to the skin and eyes. A face protection device will stop the radiation that is prevalent in arc welding. The helmet face shield and the hand-held face shield are commonly used in arc welding. The face shield generally covers the entire face, down to the lower throat. All that is exposed is a small 2 × 4½-in. opening that has a shaded lens to stop the infrared and ultraviolet rays. The lens is capable of stopping 99.5 percent of these rays. The color density of the lens depends upon the type of welding amperage being used. For standard stick electrode welding with the elec-

Fig. 10-13 Electrode holders. (Courtesy of Lenco Inc.)

Fig. 10-14   Welding ground clamp.

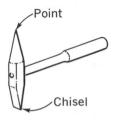

Point

Chisel

Chipping hammer

Wire brush

Fig. 10-15   Chipping hammer and
wire brush.

trode diameter size ranging from $1/16$ to $3/8$ in., the shade number for the colored lens would be between a no. 10 and a no. 14 lens. However, no. 9, no. 10, and no. 11 shade lenses are the most popular with welders. Even with the eyes protected by shaded welding lenses, ultraviolet rays from arc welding may cause eye pain if welding is performed for periods that exceed tolerance, which may be 8 to 18 hr, depending on the sensitivity of the individual.

The face shield may be hand held, or it may be a head helmet with either a solid lens or a flip front (Figure 10-16). The solid lens is stationary. The flip-front lens shading revolves upwards so that the welder can see the weld bead in order to clean it. It also allows maximum eye protection from hot slag while chipping. Long gauntlet gloves should also be used to protect the hands from the ultraviolet and infrared radiation as well as the heat that is given off by the arc column. Long gauntlet gloves also protect the hands from minor burns during the chipping operation. The gloves should be worn at all times during the welding process. Even though these gloves are excellent protection from the heat of the arc column, they should never be used to pick up hot metal or be held in the arc column because they would deteriorate rapidly as well as transmit heat into the welder's hand. The work clothes that should be worn by the welder consist mainly of long-sleeved cotton shirts and long pants with either no pant cuffs or cuffs that have been sewn shut. Shoes should be high topped. Street shoes will not prevent small globules of molten metal from dropping into the shoe. Some

Fig. 10-16   Face shields. (a) Solid
lens. (b) Flip-front.

(a)

(b)

mechanical means should be provided to remove the fumes produced by arc welding when welding indoors, such as some type of fan or air ducts to draw out the semitoxic fumes.

QUESTIONS

1. What are the major power supplies used for arc welding?

2. What is DAV? CAV? RAV?

3. What are the major ways to control current in an arc welding machine?

4. What are the advantages of a dc power supply?

5. What are the advantages of the ac power supply?

6. What are the advantages of the ac rectified power supply?

7. Why is the power output greater in three-phase machines than in single-phase machines?

8. Why must the welder be protected from the arc column?

9. What are the tools used to clean the slag from the weld bead?

10. Why are long gauntlet gloves mandatory for welder protection?

PROBLEMS AND DEMONSTRATIONS

1. Demonstrate how to adjust arc welding machines.

2. Demonstrate the proper manner to insert the electrode into the electrode holder.

3. Demonstrate chipping and brushing a weld bead.

SUGGESTED FURTHER READINGS

Giachino, Joseph W., William Weeks, and Elmer Brune: *Welding Skills and Practices,* 2d ed., American Technical Society, Chicago, 1965.

Morris, Joe L.: "Welding Principles for Engineers," Prentice-Hall, Inc., Englewood Cliffs, N. J., 1951.

————: "Welding Process and Procedures," Prentice-Hall, Inc., Englewood Cliffs, N.J., 1954.

Sacks, Raymond J.: "Theory and Practice of Arc Welding," 3d ed., Van Nostrand Reinhold Company, New York, 1943.

11 OPERATION

The arc welding circuit consists of the power supply, an electrode cable, a ground cable, a ground clamp, an electrode holder, and the electrode rod. The only two items in the welding circuit that can be adjusted are the amperage or current setting and the arc length between the tip of the electrode and the base metal. Before arc welding is begun, a few basic ideas must be understood. The arc stream is formed when the electrode is brought into contact with the work and then withdrawn slightly. The current jumps across this gap creating light and heat energy, which melts the tip of the electrode and that portion of the base metal directly under it. To do an adequate job, the welder must be able to hold the arc length at a set distance so that the base metal receives steady heat and there is an even transfer of the electrode filler metal. At first, it is quite a problem to hold this arc at a consistent length. The arc may either become too short, touching the base metal, causing a complete short circuiting of the arc, or the arc length may become so great that it will break contact. The control of this arc length can be mastered with practice.

The current requirements that are set on the machine before striking the arc depend on the diameter of the electrode. Electrode amperage settings are classified by the diameter of the work and the type of flux coating on the wire. Manufacturers of electrodes usually recommend a range of current settings that take into consideration the thickness or the mass of the metal to be welded (Chart 11-1). However, there are starting points from which a current is adjusted after the welding operation has begun. For example, if a ⅛-in. slightly dusted electrode is used, the welding machine is generally set at approximately 90 amps for a standard welding operation. Then, according to the amount of heat that is absorbed by the metal and different operator characteristics, the amperage adjustment is either increased or decreased, usually in steps of 5 amps.

STRIKING THE ARC   Two methods are used to start or strike an arc: tapping and scratching (Figure 11-1). With the tap method, the electrode is brought straight down to the base metal and withdrawn as soon as contact is made. The electrode is withdrawn to only approximately the diameter of the electrode wire. This method is the more difficult, but it is preferred by experienced welders because the beginning arc column will not deface or destroy the base metal. However, it

CHART 11-1   Current requirements for arc welding electrodes

ELECTRODE WIRE DIAMETER, IN.	DUSTED ELECTRODES		SEMICOATED ELECTRODES	
	CURRENT, AMPS	STARTING POINT*	CURRENT, AMPS	STARTING POINT*
¹/₁₆	20–40	30	—	—
⅛	70–120	90	100–150	135
⁵/₃₂	120–180	150	120–200	165
³/₁₆	140–250	180	175–250	195
¼	200–400	275	250–400	285

*Standard starting point before current adjustment for base-metal requirement.

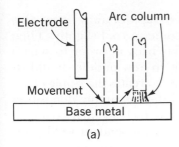

(a)

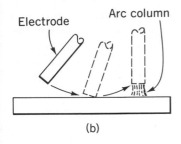

(b)

Fig. 11–1 Striking an arc. (a) Tap method. (b) Scratch method.

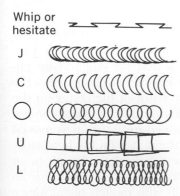

Fig. 11–2 Types of weld movements.

represents a more difficult technique, and there is greater possibility that the electrode or rod will adhere to the base metal because of the surge of current during the initial contact of the electrode. The scratch method is usually preferred by beginning students because it is an easier technique. It is comparable to striking a match. After the arc is started with the scratch method, the distance it is held from the work is the same as with the previous method, approximately the diameter of the electrode. The major hindrance in using this method is that if the electrode is not immediately retracted upon contact with the base metal, it will deposit molten metal on the base metal in a random fashion or it may adhere to the metal because of the surge of high current.

After the arc length has been established, it should be maintained at this distance. A very short arc length results in poor penetration and a drop in the current and voltage settings. An arc length that is greater than the diameter of the electrode will result in the molten metal splattering on the base metal. The correct electrode distance and amperage and current settings are indicated when the sound emitted from the arc column sounds like bacon frying in a skillet.

The manipulation of the electrode falls into two categories according to the thickness of the flux coating on the electrode. Electrodes with a light flux coating require more physical manipulation of the electrode tip than do semi-coated electrodes, because of the manner in which the arc length is established. The lightly dusted electrodes are kept from touching the base metal, whereas the heavier coated electrode can be allowed to touch the base metal, oftentimes referred to as a drag technique in welding. The basic weld movements depend upon the operator of the welding machine (Figure 11–2). Popular weld movements are the J type of movement, the C type of movement, the circle, and the U-shaped movement. All these movements are designed to control the weld puddle so that the crater can create a penetration depth that the filler material will completely fill. The most common mistake made by the beginning welder in running beads is that he does not allow the arc crater to fill sufficiently. Optimum movement allows for a hesitation period, when the arc is standing still. This hesitation allows part of the crater to fill with molten metal. The arc is then moved rapidly away from this puddle to allow the molten puddle to cool slightly. Then the arc column is moved quickly back in order to start a new weld puddle. From puddle to new puddle should be approximately $\frac{1}{16}$ in. This distance creates the ripple effect of the weld bead. Arc welding then is simply depositing one small puddle upon another which in turn creates a bead. During the time that these independent weld puddles are being stacked upon one another, the arc column is never moved far enough away to allow complete solidification of each independent puddle. If the arc is withdrawn from the puddle so that the metal begins to solidify, the gaseous slag portion of the puddle will rise to the surface of that puddle and take a solid form. Under this condition, when the arc is then moved back over the previous puddle, the slag will be entrapped in the molten bead resulting in a porous weld. It is important to maintain a constant arc length during the puddling operation. The short arc, or an arc length of approximately the diameter of the electrode wire, must be maintained. If the arc length is too great, the gaseous shield that results from the flux coating will not be sufficient to protect the molten puddle from the atmosphere, resulting in a poor weld.

RESTRIKING AN ARC

Many times a welder will encounter a requirement that he cannot meet if he uses a single electrode, or a weld bead that he must interrupt for some purpose and then restart. The operation of restarting a weld bead must also be mastered by the welder in order to insure sound welding techniques. A weld bead, after it has once been interrupted, is restarted by first chipping and brushing the solidified slag from the weld bead and approximately 1 in. from the crater, which insures that no slag will be entrapped in the restarted arc puddle. The welder strikes the arc approximately $\frac{1}{2}$ to 1 in. in front of the crater on the side opposite the weld bead (Figure 11-3). After the arc is struck, the long arc is maintained until the arc is moved to slightly above the center of the crater on the bead side. The arc is then shortened to a standard arc length or to slightly shorter than a standard arc length and held for an instant, until the cool crater fills with molten metal. Once this crater is full of molten metal, the electrode then must

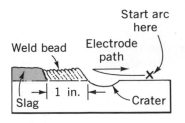

Fig. 11-3 Restarting a weld bead.

be quickly moved in the direction of travel. The greatest problem facing a welder when restarting a bead is hesitating too long in filling the cooled crater with molten metal, which allows the molten metal to overflow, resulting in a bead starting point larger than the previous bead width or height.

THE FLAT POSITION The flat position is the most-used for all shield arc welding. Arc welding not done in the flat position is generally referred to as out-of-position welding. Most shop welding is usually done in the flat position. Elaborate fixtures have been designed that will rotate the work so that the welding can be accomplished in the flat position. This favored position is the most popular because it requires the least amount of skill in order to produce a sound bead. The semicoated electrode, such as the iron-powder electrode, can be simply pulled across the metal and the metal deposited can be adequate or even superior with little operator skill involved.

The electrode in the flat position is held usually at a 90° angle to the work, with the electrode slanted from 10 to 25° in the direction of travel (Figure 11–4a). There are no absolute rules for the angle at which the electrode should be held. This angle is dependent upon the voltage and amperage settings of the machine and the thickness of the metal. With a little practice, the novice welder can determine approximate rates of travel for his own particular mode of operation. The typical rod movement for the flat position is either a simple movement of the electrode while maintaining the correct arc length or the hesitate-move type of movement in which the end of the electrode moves ahead approximately ¼ in., hesitates for an instant allowing the molten puddle to fill, sweeps ahead slightly, maybe ⅜ in. to allow the molten puddle to cool, then back to approximately ¹⁄₁₆ in. ahead of the previous molten puddle where it hesitates again.

Weld beads that are exceedingly smooth and free of slag spots can be accomplished easily in the flat position. The flat position is most adaptable for the welding of all metals, whether nonferrous or ferrous. Many metals, such as cast iron, almost require that the flat position be used. The importance of the flat position is indicated by the emphasis

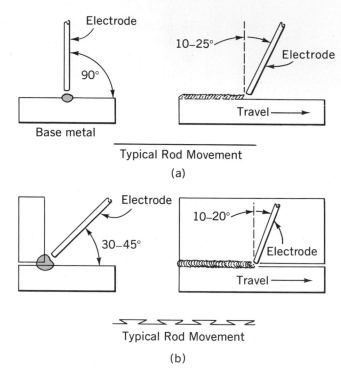

Typical Rod Movement

Fig. 11-4   The flat position. (a) Bead.
(b) Fillet.

placed on it by the electrode manufacturers. The electrodes that are in the Exx 20 or 30 class have been designed specifically for use in the flat position. These electrodes, especially in the 30 series, are high-production, high-deposit-rate electrodes, suitable for single-pass welding procedures.

THE HORIZONTAL POSITION   The horizontal arc welding position is the second most popular position because it enables the welder to deposit a large amount of weld metal although not as much as in the flat welding position. The manipulation of the electrode can be made with the same movements that are used in the flat position. Preferred movements are the C, the J, and the O; the hesitate-move method can also be used (Figure 11-2). The angle of the electrode should be from approximately 5 to 25°, tipped up in order to help alleviate the influence of gravity upon the molten pool, inclined from 10 to 25° in the direction of travel (Figure 11-5). The major errors to watch for when welding in the horizontal

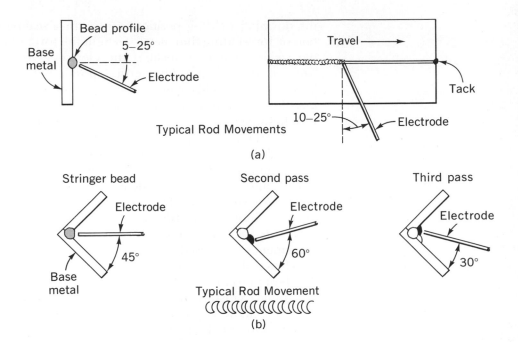

(a)

(b)

Fig. 11–5 The horizontal position.
(a) Bead. (b) Fillet multiple pass.

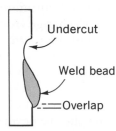

Fig. 11–6 Sagging of the horizontal weld bead.

position are undercutting and overlapping of the weld zone (Figure 11–6). These problems are caused by the force of gravity on the molten puddle which makes the puddle more difficult to control in any out-of-position welding. The sagging of the molten puddle may be prevented by maintaining a shorter arc than in the flat position and by making quicker movements instead of the slow and easy movements of the flat position. Out-of-position welding requires that the movement of the electrode be accelerated in order to help the cooling of the weld puddle. To delay in moving the electrode will cause the metal to sag. The 5 to 25° inclination of the electrode also helps fight the force of gravity upon the weld bead and allows the electrode force being emitted in the arc column to impinge upon the base metal with an upward direction.

THE VERTICAL WELDING POSITION There are two types of vertical welds: uphill and downhill. The uphill weld is most used because the heat of the electrode goes deeper into the metal, thus allowing deeper penetration of the weld. It is also the stronger of the two welds and is used when strength is a major consideration;

the downhill welding position is used for a sealing operation or for welding thin metal. To fight the force of gravity, the welder generally inclines the electrode to an angle between 10 and 25°. The vertical position is simply the laying of one weld puddle directly upon the next weld puddle. The typical rod movements are the oval, the C with hesitations at the end of the C, or the whip movement, which is the most popular (Figure 11–2). The whip movement allows the electrode to establish a weld puddle; it is then whipped vertically ½ in. to allow the molten puddle to cool, and then the electrode is brought back to approximately ¹⁄₁₆ in. above the previous puddle where it is allowed to hesitate in order to start a new pool. This procedure is the basic movement for vertical welding. Any movement can be used in all of the positions, depending on the choice of the operator, and there is no best movement. The best movement for an individual is the one that he can manipulate satisfactorily for a length of time and one that does not allow the entrapment of slag within the weld puddle. One of the major things to avoid with any of the movements in any position is breaking the arc, losing the arc column, and restarting it without cleaning the deposited weld metal. The breaking of the arc column can occur often in the vertical position because of the upward whipping motion of the electrodes. Many welders break contact and lose the arc column every time the electrode is whipped up, which allows too much solidification to take place in the weld pool, causing slag to change from a fluid to a solid material that is entrapped in the puddle (Figure 11–7a and b). This problem must be avoided. The same motions that are used in uphill vertical welding may also be used in downhill welding. The major drawback to downhill welding is that the slag often runs in front of the bead and is entrapped. Downhill welding should never be used for a strength weld.

Arc welding done in a flat or horizontal position often does not necessitate that the operator wear protective clothing. When welding in the vertical or overhead position, the operator must be protected from the flying sparks and heat. For vertical welding the minimum amount of protective clothing should include long gauntlet gloves and welding sleeves, especially sleeves that cover the

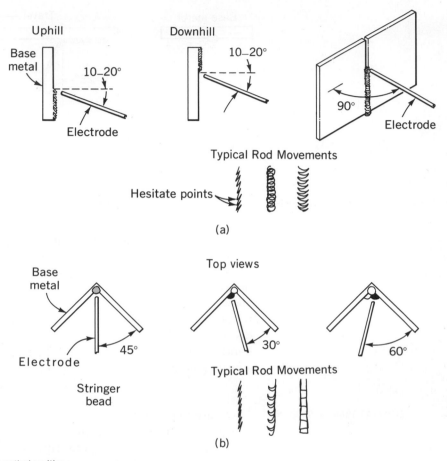

Fig. 11–7   The vertical position.
(a) Bead. (b) Fillet.

welder's arm at the elbow because sparks fall in the crease at the elbow and can burn through clothing.

THE OVERHEAD POSITION

The overhead position, while the most hazardous because of flying sparks and the possibility of the molten puddle dripping onto the operator, is not the most difficult of the four positions. The manipulations and the weld angles are the same as those used in the flat and the horizontal positions. The only difference in overhead welding is that the weld bead is in a more awkward position (Figure 11–8*a* and *b*). Instead of the weld puddle beginning to sag the puddle will have a tendency to drop from the work. How-

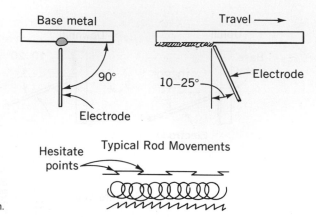

Fig. 11–8 The overhead bead position.

ever, by maintaining a short arc length and with rapid electrode manipulation, this tendency may be overcome. In all of the welding positions, the welder may find it necessary to drape the electrode cable over a shoulder and allow an arm's length of cable to lie over the shoulder, which prevents the electrodes from being deflected because of contact with the cable. It also keeps the arm and hand from tiring because the welder will not be supporting a large amount of welding cable by hand.

WELD BEAD TROUBLE SHOOTING

There are many reasons for weld bead failures. Many weld failures are largely due to such operator errors as porous welds, poor penetration, poor fusion, and undercutting.

Porous welds are impurities in the weld beads which result from improperly prepared base metal, insufficient puddling time, old or wet electrodes, or an arc column length that is too short (Figure 11–9). A short arc must be maintained when welding with a low-hydrogen or stainless electrode, but not when welding with mild-steel electrodes. When inclusions or impurities do appear or are noticeable in the weld, there are many things that the operator can do to identify and stop the conditions that result in a porous weld bead. He must first check for impurities in the base metal. If the base metal is contaminated, no further checking is necessary; the porous weld cannot be corrected when the metal is defective. However, if the porous weld condition is not the fault of the base metal, the welder should first check the electrode to be sure it is proper for the weld-

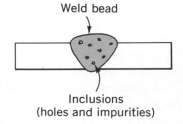

Fig. 11–9 Porous welds.

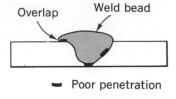

Fig. 11-10 Poor penetration or fusion.

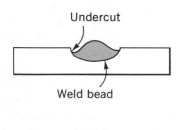

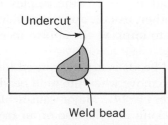

Fig. 11-11 Undercutting.

ing operation. He can then reset the current for the electrode; maintain a longer arc length; change the welding pattern manipulation, possibly to a weave configuration; allow more puddling time for gases to escape; and keep the weld area clean with the chipping hammer and brush.

Poor penetration or poor fusion results if the travel speed of the bead is the wrong feed; generally it will be too fast. Poor penetration or poor fusion occurs also if the electrode is too large and if the current setting is improper for the electrode size, generally too low (Figure 11-10). Poor penetration occurs most readily in the backside of a butt weld when the two pieces being joined do not have sufficient root clearance and the electrode has not been allowed to burn through, common faults of most beginning welders. Poor fusion is commonly caused by allowing the electrode filler metal to flow uncontrolled out of the arc column and simply lie upon the metal. This condition is called an overlap. To avoid these errors, the welder should always leave an adequate root spacing for proper penetration, weld just fast enough to fill the crater that is maintained by the arc column, and use an electrode size that can be manipulated readily.

The opposite of overlapping is undercutting (Figure 11-11), which is caused primarily by incorrect manipulation of the electrode, sometimes compounded by current settings that are too high. The major problem in undercutting is that the weld bead may be deposited correctly and have sufficient strength, but the reduction in area along the side of the weld bead creates a definite weak point in the weld. Undercutting is caused by moving the electrode away from the crater before the filler material has had a sufficient time to transfer to the base metal, and generally a check of equipment will reveal oversized electrodes and excessive amperage settings. To prevent undercutting, the welder should avoid excessive movement of the electrode. He should slow down and try to maintain a uniform motion. Undercutting occurs frequently when a beginner is welding fillet welds, especially on the top of the bead. To avoid undercutting, welders many times shorten the arc, slow their speed, or lower the current settings.

Warping or distortion conditions are caused by uneven heating and cooling, which involve the expansion and contraction of the base metal (Figure 11-12). The base metal

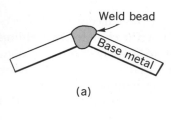

(a)

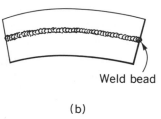

Weld bead

(b)

Fig. 11–12   Warping and distortion.
(a) Distortion. (b) Warping.

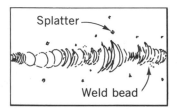

Fig. 11–13   Poor appearance of the
weld bead.

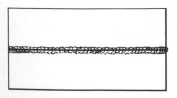

Fig. 11–14   Cracked weld bead.

is pulled out of original alignment. Shrinkage cannot be controlled to any great extent; however, the amount of heat injected into the base metal can be. The heat can be controlled by back-step welding sequences, by clamping the parts into their original position in a special jig or fixture, and by single bead welding, which means that instead of making two or three passes with a small-diameter electrode, one pass is made with a large-diameter electrode. When warping or distortion are encountered, the conditions may be eliminated or controlled by increasing the welding speed, by closing the distance between the parts to be welded, by clamping in a jig or fixture, by tack welding the joint in a number of spots, by presetting the parts to be welded so that after distortion or warping has taken place, they will be pulled into alignment. This presetting, however, means that an estimate must be made as to how far the parts will move after the welding process. Many welders can do this, but it is a skill that comes only with practice.

Much importance is sometimes placed on the appearance of a weld which does not necessarily indicate its inherent strength. The appearance can indicate the electrode manipulation used, the arc length, and the voltage and amperage setting of the welding machine. The major causes for poor appearance of a weld bead are an irregular welding technique; an arc column that is either too long or too short; inconsistent movement across the base metal; faulty, old, or wet electrodes; and incorrect polarity settings for the electrode (Figure 11–13). An arc length that is too long is indicated by the deposit of small spots or particles of metal on the base metal alongside the weld bead. These particles generally can be removed with a chipping hammer, but they are troublesome. Inconsistent speed across the base metal is indicated by the width of the weld puddle. A good weld bead always has the ripples a consistent distance from each other, usually approximately 1/16 in. Practice is the only way to improve the appearance of the weld bead.

Center-line cracking of weld beads is usually caused by the inability of the base metal to move when the weld bead solidifies and contracts (Figure 11–14). Major causes of center-line or under-bead cracking are the use of an incorrect electrode; the imbalance of the base metal masses, such as when one piece has a large mass and the other a small mass; or a too-high carbon content in the base metal. Center-line cracks are common in medium- and high-carbon steel weld beads that have been applied incorrectly or without preheating. Weld bead cracks can be completely eliminated by designing the joint correctly, by preheating the parts to be welded prior to welding, by maintaining the preheat temperature in the base metal during the welding process, by allowing the base metal to move freely as the welding process takes place, or by stress relieving the complete welded structure as soon as the welding operation has been finished.

QUESTIONS

1. Why does the diameter of the electrode determine the current requirements?

2. What is the proper arc length for any electrode?

3. What are the methods of striking an arc?

4. What are the advantages of the flat position?

5. What are the basic movements in manipulating the electrode and the weld puddle?

6. What is the proper sequence for restarting the arc for a continuous weld bead?

7. How does the angle of the electrode affect the weld bead?

8. What is the most difficult welding position? Why?

9. What is undercut? Overlap?

10. What are the major causes of poor penetration? How can it be corrected?

11. What causes impurities in the weld zone? How can they be avoided?

12. How can distortion be controlled in the base metal?

PROBLEMS AND DEMONSTRATIONS

1. Demonstrate the flat, horizontal, vertical, and overhead welding positions.

2. Select a piece of ferrous scrap metal and a ⅛ E6011 electrode. Set the machine amperage at 70 and weld a short bead. Stop welding after an inch or so and raise the amperage setting 10 or 15 amps. Weld another short bead and then repeat this process

until the electrode fails to function properly. Pay particular attention to the sound of the electrode at the various amperage settings.

3. Select a piece of ferrous metal approximately 2 in. wide and 6 in. long. Weld a bead down its complete length at the center of the piece. Notice that the amount of heat absorbed into the base metal determines the amount of distortion of the base metal.

SUGGESTED FURTHER READINGS

Atchison, R. B.: Expansion and Contraction, *Journal of the American Welding Society,* vol. 25, no. 12, p. 1195, December 1946.

Carpenter, S. T., and R. F. Linsenmeyer: Weld Flaw Evaluation, *Welding Research Council Bulletin,* vol. 42, September 1958.

Pender, James: "Welding," McGraw-Hill Book Company, New York, 1968.

Sack, Raymond J.: "Theory and Practice of Arc Welding," Van Nostrand Reinhold Company, New York, 1943.

12 WELD SYMBOLS

Three types of weld configurations can be performed by any shield arc welding process: the butt weld, the fillet weld, and the slot or plug weld (Figure 12–1). The butt weld varies from a square joint to a groove joint design, depending on the thickness of the metal to be welded. The fillet weld generally requires no special preparation. It is an angular weld bead that joins metals usually at a 90° angle and may be a traditional fillet weld or a weld such as that used in the lap joint. The slot or plug weld is designed to join one piece of metal on top of another. Generally there will be more than one slot or plug located on the piece of metal.

The terminology used helps in distinguishing the many variations in the groove weld (Figure 12–2). The root face and the distance between the root faces are dependent

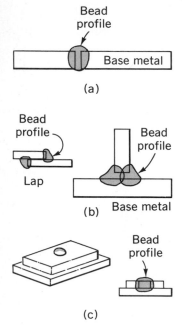

Fig. 12-1 Basic joints. (a) Butt. (b) Fillet. (c) Slot or plug.

solely upon the thickness of the metal to be welded. Square butt joints or groove joints are generally used when the thickness of the base metal is ⅜ in. or less; if the metal thickness is greater than ⅜ in., complete penetration will not take place through the base metal. For example, when a ⅜-in. plate is butt welded with a square joint, the root distance should be maintained at approximately one-half the plate thickness, or 3/16 in. (Chart 12-1). For the single-bevel, single-U, or single-J designs, the plate thickness ranges from ¼ to 2½ in. The double-joint designs are primarily used for heavy metal that can be welded from both sides. The double-joint designs, which are the double V, the double bevel, the double U, and the double J; or the flare adaptations of these designs, which include the U and J, are for plate thicknesses of approximately 1 in. or more.

The position in which the weld must be accomplished determines the angle of type of groove joint design that must be machined into the metal before welding. The joint preparation is usually performed by someone other than the welder, and the welder simply makes the weld; therefore, information concerning the preparation and the weld must be transmitted both to the person preparing the weld joint and to the welder. This is done by a standardized system of welding symbols (Figure 12-3). Standard weld

CHART 12-1 Butt joint welding limits

JOINT TYPE	WELD BOTH SIDES			WELD ONE SIDE	
	ROOT, IN.	ROOT FACE, IN.	GROOVE ANGLE*	ROOT, IN.	GROOVE ANGLE*
Square edge	One-half of plate thickness	—	—	One-half of plate thickness; 3/16 max	—
Single V	1/16	⅛	60°	3/16	45°
Single bevel	1/16	⅛	45°	3/16	45°
Single U	0-⅛	1/16	45° H 20° FVO	3/16	45° H 20° FVO
Single J	0-⅛	1/16	45° H 20° FVO	3/16	45° H 20° FVO
Double V	⅛	1/16	60°	Poor	Practice
Double bevel	⅛	1/16	45°	Poor	Practice
Double U and double J	0-⅛	1/16	20° FVO	Poor	Practice

*Welding positions: F, flat; H, horizontal; V, vertical; O, overhead.

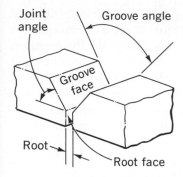

Fig. 12-2 Groove or butt weld joint terminology.

Fig. 12-3 The welding symbol. (a) The welding symbol. (b) Example.

symbols have been assigned and endorsed by the American Welding Society and the American Society of Testing Materials. The weld symbol appears complex, but the entire symbol is not used in most situations. It is generally used on blueprints where it indicates the process to be used and the amount of weld deposit that is necessary in order to support the design of the weldment (Figure 12-3).

The weld symbol points to the location at which the weld is to be done with the arrow. If the weld symbol has a black dot where the arrow breaks the main bar, it means that the weld is to be accomplished in the field. In other words, the weld would be an assembly technique such as might occur at the job site rather than at the shop where the steel structure was fabricated. If there is a circle around the black dot at this junction of the arrow head to the bar, the weld is to be made 360° around the joint. In Figure 12-3a, LP indicates the location at which information about the length and the pitch of the weld may be found. The pitch is the spacing between the weld beads measured from the center of one bead to the center of the next bead. For example, the length of the weld bead might be 2 in. and the pitch 6 in., which indicates that every 6 in., measured from one center to the next center of the weld beads, there is to be 2 in. of weld along the entire length of the joint. In Figure 12-3a, the brackets indicate the places in the symbol where information about joint design will be

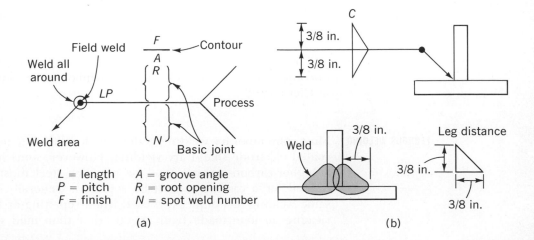

(a)   (b)

found. If the information is found above the bar, the weld is to be performed on the far side of the joint away from the direction that the arrow points; and if the information appears below the bar, the weld is to be made on the near side, or the side where the arrowhead is pointing. Directly above the area where the basic joint design information is found, may be found information (if it applies) about the root opening; the groove angle; and the contour of the finished weld, which can be flush, convex, or concave. The finish symbol may also be placed in this area. Usually, for most shield arc welding, there is no need for a finish symbol because it is always good practice to chip the slag from all finished welds to make them easier to inspect. Finish symbols are G for grind, C for chip, F for file, or M for machine finish. Machine finishes are generally performed in the shop. Directly beneath the "near side" basic joint symbol is the space reserved for information about the resistance welding process, indicating the number of spot welds to be made on a particular joint or piece. Next to the information about the basic joint designs, in the direction toward the end of the arrow and on the top of the bar, is another space that is reserved for resistance welds. The last area of the weld symbol, the tail, is generally absent when the shield arc welding process is used. When processes other than the shield arc process are indicated, the tail is used. The process may be submerged arc, TIG (tungsten inert gas), MIG (metal inert gas), or any of a number of other welding processes. In Figure 12–3b, the symbol indicates that the fillet weld with a leg distance of $3/8$ in. should be used on both sides of the joint and it is to be welded on the job site. The finished weld is to be chipped free from slag for inspection.

The symbols used for the basic joints which can be used in conjunction with the overall weld symbol are presented in Chart 12–2.

FERROUS WELDING   Most contemporary welding is done with mild steel, using coated electrode shield arc welding. However, sometimes a higher-carbon-content steel or an alloy steel must be welded, or a cast item must be repaired. Generally, the same welding techniques are used when repairing or fabricating an item made from metal other than mild steel

CHART 12-2 Weld symbols

WELD TYPE	ARROW SIDE	FAR SIDE	BOTH SIDES
Fillet			
Plug or slot			Not used
Arc seam, arc spot			Not used
Butt or groove Square			
V			
Bevel			
U			
J			
Flare V			
Flare level			

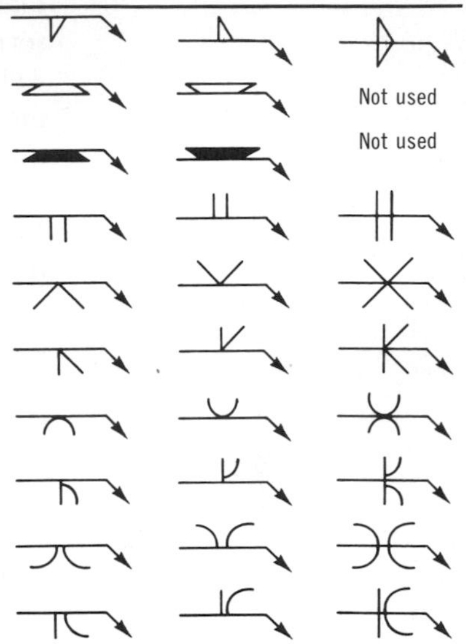

SUPPLEMENTARY SYMBOLS			
WELD ALL AROUND	FIELD WELD	FLUSH CONTOUR	CONVEX CONTOUR
○	●	—	⌒

as when welding items of mild steel, with a few exceptions. If these exceptions are not observed, a failure of the weld joint will result. In most cases, welding mild steel or low-carbon steel presents no real problem, because this steel has good ductility so there is little danger of the weld or parent metal cracking. Metal is ductile or has ductility when it has the ability to stretch, bend, or compress without breaking. Low-carbon steel is ductile because of the amount of carbon it contains; it has a 0.30 percent or lower carbon content. Low-carbon steel, or mild steel, is used to build most of the steel-structured buildings, ships, and equipment. More mild steel is in use today than all other types of steel, aluminum, and magnesium combined. Low-carbon steels are those that contain from 0.30 to

0.45 percent carbon; high-carbon steels contain more than 0.45 percent carbon. Mild steel will support approximately 60,000 psi, which is why electrodes of the E60 series are used predominantly in welding mild steel. The weld should be as strong as, if not stronger than, the base metal because the electrode metal equals or exceeds the quality of the base metal.

Factors that are of only little concern when welding mild steel become very important when welding medium-carbon, high-carbon, or alloy steels. The biggest problem in the welding of medium- and high-carbon steels is that ductility decreases as the carbon content increases or as the amount of alloying agents in a steel increase. Higher-carbon steels and alloy steels also develop stresses as a result of the heat more readily than do mild steels. Stresses are generally caused by the temperature difference between the arc and the cool mass of metal. The great stresses in the metal can lead to cracking of the weld bead itself, the base metal adjacent to the weld bead, or the base metal directly beneath the weld bead.

Primarily, a high-carbon steel cracks because it is allowed to cool too rapidly or because it has not been preheated in order to expand the structure so that it cools evenly. When a metal cools too rapidly, it does not have time to go through its normal transformations. As a result, an extremely hard, brittle structure is formed. This structure, called martensite, the hardest form ferrous metal can be, is susceptible to cracking. If the metal is allowed to cool slowly, it will not transform into martensite completely and will therefore be a softer, ductile material.

Several precautions can be taken to alleviate stresses resulting from the welding process (Chart 12–3). For example, when medium-carbon steel is welded, an economical way to solve the problem is to preheat the base metal from 350 to 700°F, perform the weld in a single pass if possible, and then allow the steel to cool slowly. Following these steps will not only prevent many of the harder crystals from forming, but will also allow many of the locked-in stresses resulting from the hot arc column to be dissipated through the metal.

A problem associated with welding high-carbon steel is porosity in the weld bead. Porosity is principally caused by hydrogen bubbles being caught in the weld metal. Hydrogen is highly soluble in the molten metal. The en-

CHART 12-3 Shield arc ferrous welding

METAL	ELECTRODE CLASSIFICATION	SPECIAL INSTRUCTIONS
Medium-carbon steel	Exx15, Exx16	Preheat to 350-700°F
		Single pass preferred
High-carbon steel	Exx15, Exx16	Preheat to 350-700°F
		Single pass preferred
		Low-amp setting
Manganese steel		
Low	E7010, E7020, E6015, E6016	2-in. bead length
		Keep cool to touch of hand
High	18-8 Cr-Ni	or preheat to 350-700°F
		2-in. bead length
		Keep cool to touch of hand
		Peen-stress relieve
Stainless steel	Same as base metal, 1/8-in. diameter used most	Very low amp settings
		Short arc
Molybdenum steel		Preheat to 350°F
Thin	E7012, E7020	Stress-relieve 1200°F, cool slowly
Thick	E7014, E7016	Peen-stress relieve
Nickel steels		Preheat to dull red
Thin	E6012	Single-pass bead preferred
Thick	E7010, E7020, E7015, E7016	Shallow penetration
Cast iron	Machinable (Ni or Cu-Ni)	Cool to touch
	Nonmachinable (Fe) 1/8-in. diameter used most	Peen-stress relieve
		Low-amp setting
		Short weld bead, 1-2 in.

trapped water in the electrode coating is superheated by the electric arc, which causes its separation into its constituent elements of oxygen and hydrogen. The hydrogen then enters the weld, becoming entrapped as the weld cools and the hydrogen tries to escape. During the escape process, bubbles, which turn into a metal void, are formed. If the weld metal cools too rapidly, the hydrogen cannot escape; and the resulting weld is both porous and weak. When the hydrogen is caught between the weld metal and the base metal, it is a contributing factor in underbead cracking. The E60, E7015, or E7016 electrodes, called the low-hydrogen electrodes, are designed to eliminate hydrogen inclusions because the flux coating has a low hydrogen content. These electrodes must be stored in a dry, warm place or the electrode coating will absorb water and will no longer be low in hydrogen.

Once the weld has been completed on high-carbon or medium-carbon steel, the weld deposit will not have the same characteristics as the base metal, but it will be suitable.

If for some reason the base metal must be restored to its original condition, the metal can be heat treated after the welding process has been completed.

Steel alloys can be welded readily with a few simple adaptations of the traditional welding process. Before the welding of an alloy steel, such as manganese, molybdenum, or nickel alloyed steel, the base metal is preheated. For example, on low-manganese steel the base metal is preheated to 350 to 700°F before welding. The same is true of molybdenum alloyed steels. Nickel steels, however, are preheated to approximately a dull red before welding (Chart 12–3). Preheating the base metal helps to eliminate the possibility of underbead cracking in alloy steels as well as in carbon steels. When most alloy steels are welded, an electrode that has a shallow penetration rating is preferred. Minimal manipulation of the electrode is also advisable because it insures that a minimal amount of base metal will mix in the weld bead. On alloy steel, weld beads should be kept as small as possible and joint designs that allow the complete weld to be performed in a single pass are preferred.

Shield arc welding of stainless steel is a useful welding process. Stainless steel is welded in the same manner as mild steel, except that out-of-position welding may produce unacceptable weld bead porosity because of the slag characteristics of the electrodes that are used. The E308–15 electrode is most used; however, other electrodes that can be used, depending upon the requirements, are the E309, E310, E316, E317, E330, and E347. These electrodes are matched to the composition of the base metal (Unit 9, Electrodes). As in the welding of alloyed steels and high-carbon steels, the narrow stringer bead is preferred, and a weaving motion should be avoided if at all possible. Also, low-ampere settings are usually required for the welding of stainless steel, and a short arc is maintained to eliminate formation of splatter on the base metal, which is extremely difficult to clean. A disadvantage of shield arc welding on stainless steel is that discolorations ranging from the shades of purple to light blue generally surround the weld bead. Since it is extremely noticeable, the appearance of the bead suffers. If the appearance of the weld is important, this discoloration must be removed by a refinishing process.

Shield arc welding of cast iron requires a departure

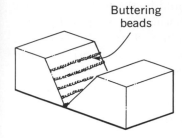

Fig. 12-4 Buttering.

from standard procedures. Cast iron may be welded by the shield arc process only while it is cool to the touch. Two basic types of electrodes are used for the shield arc welding of cast iron: the machinable electrode and the nonmachinable electrode. Machinable electrodes contain nickel or a combination of copper and nickel and yield a deposit that can be worked after the weld is accomplished. The nonmachinable electrode is basically a mild-steel wire electrode that deposits a steel material into the weld zone which is nonmachinable. In general, the weld deposits of both the machinable and nonmachinable electrodes present a weld bead that is 3 to 4 times as strong as the cast-iron base metal. Many times, on cast-iron pieces that are contaminated through usage, such as oil-soaked engine blocks, a stainless steel electrode containing 18 percent chromium and 8 percent nickel with a low carbon content is applied to the cast iron in a buttering technique which will support the standard cast-iron shield arc electrode (Figure 12-4). After the cast iron has been prepared if it needs to be by the buttering process, the weld beads are then produced. Because of the cumulative effects of internal strain that is consistent with the structure of cast iron, short weld beads must be used. Weld beads that range in length from 1 to 2 in. are used most often. After 1 to 2 in. has been deposited on the weld zone, the welder should test the base metal to insure that it is cool to the touch. After each weld bead has been deposited, it is generally necessary to stress-relieve the weld by peening (Figure 12-5). To peen the weld means to strike the weld with a ball-peen or similar hammer, resulting in the structure of the weld being stressed, thus relieving the internal strain. Weld beads may also be stress-relieved by heat treating.

In preparation for the shield arc welding of cast iron, four separate steps must be followed in order to insure a sound cast-iron weld. First, the casting skin must be removed in the area that is to be welded. This skin is usually removed by grinding the strip that encompasses the weld area. The grinding of the surface of the cast iron should remove many of the impurities that were left at the time the metal was cast. If these impurities are not removed, they may well become inclusions within the weld. Second, if the cast iron is more than $\frac{3}{16}$ in. thick, a joint design must be ground into the weld zone (Figure 12-6). The

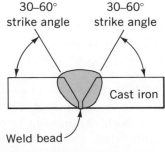

Fig. 12-5 Peening a weld.

included angle of the joint should be 60° to insure proper fusion. Third, all rust, dirt, oil, scale, grease, or other contaminating materials must be removed from the weld area. Many times the cast-iron casting is soaked with oil, such as with an engine block. Kerosene is an excellent cleaner to remove much of the oil. When cast iron is welded, because of its lack of ductility, small cracks become larger. It is mandatory, therefore, to drill a small hole, usually 1/8 in. in diameter, just beyond the end of each crack to stop the cracking. The weld bead will then extend beyond each drilled hole.

NONFERROUS SHIELD ARC WELDING

Stick electrode shield arc welding of nonferrous metal represents only a minor portion of stick electrode welding because other processes are more adaptable to the welding of nonferrous metals. The major types of electrodes, however, that are used for stick electrode welding of nonferrous metals are aluminum electrodes for welding aluminum; aluminum-bronze electrodes for welding some ferrous metals and some nonferrous metals as well as for hard surfacing ferrous metals; phosphorus-bronze electrodes for welding copper, brass, bronze, steel, cast iron, and malleable iron; and the nickel electrodes (primarily chromium nickel) for welding nickel-based metals (Chart 12-4).

Two basic electrodes are used for the shield arc welding of aluminum. They may be used for welding rod in the oxyacetylene process as well, for they are highly adaptable. The two electrodes have approximately the same chemical composition; however, one electrode is pure 100 percent aluminum, while the other has 5 percent silicon added to the aluminum. The aluminum electrode with the silicon controls much of the grain growth and produces a harder bead than the pure-aluminum electrode. However, the pure-aluminum electrode is more adaptable to corrosion-resistant welds. If the weld bead must have the same appearance as the base metal, the 100 percent aluminum electrode must be used. If the color match, however, is not essential and strength is more important, the 95 percent aluminum and 5 percent silicon electrode is used. The welding procedure is the same for both. Aluminum has been found to transfer more readily when a short-circuit

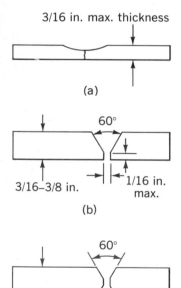

Fig. 12–6   Cast-iron joint preparation.
(a) Grinding method. (b) Single V.
(c) Double V.

CHART 12-4 Shield arc nonferrous welding

METAL	ELECTRODE COMPOSITION	SPECIAL INSTRUCTIONS
Aluminum (all)	95% aluminum, 5% silicon or 100% aluminum	Short arc length DCR Preheat to 500–800°F.
Deoxidized copper	Phosphorus bronze (heavy flux coating)	1/4 in. thickness or less Preheat to 200–400°F
Brass	Phosphorus bronze	1/4 in. recommended as maximum thickness Small beads High amp
Bronze	Phosphorus bronze	1/4 in. recommended as maximum thickness Small beads Medium amp
Nickel base	18–8 chromium nickel or 50% nickel	Narrow beads 18-gage metal minimum
Dissimilar (Fe to non-Fe)	Aluminum bronze	Preheat to 350–600°F Weaving motion Peen stress-relieve

arc is used rather than when the spray arc technique is used. Using this short-circuit arc effect means that the voltage must be controlled. The voltage requirements for aluminum electrodes range from 14 to 32 volts maximum. Greater voltage would require that the metal transfer be effected by the spray technique, which is hard to control for good welding. Arc column control is difficult in the welding of aluminum because the arc has a tendency to wander. The restarting of a weld bead is also difficult and requires much practice. The common method of striking an arc, using the scratch technique, has been found to be the most advantageous, mainly because of the heavy flux that forms over the bead and the heavy flux that surrounds the electrode wire. In general, the highest current amperage should be the highest that can be easily used. The highest current settings possible are determined by watching the edges of the joint, which should melt before the backing, the joints, or the surrounding area melts through. This high amperage rating is necessary because of the heat-dissipating qualities of aluminum which produce a chilling effect upon the arc column. Because of this chilling effect from the heat-dissipating qualities, the base metal should be preheated to between 500 and 800°F before welding. If this preheated condition can be maintained throughout the welding process, the weld will be less porous than if it were allowed to cool during the welding process. The operator may use almost any type of manipulation movements. The major thing to watch out for is the travel speed which has to be far greater than the speed used when welding mild steel. Flat welding of aluminum is the preferred method because the electrode is easier to manipulate, and because of the burn-through characteristics of aluminum. Out-of-position welding proves difficult. Vertical welding is best accomplished by using the downhill method; overhead welding should be avoided if at all possible because of its extremely difficulty; and the horizontal position is found to work best if the work is rotated into a slightly downhill, horizontal weld position.

All aluminum welding that uses a flux-coated electrode or a solid fluxing agent requires that the flux be removed from the base metal and from the weld bead. If it is not removed it will continue its corrosive action and will eventually destroy the aluminum. Flux may be removed by scrubbing with a wire brush and either a solution of hot water and 5% nitric acid or a 10% sulphuric acid solution. The acid helps to remove the flux, but it must be removed from the surface or it will continue its corrosive action. The acid, however, is easily removed by rinsing the metal with clear hot or warm water. The acids that are used to remove the aluminum flux are harmful to the skin and care should be taken to avoid contact with them.

Phosphorus-bronze electrodes are used with a wide range of metals, usually to stick electrode shield arc weld copper, brass, bronze, steel, cast iron, and malleable iron. They are capable of joining dissimilar metals, such as cast iron to steel, or steel to brass. The electrodes can be used in all four welding positions. They are either flux coated or bare, depending upon the amount of phosphorus in the wire. Those that contain a small amount of phosphorus, such as 0.20 percent, require additional shielding agents; therefore, flux must be used in conjunction with the electrode. However, if a large amount of phosphorus is used, there is no need for any additional fluxing agent on the electrode. The standard sizes for the coated electrodes are $1/8$ and $5/32$ in. in diameter. The larger electrode requires higher amperage settings, which range from 175 to 350 amps for the bare phosphorus-bronze electrodes. The amperage range for the $\frac{1}{8}$-in. covered electrode is 80 to 120 amps. The application of the phosphorus-bronze electrodes determines the amperage settings for the electrode. For example, if cast iron, cast steel, or malleable iron is to be used, the lower side of the amperage range, or approximately 80 amps, should be used for a $\frac{1}{8}$-in. electrode and 130 amps for a $\frac{5}{32}$-in. electrode. If, however, the base metal is bronze, the amperage should be set more toward the medium range for the electrode, or around 100 amps. If the base metal is one of the brasses, especially zinc-brasses, the amperage setting should be in the high range, or 120 amps for the $\frac{1}{8}$-in. electrode and 180 amps for the $\frac{5}{32}$-in. electrode.

The welding procedure is basically the same as that used when welding mild steel; however, a short arc length must be held at all times in order to control the arc column. The heavy electrode, though, presents a problem to the beginning welder because the flux coating on the electrode should not be allowed to touch the molten puddle that is created by the arc column. If the slag coating of the electrode is allowed to touch the molten puddle, the flux will float in the puddle and possibly will remain as an inclusion in the finished weld bead. Copper base metal that is less than $\frac{1}{4}$ in. thick needs no preheating, but copper that is thicker than $\frac{1}{4}$ in. should be preheated to 200 to 400°F to combat the heat-dissipation qualities of the copper. Because of the chill effect, small-width beads or stringer beads are recommended for most welding.

Many electrode manufacturers have also added lead to a special class of phosphorus-bronze electrodes. This lead enables the welder to deposit a bearing type of surface on worn bronze parts, making it possible to repair bearing plates rather than to discard them. The technique used for welding with lead-phosphorus-bronze electrodes is basically the same as when welding with a phosphorus-bronze electrode. The flat position is required for lead-phosphorus-bronze, and DCRP is generally recommended. The flat position is required because of the high density and low melting temperature of lead. Caution must be employed when using this lead-phosphorus-bronze electrode because lead fumes are poisonous, and the operator must be protected from the toxic effect of the fumes.

Lead-phosphorus-bronze welding should always be done in a well-ventilated area where forced air can be blown across the operator area.

ALUMINUM BRONZE   Aluminum-bronze electrodes can be used for welding the same metals that phosphorus-bronze electrodes are used for. Aluminum-bronze electrodes can weld malleable iron, cast iron, steel, aluminum bronze, most of the phosphorus bronzes, copper, tin, phosphorus alloys, and brasses. The aluminum bronzes, however, can be used also to increase the resistance of metals to wear. It can be used as a hard-facing electrode as well as the method to join such metals as manganese bronze, Muntz metals, and brasses. Aluminum bronze can be used as an overlay to extend the normal lifetime of a moving part, to increase the wearing qualities of metal, and to improve shock or the impact ability of metal. Aluminum-bronze electrodes range in diameters from $5/64$ to $1/4$ in. The $1/8$-in. diameter and the $5/32$-in. diameter electrodes are most popular. The method of transfer used is the short arc method which requires low voltages. The amperage range for the $1/8$-in. electrode is from 90 to 130 amps and for the $5/32$-in. electrode is from 130 to 150 amps. The welding movement with the aluminum-bronze electrode should be a weaving motion with the width of the deposit ranging from 3 to 5 times the diameter of the electrode. The aluminum-bronze electrodes are designed specifically for a wide bead or for an overlay type of deposit. All base metals welded with the aluminum-bronze electrode should be preheated. The preheating prevents hydrogen pickup because of the high hydrogen content of the flux which in turn would create porosity in the deposit metal; therefore, a minimum heat of 350°F should be used for the first pass with an aluminum-bronze electrode. If there are to be successive passes, the weld bead must be cleaned thoroughly of flux and peened to stress-relieve the deposit in the surrounding area.

Aluminum-bronze electrodes are classified into five or six major categories, depending on the hardness of the metal that they deposit. The hardnesses range from a deposit with a hardness similar to that of the base metal to a deposit used uniquely for its wear-resistant or hardness characteristics. Some aluminum-bronze electrodes have a Brinell hardness that ranges as high as 300 on the Brinell hardness scale. A hardness of 300 would indicate a deposit that would be used for overlays on forming or drawing dies. At the other extreme, aluminum-bronze filler material can be used for welding repair work such as with cast iron, where the weld bead has to be machinable. The weld bead will be slightly harder than the cast iron surrounding it, but it will be machinable.

QUESTIONS

1. What are the three types of weld joints?

2. What is a fillet weld?

3. What is the function of the root?

4. Why is the square butt joint limited by metal thickness?

5. What information is given by the basic weld symbol?

6. Why is it poor practice to weld only one side of a double bevel joint?

7. How is the size of the weld bead indicated by the welding symbol?

8. Why is the tail of the welding symbol deleted in the shield arc welding process?

9. What is the biggest problem when shield arc welding medium- and high-carbon steel?

10. What special preparation must be performed prior to welding medium-carbon steel? Nickel steel? Molybdenum steel?

11. Why must cast iron be kept cool to the touch during the welding process?

12. What is the correct way to peen a weld?

13. What is buttering?

14. What nonferrous metals can be arc welded?

15. What are the special instructions for preparing aluminum base metal prior to welding?

PROBLEMS AND DEMONSTRATIONS

1. Demonstrate preheating the base metal with an oxyacetylene flame and with a standard heat-treating furnace.

2. Weld a piece of medium-carbon steel using proper preheating methods and weld another piece of medium-carbon steel without preheating. Allow the two pieces to cool to room temperature. Check the pieces for possible cracks along the weld bead and the back side of the base metal.

3. Weld a piece of cast iron using the correct welding and peening procedures. Weld a piece using the methods used when welding mild steel. Compare the results.

4. Shield arc weld a piece of aluminum using DCRP. Run another bead with DCSP. Compare the results during the welding process and the different effects of polarity on the aluminum.

SUGGESTED FURTHER READINGS

Barnett, O. T.: "Filler Metals for Joining," Van Nostrand Reinhold Company, New York, 1959.

Rossi, Boniface E.: "Welding Engineering," McGraw-Hill Book Company, New York, 1954.

Spencer, Lester F.: Hardfacing. Picking the Proper Alloy. *Welding Engineer*, vol. 55, pp. 39–48, November 1970.

Stieri, Emanuel: "Basic Welding Principles," Prentice-Hall, Inc., Englewood Cliffs, N.J., 1953.

13 SUBMERGED ARC WELDING

Welding through or under a blanket of flux has been practiced by various methods since the early 1920s. Both carbon and metallic electrodes have been used with varying degrees of success for hand and automatic processes. In recent years perfection of fully automatic and semi-automatic submerged arc welding equipment has resulted in an increase in welding speed considerably above anything previously possible. This increased speed has greatly reduced the cost of individual welds.

The submerged arc process, which may be done either automatically or manually, creates an arc column between a bare metallic electrode and the workpiece. The arc, the end of the electrode, and the molten weld pool are submerged in a finely divided granulated powder that contains appropriate deoxidizers, cleansers, and any other necessary fluxing element. The fluxing powder is fed from a hopper that is carried on the welding head. The tube from the hopper spreads the powder in a continuous mound in front of the electrode along the line of the weld. This flux mound is of sufficient depth to submerge completely the arc column so that there is no splatter or smoke, and the weld is shielded from all effects of atmospheric gases. As a result of this unique protection, the weld beads are exceptionally smooth (Figure 13–1). The flux adjacent to the arc column melts and floats to the surface of the molten pool; then it solidifies to form a slag on top of the welded metal. The rest of the flux is simply an insulator that can be reclaimed easily. The slag that is formed by the molten flux solidifies and is easy to remove. In fact, in many applications, the slag will crack off by itself as it cools. The unused flux is removed and placed back into the original hopper for use the next time. Granulated flux

Fig. 13–1 Submerged arc process.

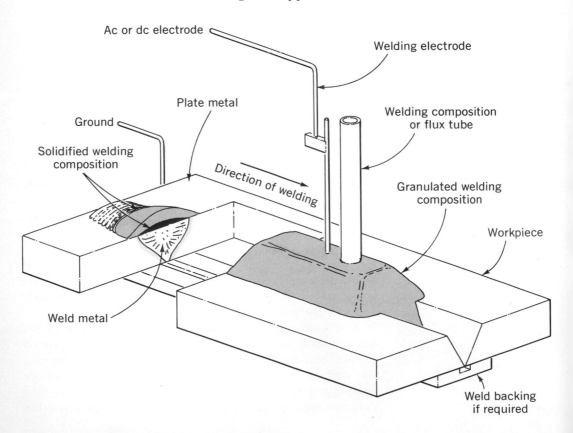

is a complex, metallic silicate that can be used over a wide range of metals. A typical flux, no. 660, can be used for welding mild steel; low-alloy, high-tensile steels; straight chrome; and chromium-nickel steels.

In either automatic or semiautomatic submerged arc welding, the bare electrode is mechanically fed through an electrical contacting jar, nozzle, or collet. A welding current passes through the collet to the electrode and the arc is maintained between the base metal and the electrode. The heat of the arc causes a melting of the base metal, electrode, and the adjacent flux to accomplish the tasks of penetration, filling the crater, and fusion of the joint. The electrode is fed into the arc automatically so as to maintain a preset arc voltage and arc length. Direct current supplied by a motor generator welder produces the arc between the electrode and the base metal; however, ac power supplies may be used even though they are not as popular. Use of a dc power supply assures a simplified, positive control of the submerged arc welding process.

The bare wire electrodes can be purchased either as bare electrodes or with a slight mist of copper coated over the wire to prevent oxidation during the shelf life of the electrode. The coiled electrode comes on a standard reel and is available in diameters ranging from $3/32$ to $1/2$ in. The most popular electrode is the copper-coated one that has an extended shelf life because of its coating that prevents rusting. The coating also increases the electrical conductivity of the electrode wire.

The submerged arc process is characterized by high welding currents (Figure 13-2). The current density in the electrode is 5 or 6 times that used in ordinary manual stick electrode arc welding. As a result the melting rate of the electrode as well as the speed of welding is much higher than in the manual stick electrode process. The high current density also results in deep penetration so that plates $5/8$-in. thick can be square butt welded without special edge preparation. Currents as high as 1000 amps may be used with a $3/16$-in. filler wire. In a sense, the welding wire carries with it a "container" full of molten metal along the weld line. The walls of this container consist of the unmelted flux ahead of the wire, solidified weld metal, slag behind the weld electrode, and base metal on either side or below the container. Because of the effect of this

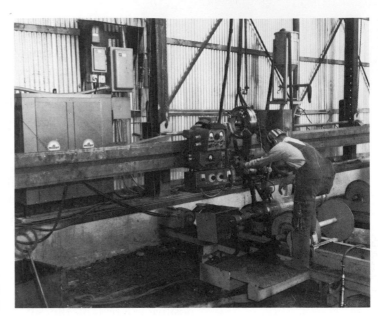

Fig. 13-2 Submerged arc welder.
(Courtesy of Miller Electric Manufacturing
Company.)

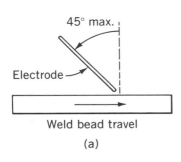

(a)

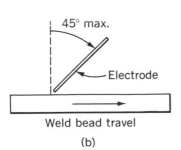

(b)

Fig. 13-3 Wire slant. (a) Forward
point. (b) Backward point.

container image, base metal as thick as 1½ in. can be welded in a single pass. The use of higher welding currents generally means deeper penetration and greater use of base metal for making many types of welds. Faster welding speeds result, and faster welding speeds minimize distortion and warpage.

The submerged arc process is also capable of welding fairly thin gage material. However, because of the greater welding current densities, the electrode many times is pointed or slanted in a particular direction. This wire slant determines to a great extent the bead appearance (Figure 13-3). The forward pointing wire points in the direction of the weld travel and is similar to forehand welding in the oxyacetylene welding process, which deflects much of the heat of the arc column, yielding a decrease in the penetration, a wider and flatter crown to the bead, and a smaller heat-affected zone. The backward-slanting point, however, does just the opposite and is comparable to the backhand welding technique in the oxyacetylene process. The backward-pointing electrode yields greater penetration, a larger heat-affected zone, and a narrow weld bead.

Submerged arc welding can be done manually (Figure

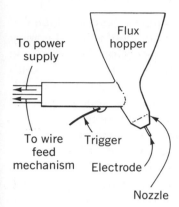

To power supply

Flux hopper

To wire feed mechanism

Trigger

Electrode

Nozzle

Fig. 13-4 Manual submerged arc. Nozzle construction is comparable to a MIG nozzle.

13-4). A small hopper for the granular flux is built into a special electrode holder. The combination electrode feed and cable is attached to the electrode holder. The operator guides the electrode and is able to make fast welds in places where automatic submerged arc welding is not suitable, such as on curved lines or irregular joints. Manual submerged arc welding can also be used whenever normal, manual shield arc stick electrode welding is used. The manual and the automatic submerged arc welding processes are most suited to the flat welding position or the slightly vertical, downhill welding position.

Welds made by the submerged arc welding process have high strength and ductility with low hydrogen or nitrogen content. This process is suitable for welding low-alloy, high-tensile steels as well as the mild, low-carbon steels. Whatever type of steel is used, the principles governing the mechanism of metal transfer remain the same in the submerged arc process as in the other shield arc processes. The voltage requirements of the short arc used are fairly low, ranging from 20 to 40 volts; 40 volts is a sufficient amount of potential for a 1000-amp demand. The submerged arc process is also capable of joining medium-carbon steel, heat-resistant steels, corrosion-resistant steels, and many of the high-strength steels. Also, the process is adaptable to nickel, Monel, Everdur, and many other nonferrous metals.

The submerged arc welding process has many industrial applications. It is used for fabricating pipe, boiler pressure vessels, railroad tank cars, structural shapes, practically anything that demands welding in a straight line. The many applications of the submerged arc welding process have created three general types of equipment: the stationary electrode with the work moving, the electrode mounted on a carriage with the work stationary, and the submerged arc mounted on a self-propelled tractor with the work stationary. The last type, of course, is the manual process, and in this case the special flux-hopper electrode holder is hand held.

All of the applications of submerged arc welding process demand that the heavier sections of the base metal to be welded be prepared in order to insure optimum penetration. Many times in metal plates that are more than $1/2$ in. thick, the joints are prepared. The major joint de-

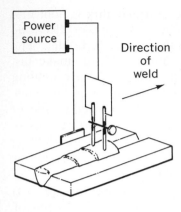

Fig. 13–5 Twin arc welding.

signs used are the single-V, the single-U, and the double-V butt joints, with the single-V butt joint being the most popular. The square butt joint, fillet joint, and plug weld joint designs can also be used in the submerged arc welding process. A major advantage of the submerged arc welding process is that joint designs can be kept simple, yet the process can insure adequate penetration. An oxygen-fuel cutting process is a satisfactory method to use for preparing a joint, such as a single V; the carbon arc and the plasma arc cutting processes are also useful. Generally, in all except the lap of fillet type of weld, a backing strip is used to support the metal directly beneath the weld zone because of the size of the molten weld pool and the possibility that the molten metal will sag or that the arc crater will burn through the base metal. Usually, the backing plate becomes an integral part of the weld and remains with the weldment.

Fig. 13–6 Multiple-electrode submerged arc. (a) Multiple power tandem electrodes. (b) Single power supply, parallel electrodes. (c) Series connection, parallel electrodes.

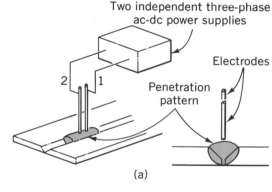

(a)

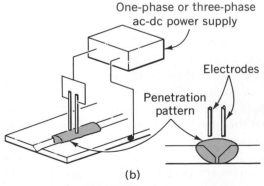

(b)

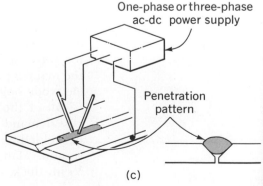

(c)

Submerged arc welding may be done also with more than one metal electrode. This process is called twin submerged arc or twin arc welding and is shown in Figure 13−5. This adaptation changes the weld deposit size considerably. The multiple electrode, submerged arc process feeds two electrodes simultaneously, instead of one, into a dual welding head, which permits use of higher currents. Current demands as high as 1500 amps can be sustained with the multiple electrodes submerged arc process. There are three basic ways that the multiple electrode process can be achieved (Figure 13−6). The multiple power supply units with the electrodes in tandem (Figure 13−6a) yield the heaviest, deepest weld penetration pattern deposits. The electrodes are generally in line with the weld, which creates a deep, narrow weld bead. This welding setup is used for the heavier, thicker base metals. If a wide bead is required, the head is rotated 90° so that the electrodes are across the weld. The spacing between the electrodes can be varied by lengthening the area between the two electrode holders. The multiple power support can be used with tandem electrodes, but the parallel electrodes usually have a single power supply. The parallel electrodes (Figure 13−6b) will not deposit the metal as deeply as tandem electrodes will; however, the weld penetration pattern can be widened to encompass completely a number of narrow weld beads. Another possibility in using multiple electrodes is the multiple electrodes connected in a series so that one electrode is the cathode and the other the anode. The electrodes are parallel to each other across the weld work line. This three o'clock welding yields the least amount of penetration, but permits welding thin-gage metal at a high rate of speed. The penetration will be thin; therefore, the metal to be welded is usually under $\frac{1}{2}$ in. thick.

Normally, the submerged arc welding process is known as a process that achieves deep penetration. However, penetration of the series dual electrode is based on the magnetic lines of force opposing each other, resulting in a force being exerted at an angle of 90° to the electron flow which accounts for the shallow penetration (Figure 13−7). This process is applicable to the surfacing of a metal with a dissimilar metal, such as in hard surfacing. The submerged arc process lends itself to the production of high-

Magnetic lines of force (flux lines)

Electrode

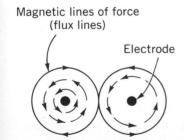

Resultant tangent vector force exerted upon electron flow of arc columns

Fig. 13−7 Series-dual electrode penetration theory.

quality welds, where the weld surface will show a uniform-
ity not obtainable with manual welding. The elimination
of fumes, smoke, and any visible arc column adds to its
ease of operation and efficiency, which, in turn, encourages
its application in the welding industry.

QUESTIONS

1. What is a submerged arc?

2. How is fluxing powder added in the submerged arc welding process?

3. What is the function of the flux?

4. What are the types of flux?

5. Why are the electrodes coated with a copper mist?

6. What are the advantages of the submerged arc process?

7. What effect does wire slant have upon the weld bead?

8. What are the types of submerged arc welding machines?

9. Why is submerged arc generally limited to welding in the flat position?

10. What are the differences between the tandem and parallel applications of the multiple electrode submerged arc welding process?

11. How does submerged arc electrode placement determine the penetration pattern of
the weld bead?

SUGGESTED FURTHER READINGS

Morris, Joe Lawrence: "Welding Processes and Procedures," Prentice-Hall, Inc., Englewood
Cliffs, N.J., 1955.

Phillips, Arthur L.: *Current Welding Processes,* American Welding Society, New York, 1964.

Rossi, Boniface E.: "Welding Engineering," McGraw-Hill Book Company, New York, 1954.

Séférian, D.: "The Metallurgy of Welding," John Wiley & Sons, Inc., New York, 1962.

14 CARBON ARC WELDING

Carbon arc welding was probably the first electric welding process developed. The first recorded attempt to electric weld was made in 1881 when Auguste de Meritens, a French inventor, used the carbon arc as a welding source with storage batteries as a power source. In his experiment, he connected the work to the positive pole of a battery and attached a carbon rod to the negative terminal. Although this was an inefficient method, it did accomplish the fusion of lead to lead, and thus fusion welded several parts together with an electric arc.

Carbon arc welding is a process in which the fusion of metal is accomplished by the heat of an electric arc. Generally, no shielding atmosphere is used. Pressure is not used and filler rod is used only when it is necessary. The carbon arc welding process differs from the more common

forms of shield arc welding in that a carbon or graphite
rod is used instead of the consumable flux-coated electrode. The only function of the arc column in carbon arc
welding is to furnish heat to the base metal. The heat
that results from the arc between the base metal and the
carbon rod or between two carbon rods is used in much the
same manner as the flame from the oxyacetylene torch in
that it is used to melt the base metal or filler rod, allowing
a fusion weld to take place between base metals. Carbon
arc welding was commonly used in the fabrication of aluminum and steel parts before introduction of the inert-gas
welding processes. The carbon arc process may also be
used for the welding of copper, nickel, Monel, and other
nonferrous, as well as ferrous, metals. Carbon arc or
carbon electrode welding can also be used as an excellent
heat source for brazing, braze welding, soldering, and heat
treating.

The carbon arc processes use either dc or ac power
sources. Either of these two power sources can have a
manually operated or an automatically operated carbon
arc electrode holder. Either, depending upon the amperage, can be either air-cooled or water-cooled. Generally,
electrodes that operate in excess of 200 amps are water-
cooled, while those that operate at approximately 200 amps
or below are air-cooled.

SINGLE-ELECTRODE CARBON ARC The single-electrode carbon arc process can be used for
either welding or cutting. The power supply for releasing
the electrode carbon arc process is dc, where the electrode
is always negative (Figure 14-1). The electric arc is created

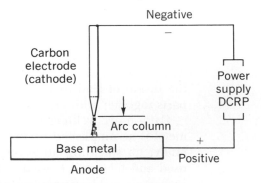

Fig. 14-1 Single carbon arc.   Arc length average, 1 1/2 in.

between the cathode, or the negative end of the electrode, and the anode, which is always the base metal. The electric arc is simply a sustained electrical discharge between the cathode and anode through a gaseous ionized field, which is composed of the gases in the atmosphere. As the gases begin to ionize, resistance builds up. As the resistance becomes stronger, a plasma state results between the cathode and anode, causing three basic heat zones in the arc: one at the cathode, one in the plasma area of the arc column, and one at the anode. The heat of the arc column is caused by negatively charged electrons, which flow off the cathode, pass through the small plasma column, and strike the anode. The mass of the electrons, although small, transmits energy, which can be converted into heat. The basic formula for this conversion of energy is $W = EIT$. W is equal to the heat, E is the voltage or the potential, I is the amperes or current, and T is time in seconds. This formula indicates the relationship between the input of electrical energy and the output of heat energy. The length of the arc column is restricted by the atmosphere surrounding it. The electrons can travel long distances in a vacuum; however, when the electrons are subjected to the atmosphere, they can travel only short distances. The distance depends upon the amount of potential energy that is crossing the arc gap, making the arc hottest at the cathode, or the negative pole. The plasma closest to the cathode will be hottest. As the voltage crosses the arc column, the voltage drop causes a corresponding decrease in the amount of heat that is released. Electron-beam welding devices were begun by placing the cathode in a vacuum.

ELECTRODES The electrode used in carbon arc welding is approximately 12 in. long. It varies in diameter from $3/32$ to $1/2$ in., and is made of carbon or graphite (Chart 14–1). Each material has its own operating characteristics. Graphite electrodes are harder and more brittle. They withstand higher current densities, lasting longer than electrodes made of carbon; however, the arc column is harder to control. On the other hand, carbon electrodes are softer and do not carry as much current as the graphite electrode, but the arc column is easier to control. In general, with the carbon electrode, higher-current density adds to arc stability, and a low-current density means longer life. Although both the

CHART 14-1 Carbon arc electrodes

SIZE (DIAMETER), IN.	AMPERAGE	
	CARBON, AMPS	GRAPHITE, AMPS
$\frac{1}{8}$	15-30	15-35
$\frac{3}{16}$	25-55	25-60
$\frac{1}{4}$	50-85	50-90
$\frac{3}{8}$	100-150	110-165
$\frac{1}{2}$	300	300
1	600	600

carbon and graphite electrodes are considered nonconsumable, the disintegration of the electrode occurs at a slow rate during the process of welding, as a result of vaporization and oxidation. The disintegration process can be greatly increased by reversing the polarity so that the workpiece is the cathode, causing the electrode to disintegrate more quickly from the higher heat. However, when DCSP is used, causing the electrode to act as the cathode, disintegration occurs more slowly so that there is little contamination of either graphite or carbon in the weld zone.

Carbon arc electrodes are easier to manipulate if a point is ground on the electrode. This point resembles the point on a lead pencil, angling back approximately $\frac{3}{4}$ to 1 in. The tip should have approximately a $\frac{1}{16}$-in. diameter (Figure 14-2). If the point is too small, it will burn off too quickly, leaving a broad face that will make the arc column uncontrollable. If the point is too sharp, it will not be able to carry the correct amperage and will burn a bright cherry red. In this situation, the amperage can be reduced or the electrode point reshaped. The electrode is generally placed in the holder so that approximately 3 in. of it extends from the electrode holder. The electrode is easily inserted into the holder and tightened in collet fashion. The electrode is started by simply adjusting the amperage on the dc welder, turning the welder on, and bringing the carbon electrode into contact with the workpiece. After contact has been made, the arc column will start. The electrode is then withdrawn from 1 to $1\frac{1}{2}$ in., and the arc is maintained at this distance. The arc is discontinued by simply removing the electrode from the workpiece completely (Figure 14-3).

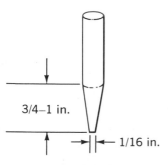

3/4-1 in.

1/16 in.

Fig. 14-2 Electrode point.

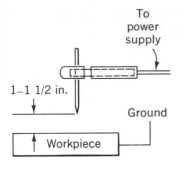

To
power
supply

1–1 1/2 in.

Ground

Workpiece

Fig. 14–3 Electrode holder.

The joint designs that can be used with carbon arc include any of the designs that are used with oxyacetylene welding or shield arc welding: butt joints, bevel joints, flange joints, lap joints, and fillet joints.

Carbon arc welding has certain advantages in that it is relatively simple to control the temperature of the molten pool by varying the arc length. The process is also easily adaptable to automation, especially if the volume of weld deposit needed is great and the materials to be fabricated are of simple geometric shapes, such as in the fabrication of water tanks. These simple shapes need only to have square joint edges because of the deep penetration characteristics of this process.

There are disadvantages, too, in using the carbon arc process. A separate filler rod must be used if any filler material is needed. The carbon arc is only a heat source and does not transfer metal to help reinforce the weld joint. Adequate joint design can overcome this disadvantage, however. The major disadvantage of this process is that blow holes occur, which is common to all dc welding processes as well as to the carbon arc process. The blow holes result from the turbulence caused by the magnetic field that surrounds the arc column, a phenomenon known as magnetic arc blow. This condition is present when the arc begins to wander and becomes erratic. Magnetic arc blow occurs most often when welding near the edge of base metal.

Fluxes and slag introduced into the carbon arc weld vary according to the metals to be welded and in what position the weld will be made. The means by which the flux or slag is introduced into the arc combine techniques from both the gas-welding process and the shield arc-welding process. Fluxes are introduced into the arc stream to control the ions of carbon and metal in the arc as well as certain substances derived from the atmosphere, such as the oxides of carbon and carbon-nitrogen mixtures. These substances decrease the control over the arc. Concentration of oxygen is greatest in the ionization area of the arc column and lowest in the plasma area of the arc column. Because of the extreme heat in the plasma area, the oxygen is consumed. A neutral atmosphere for the weld zone, so that it can be protected from oxidation caused by the atmosphere, can be provided by simply injecting a piece of paper into the arc stream. When this paper substance burns, it forms a small amount of water vapor and a large amount of carbon monoxide. Carbon monoxide is not compatible with steel. By forming a carbon monoxide atmosphere, successful fusion can be accomplished. Many of the nonferrous metals have commercial fluxes which will protect the weld zone from oxidation. Carbon arc welding requires more flux than gas welding and less flux than electric arc welding. The common method used for controlling the atmosphere surrounding the weld zone is derived from a technique used in the gas-welding processes, however. The flux is put on the filler rod, if any is used, or joints to be welded are fluxed.

TWIN CARBON ARC WELDING

The twin carbon arc welding process operates on the same principle as the single carbon arc process, except that the arc is maintained between the two carbon electrodes instead of between the carbon electrodes and the base metal. In the twin carbon arc process, the electric torch consists of two carbon electrodes that can be adjusted by means of a thumb-operated mechanism located on the handle of the electrode holder (Figure 14-4). The electrodes can be moved closer together or farther apart, making the arc length, by this movement, either hotter or colder to suit the desired heat of the weld. Another method used to control the heat radiated into the base metal is to vary the distance between the electrodes and the base metal (Figure 14-5). Low heat in the base metal is maintained by holding the carbon electrodes far enough away from the base metal so that the arc column is clearly between the two carbon electrodes. For slightly more heat, the carbon electrodes are moved closer to the base metal, which allows part of the arc column to jump to the base metal, thereby heating the base metal through direct contact with the hot gases in the arc column as well as heat radiation. The base metal can be heated to as high a point as possible with the twin carbon arc torch by holding the electrodes in close proximity to the base metal, allowing the major part of the arc column to play upon the base metal.

The power supply for the twin carbon arc torch is ac. Generally, an ac welding machine with a duty cycle of at least 20 percent is used because any shorter length of time would not make the welding process profitable. A

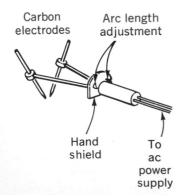

Fig. 14-4 Twin carbon arc torch.

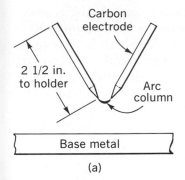

(a)

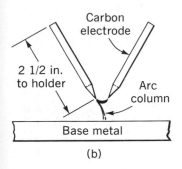

(b)

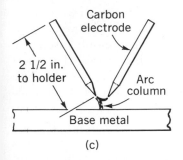

(c)

Fig. 14-5 Base metal heat control. (a) Low heat. (b) Medium heat. (c) High heat.

20 percent duty cycle means that the welding machine may be operated 2 min out of each 10-min period. More use than this will overheat the windings and cause the insulation inside the machine to break down.

The size of the electrode depends upon the thickness of the material to be welded. Generally, the diameter of the electrode approximates the thickness of the metal to be welded. After the electrodes have been selected, a point is ground on each. This point, however, is not as sharp as the point on the single carbon arc torch but is approximately ⅛ in. in diameter, resembling a blunt pencil point. The electrodes are then inserted into the torch and extended from 2 to 2½ in. beyond the clamping device. The ac welding machine is checked to be sure that it is off and then the correct amperage is set (Chart 14-2).

CHART 14-2 Amperage settings for the twin carbon arc torch

ELECTRODE SIZE, IN.	ALTERNATING CURRENT, AMPS
¼	30-60
⁵⁄₁₆	70-80
⅜	90-125
½	150-200

The correct amperage setting is determined by the amount of heat desired; example, a ¼-in. electrode needs an amperage setting ranging from 30 to 60 amps depending on the thickness of the metal to be heated. If the metal were ⅟₃₂ in. in diameter, the amperage setting would be toward the lower end of the range, or around 30 amps; if the metal were approximately ⅟₁₆ in. thick, the amperage setting would be 50 to 60 amps. The carbon rods are brought together until the shorting starts in order to strike an arc. Then the electrodes are adjusted outward until a gap, approximately ⅟₁₆ in., is established between the electrodes. When the arc is established, it creates rays that cause severe damage to exposed skin or to eyes. Therefore, an arc welding face shield, work gloves, and appropriate long-sleeved shirts must be worn. Most of the eye injuries occur in the carbon arc process when the arc is adjusted. Many welders try to make the adjustment between the electrodes without using eye protection.

The twin carbon arc torch is similar to the oxyacetylene

gas welding torch; however, carbon arc welding is easier
to use when welding dissimilar metals because the stream
of the carbon arc torch lacks the excessive carbon contami-
nators present in the oxyacetylene torch. Carbon arc
welding is particularly successful in joining copper-alloy
base metals to each other and to ferrous metals. When
dissimilar metals are joined, great care must be taken to
insure that the proper filler metal is selected before welding. For the copper-ferrous weldment, a silicon-bronze
filler rod is generally used, the same filler rod used for the
joining of galvanized steel. The technique employed is the
same as that used in oxyacetylene welding, except that an
arc stream is controlled instead of a chemical flame. A
weld with good strength and ductility will result. The carbon arc process is more versatile than the oxyacetylene
process, however, because the heat is more easily controlled. Consequently, the carbon arc process is sometimes
chosen for the welding of nonferrous metals that must be
welded at low temperature, such as aluminum, nickel,
zinc, and lead.

Both carbon arc welding processes adapt easily to inert
gas shielding of the weld. When the single carbon rod is
used, it is a simple matter to adapt the rod fixture so that
inert gas can be ejected around the rod. The adaptation
is made by placing a cup around the carbon rod and ar-
ranging a device to control the amount and velocity of
the shielding gas. The gases most commonly used are
helium and argon, or a mixture of these gases. If an inert
gas shield is used, no other type of flux is needed. The
flowing gases are used to shield the weldment from contamination. The ionized portion of the gases helps to retain the heat of the arc as well as to wash out any foreign
particles. Inert gas also cools the electrode, thus lessening
the possibility of flaking or deterioration. A weld can still
be seriously contaminated, however, by using improper
gas volume or velocity and by contaminating the weld with
the carbon rod itself if it is dipped into the weld metal.
Selecting the wrong type of current for the carbon rod also
causes contamination. DCRP causes the carbon rod to
deteriorate rapidly and to project a sooty residue over and
through the weld. Because the carbon arc process is readily
adaptable to the use of inert gases and is free of contamination, it was used extensively at one time in the welding

of aluminum and magnesium. Now, however, in the welding of these metals, the carbon arc process has been replaced by the TIG process, with which more efficient techniques are available.

AIR CARBON ARC The air carbon arc cutting torch is an adaptation of the standard single carbon arc welding torch. The major difference between the single carbon arc welding torch and the air carbon arc cutting torch is the holding device. The carbon electrodes remain the same. The electrode holder has passages so that compressed air can be ejected in line with or on the same axis as the carbon electrode (Figure 14-6). The power supply, as in single carbon arc welding, is dc; however, because of the penetration patterns resulting from straight polarity, reverse polarity is used. The carbon electrode acts as the positive pole, or the anode, because of the unequal heat distribution within the arc column in its three major areas, the cathode, the plasma, and the anode. More heat is liberated at the anode than at the negative pole, the cathode. This phenomenon, which is referred to as reverse polarity, is commonly used to cut stainless steel and steel; however, it is not suitable for cutting nickel. DCSP works better with nickel and its alloys. Special electrodes have been designed that can be used with alternating current for the power supply. These electrodes, used with ac power, have been found to

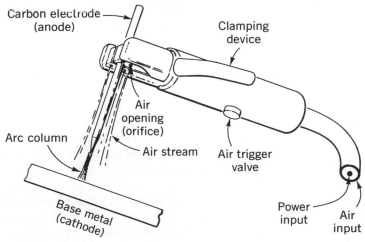

Fig. 14-6 Air carbon arc electrode holder.   (14-6)

work well with cast iron or copper base alloys. When direct current is used on cast iron or copper-base alloys, an unstable arc is produced, which is difficult to control. The diameter of the electrode depends upon the thickness of the material to be cut. Although a small-diameter electrode can be used to cut thick material, it will not cut efficiently. For efficient cutting, the diameter of the electrode should be increased as the thickness of the material to be cut is increased. Current settings for air carbon arc cutting range from approximately 80 to 1400 amps. The compressed air supply ranges from 60 to 100 psi (Chart 14-3).

CHART 14-3 Amperage settings for the air carbon arc torch

ELECTRODE DIAMETER, IN.	CURRENT, AMPS	COMPRESSED AIR SUPPLY, PSI
3/32	80-150	60-80
3/16	110-200	60-80
1/4	150-350	60-80
3/8	300-550	80-100
1/2	400-800	80-100
3/4	800-1400	80-100

The cutting procedure when using an electric cutting torch is much the same as that for oxyacetylene cutting, except that little or no preheating is necessary. As in oxyacetylene cutting, cuts can be easily made in the flat, vertical, and horizontal positions. When cutting in a level position, gouging or piecing can be accomplished. Practically any operation that is performed with the oxyacetylene cutting torch can be performed by the electric cutting torch. However, the electric cutting torch generates more heat than does the oxyacetylene torch. The electric torch generates temperatures around 10,000 to 11,000°F in the arc column; the oxyacetylene cutting torch generates a maximum of 6500°F.

In recent years, a further adaptation of the air carbon arc torch has been made, replacing the carbon electrode with a hollow steel electrode. This steel tube is covered with a flux that protects the arc column, and the hollow inside provides a passage that directs the oxygen to the cutting area. These newer steel-tubed electrodes have outside diameters ranging from 3/16 to 5/16 in. and are from 14 to 18 in. long.

QUESTIONS

1. What are the differences between the two types of carbon arc welding?

2. Why are some of the carbon arc torches air-cooled only?

3. What does the formula W = EIT explain?

4. What function does the electrode perform in carbon arc welding?

5. How is the arc column started?

6. How does the diameter of the electrode affect the amperage requirement?

7. What are the advantages of the twin carbon arc process over the single carbon arc process?

8. How is the base metal heat regulated with the twin carbon arc process?

9. What is the relationship between the thickness of the base metal to be welded and and the diameter of the electrode that is chosen?

10. Why would the carbon electrode be replaced with a tubular steel electrode?

11. Explain the operation of the air carbon arc cutting torch.

PROBLEMS AND DEMONSTRATIONS

1. Demonstrate how to weld a bead with the single carbon arc and the twin carbon arc.

2. Weld a bead with a square ended electrode. Then grind the proper point on the electrode and weld another bead. Notice the difference in controlling the arc column.

3. Weld a bead with the twin carbon arc torch varying the torch height. Notice how the arc column shifts from the electrode to the base metal as the height is decreased.

SUGGESTED FURTHER READINGS

Henry, O. H., and G. E. Clausen: *Welding Metallurgy,* American Welding Society, Inc., New York, 1962.

Kerwin, Harry: "Arc and Acetylene Welding," McGraw-Hill Book Company, New York, 1944.

Procedure Handbook of Arc Welding, The Lincoln Electric Company, Cleveland, 1957.

Rossi, Boniface E.: "Welding Engineering," McGraw-Hill Book Company, New York, 1954.

Séférian, D.: "The Metallurgy of Welding," John Wiley & Sons, Inc., New York, 1962.

Strong, C. L.: The Amateur Scientist, *Scientific American,* November, 1966, pp. 144–148.

Wulff, John, Howard F. Taylor, and Amos J. Shaler: "Metallurgy for Engineers," John Wiley & Sons, Inc., New York, 1956.

Ⅲ GAS SHIELD ARC WELDING

15 INTRODUCTION

The theoretical development of gas shielded arc welding was begun in approximately 1890, when a patent was granted for the idea of surrounding a bare electrode with a carbon dioxide gas shield. This idea eventually led to the atomic-hydrogen arc-welding process, which is the basis for the inert-gas arc-welding processes of today. In 1920, a nonconsuming tungsten electrode, capable of carrying high currents, was developed. Also in 1920, helium and argon, two inert gases, were found to protect the tungsten electrode and the weld puddle from the oxidation that occurs at the high welding temperatures created by the arc between the tungsten electrode and the base metal. Henry M. Hobart and Philip K. Bevers were granted patents in 1930, covering the arc welding of metals without flux in an atmosphere of either helium or argon with electrodes which have high melting temperatures, such as tungsten and carbon electrodes.

Because of the high price of the inert gases, they were not used for any commercial welding until 1940; however, helium was a fairly cheap gas to produce and the atomic-hydrogen welding process was used for joining many nonferrous and ferrous metals. In 1940, Russell Meredith developed the inert-gas process for the welding of aluminum and magnesium airplanes. The first inert-gas shield arc process used a shield arc electrode holder with a tube extending almost to the electrode point. The electrode used was a tungsten rod, and helium was the gas piped to the arc column. This first process was called Heliarc. From this basic process, the process of metal inert-gas (MIG) welding has logically followed. In MIG welding the tungsten electrode has been replaced by a bare wire electrode that feeds automatically through the electrode clamps of the holder, supplying the material for the arc column and the transfer of metal. This process also uses a shielding gas.

The tungsten inert-gas process (TIG) was also developed, in another logical step from the original Heliarc process, into the plasma arc welder. The plasma column that exists in the arc column has been expanded, constricted, and controlled so that it can now be used as an ultrahigh-temperature cutting and welding device.

ATOMIC-HYDROGEN WELDING

Atomic-hydrogen welding is a nonpressure fusion welding process. The atomic-hydrogen welder is a heat supply only and does not apply any pressure to the base metal. It is used only for heating base metal; however, the hydrogen arc column that results from the interaction of the arc column and the recombining of the disrupted hydrogen molecules has an even temperature and is an easily controllable heat source.

The theory behind the production of the heat derived from the atomic-hydrogen welding process is that an arc column is established between two tungsten electrodes with an ac power supply. Standard hydrogen, also called molecular hydrogen, is forced through this arc column between the two electrodes, which separates the hydrogen into its atomic form. Hydrogen is a diatomic gas, each

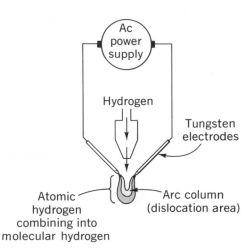

Fig. 15-1 Atomic-hydrogen flame.

molecule of which contains two hydrogen atoms. The molecular hydrogen is forced to separate because of the extreme temperature in the arc column. As the hydrogen particles separate, they become atomic hydrogen particles which accelerate because of the low pressure being exerted on the molecular hydrogen that is being injected into the arc column. The atomic hydrogen cannot maintain itself in this unnatural state, and, seeking its stable state, it will come back to recombine into molecular hydrogen. The molecular hydrogen is formed as the hydrogen atoms fuse together. This fusion of the atomic hydrogen into molecular hydrogen releases a consistent heat energy of approximately 5300°F (Figure 15-1).

The arc column established between the tungsten electrodes that are used in atomic-hydrogen welding is slightly different from present-day arc columns. The electron travel between the electrodes is not in a direct path from one electrode to another but is in the shape of a fan (Figure 15-2). The reason for this arc fan instead of a traditional arc column is that the hydrogen, being forced directly between the electrodes, causes the electrons and ions to travel in the direction of the force, thus distorting the arc column into a fan shape. The fan shapes can be manipulated by either offsetting the electrodes, varying the distance between the electrodes, or changing the amount of current being induced into the system. There are three basic arc fans (Figure 15-3). The silent arc fan results when low current is used and the electrodes, either off-

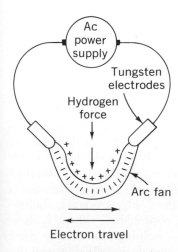

Fig. 15-2 Atomic-hydrogen arc column.

set or standard, are close together. The standard, manual arc fan and the automatic arc fan, which is used for automatic welding process, are the other two types. The tungsten electrodes used in atomic-hydrogen welding are long-lived because of the ac power supply. If the power supply were dc and the electron flow were in only one direction, one electrode would melt and become contaminated long before the other because of the electrons impinging upon the one electrode; however, with alternating current, the electron flow is alternated from one electrode to the other, allowing the tungsten electrodes to cool slightly. Also, as the electrons flow off the surface of the tungsten, the electrodes are cleansed alternately as the cycle changes in the alternating current. These factors give the tungsten electrodes long process lives.

TUNGSTEN INERT-GAS (TIG) WELDING

In all welding, the best weld is one that has the properties closest to those of the base metal; therefore, the molten puddle must be protected from the atmosphere. The atmospheric oxygen and nitrogen combine readily with molten metal which yields weak weld beads. In TIG welding, the weld zone is shielded from the atmosphere by an inert gas, which is ducted directly to the weld zone where it surrounds the tungsten. The major inert gases that are used are argon and helium.

The TIG process offers many advantages. TIG welds are stronger, more ductile, and more corrosion resistant than welds made with ordinary shield metal arc welding. In addition, because no granular flux is required, it is possible to use a wider variety of joint designs than in conventional shield arc welding or stick electrode welding. The weld bead has no corrosion because flux entrapment cannot occur, and because of the gas fluxing there is little or no postweld cleaning operation. There is also little weld metal splatter or weld sparks that damage the surface of the base metal as in traditional shield arc welding. Fusion welds can be made in nearly all commercial metals. The TIG process lends itself ably to the fusion welding of aluminum and its alloys, stainless steel, magnesium alloys, nickel-base alloys, copper-base alloys, carbon steel, and low-alloy steels. TIG welding can also be used for the combining of dissimilar metals, hardfacing, and the surfacing of metals (Figure 15–4).

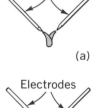

Electrodes

(a)

Electrodes

(b)

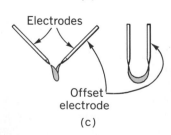

Electrodes

Offset
electrode

(c)

Fig. 15–3   Arc fan. (a) Silent arc fan.
(b) Standard manual arc fan.
(c) Automatic arc fan.

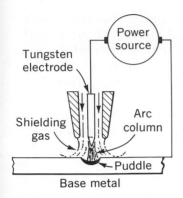

Fig. 15-4 Tungsten inert gas (TIG).

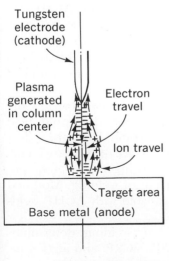

Fig. 15-5 TIG arc column.

The TIG arc column is similar to that of the standard shield arc welding arc column. The electron travel is identical to that in shield arc welding (Figure 15-5). The amount of plasma that is generated in the center of the arc column, however, is thought to be greater than that generated in the standard shield arc stick electrode welding process because the tungsten electrode has a higher thermionic rating than the mild-steel electrode does. The thermionic function or emission is the ability of a metal to release its valence electrons.

There are three basic power supplies used in the TIG welding process. They are a direct current straight polarity (DCSP) power supply, the direct current reverse polarity (DCRP) power supply, and the alternating current high-frequency (ACHF) modified power supply. In the DCSP power supply, the tungsten electrode is negative or the cathode, and the base metal is positive, or the anode. The electron flow is from the cathode to the anode. Because of the direction of this flow, approximately two-thirds of the total heat in the arc column, which is approximately 11,000°F, is released in the base metal. The tungsten electrode, then, can be of a very small diameter because the tungsten will not receive the major portion of the heat. It can have a smaller mass to dissipate the heat that it does absorb. The positive ions flow from the base metal to the tungsten electrode. When they impinge on or strike the electrode, a small amount of heat is released. The positive ions strike molecules in the atmosphere, creating an ionization layer of gas or an ionized gas, which protects the major electron flow as in shield arc welding. The DCSP power supply does not perform any cleaning action; therefore, the base metal must be clean in order to use DCSP (Figure 15-6).

In DCRP power supply, the electron flow is from the base metal, or workpiece, to the electrode. Because of this electron flow to the tungsten electrode, approximately two-thirds of the heat generated by the arc column is absorbed by the tungsten electrode, which places a requirement for a larger mass upon the electrode. The minimum diameter of the electrode mass is thought to be ¼ in. so that it can absorb excess heat. The electrons that leave the base metal and are accelerated toward the electrode carry with them some of the oxides that protect the surface of the metal, especially in the case of aluminum. The cleans-

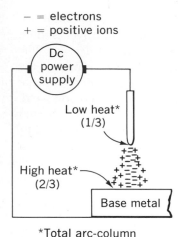

*Total arc-column
temperature = 11,000°F

Fig. 15-6   TIG DCSP.

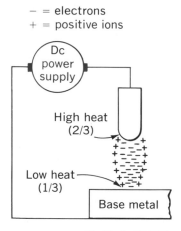

Fig. 15-7   TIG DCRP.

ing of the oxide film results from the bombardment of the surface by the positively charged ions, which are thought to cause an electrolytic disassociation of the surface oxide from the metal, and by the explosion of the electrons off the base metal, through the oxide film, toward the anode electrode (Figure 15-7).

Alternating current is also used as power supply for inert-gas welding. However, because alternating current is a combination of DCRP and DCSP, the current value is essentially zero at the instant when the current reverses direction. At this point, the arc column is extinguished and must be reignited at the zero amperage point between DCSP and DCRP (Figure 15-8a). Even though this standard alternating current does extinguish the arc column, it can still be used; but it is extremely difficult to maintain or establish an arc column. The arc column that is established is unstable and erratic. The arc can be stabilized by superimposing by a machine a high-voltage, high-frequency dc cycle. Instead of a cycle running for a given time, like 1 sec, there will be, on high frequency, 100 Hz running in this same 1-sec period of time, minimizing the effective zero point between DCRP and DCSP. The hard-starting characteristic of a standard ac cycle is alleviated, and unstable, erratic arc qualities are corrected. With ACHF, there are better arc-starting capabilities, and longer arc lengths can be maintained without extinguishing the arc column. In the standard ac cycle, one-half of the heat is absorbed into the tungsten electrode. The ACHF can be further modified by lengthening out the DCSP cycle and shortening the DCRP cycle (Figure 15-8b). This ratio between the DCSP and the DCRP may be as high as 30:1, which means that the tungsten electrode would be a cool-running electrode most of the time and the workpiece would receive two-thirds of the heat. During the short DCRP cycle, the tungsten electrode would heat slightly, but the reverse polarity would have the effect of breaking up the oxides on the surface of the base metal, performing a cleaning action, allowing the straight polarity to perform the major portion of welding. Modified alternating current is usually available with ACHF welding units.

The electron flow, as in shield arc welding, determines the penetration patterns of DCSP, DCRP, and ac (Figure

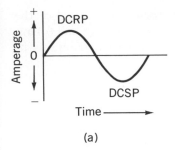

DCRP to DCSP Ratio
can equal 1 to 20

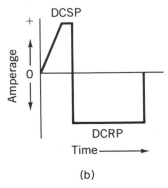

Fig. 15-8 Ac cycles. (a) Standard ac
cycle. (b) Modified ac cycle (pulsed)

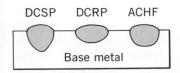

Fig. 15-9 Base-metal penetration.

15-9). The penetration of DCSP is narrow and deep be-cause the electrons impinge on the base metal. The electrode is thought of as a cool-running electrode. In DCRP, on the other hand, the electrons flow from the base metal to the electrode and the major portion of heat is absorbed by the electrode; therefore, the weld bead is relatively shallow and wide. The alternating current and the ACHF current are a combination of the DCSP and DCRP (Figure 15-9).

Three basic kinds of tungsten or tungsten alloys are used for the electrode in TIG welding: pure tungsten, zirconiated tungsten, and thoriated tungsten. Pure tungsten has a melting point of approximately 6170°F and a boiling point of 10,706°F, which gives the tungsten electrode a long life. Tungsten is also an excellent thermionic metal, or a metal that releases electrons extremely well at elevated temperature. However, slight modification of tungsten can improve the thermionic emission of electrons, which helps the starting of the arc column because the improvement results in an easier release of electrons from the tungsten alloy. Thorium or zirconium are alloyed with the tungsten in small amounts, ranging from 2 percent down to 0.001 percent of the alloying ingredient. Thorium has a melting point of 3182°F, and zirconium a melting point of 3366°F. Although there is a decrease in the melting point from the pure tungsten, the improvement in the thermionic emission of electrons aids in maintaining the life of the electrode. The mass of the electrode also helps to keep the electrode from melting, as does the inert gas surrounding the tungsten electrode, which cools it. The tip of the tungsten electrode is the only part that becomes molten. Generally, a molten droplet is formed at the end of the electrode, especially in the DCRP. With alternating current, the small ball-shaped molten droplet most definitely forms at the end of the tungsten electrode. The initial shape of the tungsten electrode can either aid or hinder the thermionic emission of electrons from the tungsten electrode. With DCSP, the tip should be ground to a conical shape (Figure 15-10). With DCRP, the tip should be ground into a conical shape but with a blunted end. With alternating current, the tip should be rounded slightly. The current passing through the tungsten electrode and the melting of the tip of the electrode slightly change the shape. With DCRP, a small ball forms at the end of the electrode, and with alternating current, a large ball forms. However, with DCSP, the tip pretty much maintains its shape.

The inert-gas consumable electrode process, or the MIG process, is a refinement of the TIG process; however, in this process, the tungsten electrode has been replaced with the consumable electrode. The electrode is driven through the same type of collet that holds a tungsten electrode by a set of drive wheels. In the TIG process, the tungsten electrode acts as the source of the arc column. The consumable electrode in the MIG process acts as the source for the arc column as well as the supply for the filler material (Figure 15-11).

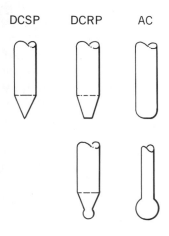

DCSP   DCRP   AC

Fig. 15-10   Tungsten electrode tips.

Three basic processes are employed in the MIG welding process: the bare-wire electrode process (Figure 15-11), the magnetic-flux process (Figure 15-12), and the flux-cored electrode process. All three processes use shielding gases. The magnetic-flux and the flux-cored MIG processes, however, yield a slightly different weld bead. The weld bead that is deposited by these two processes has a slag coating as in conventional stick electrode welding. The standard inert-gas consumable electrode process does not need any postweld cleaning. Only the magnetic-flux and the flux-cored welding processes need postweld cleaning.

The standard inert-gas consumable electrode process that just uses a bare-wire electrode and a shielding gas is capable of fusion welding aluminum, magnesium, copper base metal, nickel-base metals, titanium, carbon steel, and low-alloy steel. The deciding factor in the use of this process is the shielding gas. The shielding gases many times determine the characteristics of the arc column and the penetration of the arc column into the base metal, which is true, however, of all three MIG welding processes. A difference exists in the metals that can be joined by the three different MIG welding processes. The standard processes can join the previously mentioned metals, while the magnetic-flux MIG process lends itself more to the welding of medium- and high-carbon steels. In addition, the magnetic-flux and the flux-cored electrodes can also

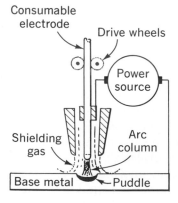

Fig. 15-11   Inert-gas consumable
electrode (MIG).

weld many of the metals that a standard consumable electrode does, except that slag removal is required. The flux components, both with the flux-cored and the magnetic-flux processes, are used to perform the usual functions of flux: to deoxidize the metal, to scavenge the metal from impurities, and to form a slag that controls post-welding heat treatment. In either the flux-cored or the magnetic-flux MIG process, carbon dioxide is the major shielding gas used (Figure 15–12).

METAL TRANSFER

Two basic mechanisms of metal transfer are used in the MIG welding processes; the shorting arc, or the dip-transfer, method, and the spray arc method. The difference is based on the reaction of the consumable electrodes to the emf or the potential voltage used to accelerate the electrons from the metal electrode to the base metal. The short arc method of metal transfer incorporates a four-step transfer mechanism. In the first stage of the short arc metal transfer (Figure 15–13a), the end of the electrode, moving at some rate per inch, heats and distorts because of capillary attraction, magnetic influences, and the force of gravity. Because of capillary attraction and electron movement, the metal touches the base metal. At this instant (Figure 15–13b), there is a short-circuiting effect, which lasts just for an instant. During this time, a magnetic field is set up at a 90° angle to the current flow, which necks down the metal that has short circuited (Figure 15–13c). This necking down is called the pinch effect. The pinch effect then takes over, separating the metal (Figure 15–13d). The ever-moving electrode starts to reheat, beginning another cycle; and the end of the electrode begins to go through the short-circuit interval again. A standard short-circuit metal transfer in a MIG welding machine occurs at a rate of approximately 100 to 200 short-circuiting cycles per second. One major reason that the short circuiting does take place is that the potential voltage or the emf used to drive the amperage or the current through the base metal is low, usually under 30 volts. The advantage of the short-circuiting arc metal transfer is that penetration can be controlled precisely; therefore, the short-circuiting arc lends itself readily to the welding of thin or light-gage metal. Because of this short-circuiting effect, poorly fit

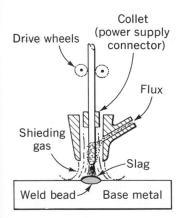

Fig. 15–12 Magnetic flux (MIG).

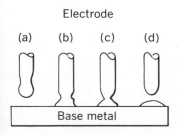

Fig. 15–13 Short arc metal transfer (MIG); 100 to 200 short-circuiting cycles per second.

joints may also be welded readily; and the welding position, whether flat, vertical, horizontal, or overhead, does not affect the quality of the weld bead. Also, because of the short-circuiting effect, temperature that is injected into the base metal is not as high as in the spray arc method so that the heat-affected zone of the base metal is smaller than the heat-affected zone when using spray arc metal transfer in MIG welding.

In spray arc metal transfer, the densities of both the current and voltage are higher; therefore, more heat energy is injected into the base metal. The process is based on the use of an electrode with a relatively small diameter, generally around $1/16$ in., and high voltage and current values. For example, a $1/16$-in. steel electrode wire requires approximately 275 amps and about 35 to 40 volts. The spray arc with standard steel electrodes operates on a DCRP power supply so that the electron movement is from the base metal to the end of the steel electrode that is being fed into the weld pool. However, the heat-liberation zones in the MIG process are opposite to those of the TIG process. When DCRP is used in the MIG process, more heat is liberated in the base metal than in the end of the electrode because the distance from the electrode to the base metal is much closer, and a larger number of metal ions are mixed into the plasma arc column. These positive ions, attracted to the base metal, strike the base metal in a much hotter condition than the positive metal ions that strike the base metal during the TIG process (Figure 15-14). The wire electrode, because of its lower melting temperature, becomes molten at a much lower temperature than the tungsten electrodes. Therefore, the ion flow creates bubbles within the molten mass of the electrode tip. The electrons impinge upon this molten mass, which has been dislocated by the small bubbles. The emf then causes the disrupted molten mass to fly off the electrode in metallic particles, creating the spray arc effect. The higher the emf, the finer the size of the particles. The distance of the filler metal from the base metal determines the effectiveness of the spray arc. Because of the metal ions and their large mass, the penetration qualities of the spray arc are confined to heavier metals and cannot successfully weld metals as thin as those that can be welded by the short arc process. The distance of the electrode from

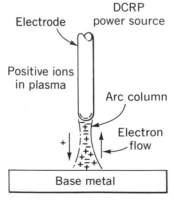

Fig. 15-14 Spray arc metal transfer (MIG).

the base metal determines the success of the spray arc. An optimum distance is approximately ½ in. from the wire electrode tip to the base metal. When the base metal is farther away than ½ in. to 1 in. maximum, the arc column becomes erratic and will wander; and the weld bead will be extremely difficult to control. One way to reduce the penetration of the spray arc is to allow the wire to stick out of the wire-feed mechanism far enough so that the wire will be preheated prior to the ignition of the arc column. This preheating will cause the burn-off rate of the electrode to be increased, thus lowering the penetration of the spray arc.

SHIELDING GAS The major shielding gases used for both the TIG nonconsumable electrode and the inert-gas consumable electrode processes are oxygen, argon, carbon dioxide, and helium or a combination of these four gases. The gases that are used in the shielding gas systems can be classified either as inert gases, which are those gases that have their outer electron subshells completely full and are stable, or the chemically reactive gases, such as carbon dioxide or oxygen, which do not have their outer electron shells filled, making it possible for the bare electrons to combine with other elements in the welding zone to create impurities.

The two inert gases that are generally used for shielding gases are argon and helium. Helium is a less dense gas than argon which means that when helium is used, more cubic feet per hour are required than with argon. Lower voltages can be used with argon than with helium because it has a lower electrical resistance. The lower voltages required in the short arc process lend themselves well to the electrical characteristics of argon. Since argon permits the operation of lower voltages at any amperage setting, it is better suited for the welding of thin metals. Also, argon does not greatly affect the heat during a manual operation in which the arc length varies as the weld operator manipulates the inert-gas arc torch.

Helium can be operated at much higher voltages. It is used often in the consumable electrode gas arc welding processes and is often used in the automatic MIG welding processes. Because a greater flow of cubic feet per hour is required with helium, greater weld speeds are possible. It is possible to weld approximately 35 to 40 percent faster with helium than when using argon for the shielding gas.

Differences in penetration are caused by the intermingling of the inert-gas molecules in the ionization column of the arc stream, creating an ionized gas, which offers more resistance to the arc column in certain locations of the weld bead (Figure 15-15). Welding speed is increased and a required penetration pattern is maintained many times by mixing argon and helium, capitalizing on the characteristics of each of the inert gases. In this way, the speed permitted by helium can be used with the penetration patterns that result from the use of argon. A combination that is used many times is a mixture of 25 percent argon and 75 percent helium or a 20 percent argon and an 80 percent helium mixture for the welding of aluminum. Most nonferrous metals can be welded with either argon or helium or a mixture of argon and helium. All ferrous metals can be welded with these gases.

Carbon dioxide is a gas that is used only for the welding of ferrous metals because of its inexpensiveness. Carbon dioxide costs approximately one-tenth as much as argon or helium and is capable of producing a high-quality weld when used as a shielding gas. The penetration characteristics of carbon dioxide are similar to the penetration characteristics of helium because of the similarities in the weights of the gases. The carbon dioxide that is used for welding must be free of all moisture because moisture releases hydrogen which will in turn produce porosity in the weld metal. Because carbon dioxide has greater electrical resistant qualities, the current setting must be 20 to 30 percent higher than those used with argon or helium. One small negative characteristic in the use of carbon dioxide as a shielding gas in a welding process is the formation of a small amount of oxygen in the metal, resulting from the heat in the arc column. This

Fig. 15-15 Shielding-gas penetration effects.

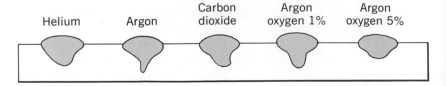

| Helium | Argon | Carbon dioxide | Argon oxygen 1% | Argon oxygen 5% |

small amount of oxygen can reduce the nominal test strength of the metal. When the oxygen molecule is released from CO_2, the gas that is formed in the outer shielding portion of the shielding gas is carbon monoxide, a toxic gas. Another major disadvantage of CO_2 is its extreme resistance to current. Because of this resistance, the arc length is sensitive. When the arc length is too great, it will extinguish more readily than when an inert gas such as argon or helium is used.

Other inert gases used for gas shielding are nitrogen, oxygen, and hydrogen. Nitrogen is used for a mixing or carrier gas with a very small percentage of nitrogen used. Oxygen can be used up to the amount of 10 percent. Oxygen used above this will yield porosity in the weld bead. Mixtures of oxygen as CO_2, or mixtures of oxygen, CO_2, and argon, are generally used for the welding of mild steel. Mixtures of nitrogen and argon, or hydrogen and argon, are used for the welding of stainless steels. The nitrogen content for the nitrogen and argon mixture can be as high as 20 percent and the hydrogen in the hydrogen and argon mixture can be as high as 15 percent. Many times the mixtures that contain nitrogen or hydrogen are used as backup gases for the far side of the weld, which protect the root of the weld bead.

QUESTIONS

1. Why is the atomic-hydrogen welding process thought to be the forerunner of the inert-gas welding processes?

2. What causes the hydrogen to disassociate in the atomic-hydrogen welding process?

3. How is the shape of the arc controlled in the atomic-hydrogen process?

4. How is heat liberated in the TIG process? The MIG process? The atomic-hydrogen process?

5. What are the current requirements of the atomic-hydrogen process? the TIG process? the MIG process?

6. Why is the heat released by the atomic-hydrogen flame of such an even nature?

7. How does the penetration differ between the TIG and MIG processes?

8. What is the function of the shielding gases?

9. How do the shielding gases affect the penetration of the arc column?

10. What are the effects of DCRP, DCSP, and alternating current upon the deposited weld bead?

11. How do thorium and zirconium affect the tungsten electrode in the TIG process?

12. Why is TIG electrode preparation so crucial to proper arc column performance?

13. How does the standard MIG process differ from the magnetic-flux MIG welding process?

14. What are the major ways that metal is transferred to the inert-gas processes? How do they differ?

15. Why do different shielding gases yield different weld bead shapes?

PROBLEMS AND DEMONSTRATIONS

1. Demonstrate the silent arc fan and the standard atomic-hydrogen arc fan.

2. With the TIG torch, weld a bead with DCSP and an appropriate current setting for the base metal. After one pass has been performed, change the polarity to DCRP and weld another bead. Notice the arc effects upon the tungsten electrode and the effects upon the base metal.

3. Select two tungsten electrodes of the same size. Prepare one electrode by carefully grinding the correct point. Leave the other point of the electrode blunted. TIG weld a bead with each electrode, using correct current settings for the base metal. Compare the stability of the arc column.

4. TIG weld a bead using argon as a shielding gas. After a bead is completed, substitute helium for the argon shielding gas. Weld another bead prior to readjusting the welding machine. Compare the effects of the different shielding gases on the penetration of the arc column and the voltage and amperage requirements.

SUGGESTED FURTHER READINGS

Bruckner, Walter H.: "Metallurgy of Welding," Sir Isaac Pittman and Sons, Ltd., London, 1954.

Giachino, J. W., et al.: *Welding Skills and Practices,* American Technical Society, Chicago, 1965.

Griffin, Ivan H., and Edward M. Roden: "Basic TIG Welding," Delmar Publishers, Inc., New York, 1962.

Simonson, R. D.: "The History of Welding," Monticello Books, Inc., Morton Grove, Ill., 1969.

The Welding of Aluminum and Its Alloys, The Aluminum Company of America, Pittsburgh, 1960.

16 EQUIPMENT

The atomic-hydrogen welding system includes a hydrogen cylinder and appropriate regulator, an ac power supply, and a welding torch. The atomic-hydrogen torch may or may not have a remote on-off switch. The trigger of the welding torch is used for moving the tungsten electrodes closer together to maintain the arc column (see Figure 16-1). The ac power supply requirement for the atomic-hydrogen welding process is the standard ac welding transformer. Almost any of the heavy-duty or medium-duty ac transformer type of power supplies may be used. The hydrogen supply can be purchased in cylinders from the local gas company or from the local oxyacetylene supplier. When the special hydrogen regulator is not available, a standard new oxygen regulator will perform as well. However, before an oxygen regulator that has been used

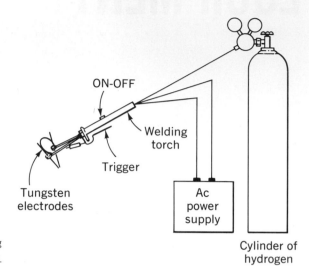

ON-OFF

Welding
torch

Trigger

Tungsten
electrodes

Ac
power
supply

Cylinder of
hydrogen

Fig. 16-1 Atomic-hydrogen welding
system.

for hydrogen gas is placed back on an oxygen cylinder, the complete regulator must be purged of all hydrogen because hydrogen is combustible, and when it is mixed with oxygen, an explosion may result. The standard hose used to pipe oxygen may also be used to pipe hydrogen if the same precautions are followed about not mixing the gases. The special piece of equipment used in atomic-hydrogen welding is the atomic-hydrogen torch (Figure 16-2).

The atomic-hydrogen torch has collets for two tungsten electrodes. The length of the electrodes can vary from approximately 2 in. to the maximum length of the tungsten electrode. In the atomic-hydrogen torch, one tungsten electrode is stationary. The movement of the other elec-

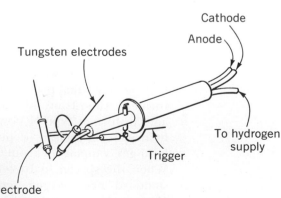

Cathode

Anode

Tungsten electrodes

To hydrogen
supply

Trigger

Fig. 16-2 Atomic-hydrogen electrode   Electrode
holder.   collets

trode is controlled by the trigger mechanism which adjusts the distance between the electrodes, called the arc or air-gap distance. The adjustment of the air-gap distance is generally the only adjustment for the tungsten electrodes. Often an on-off switch for the power supply is located on or fastened to the atomic-hydrogen torch. However, this switch may also be portable so that the atomic-hydrogen welding machine operator can simply push a button to start the current flow in the torch with the electrodes set wide, and then he can move the electrodes closer together with the trigger until the arc column is established. Generally, the flow of the hydrogen is controlled automatically. When the trigger is adjusted, the tungsten electrode moves and the orifice for the flow of hydrogen across the arc column is automatically changed, too. Common amperage ranges in atomic-hydrogen welding are from 15 amps minimum to 150 amps maximum, alternating current. This amperage range uses tungsten electrodes that range in diameter from 0.040 to 0.187 in. The voltage range for the ac power supply is from 50 volts minimum to 75 volts maximum. Standard arc welding protective clothing is used for operator protection. The lens shielding numbers used for atomic-hydrogen welding range from a standard no. 8 lens to a standard no. 11 lens, depending upon the preference of the operator. The lens used most often for atomic-hydrogen welding, as in shield arc welding, is a no. 10 lens.

TUNGSTEN INERT-GAS (TIG) WELDING The standard inert-gas welding system includes a gas supply, a gas regulator, a power supply, and the TIG torch (Figure 16-3). There are also optional machine accessories that may be purchased in order to economize on inert-gas usage or to control amperage output.

Any standard shield arc welding machine may be used to weld with a TIG torch. However, certain types of machines are more suitable than others in the welding of certain metals with the TIG welding process. The major types of machines that are used in TIG welding are the motor generator, dc only, the ac transformer, and the combination of the ac transformer that is rectified to direct current. When the ac transformer is used, the cycles that the transformer is capable of producing must be modified

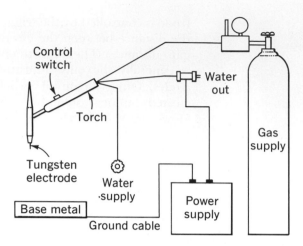

Control
switch

Water
out

Torch

Gas
supply

Tungsten
electrode

Water
·supply

Base metal

Ground cable

Power
supply

Fig. 16-3 TIG welding system.

to control the reestablishment of the arc column as the polarity of the direct current swings from positive to negative, for the arc is extinguished when the current is 0. The arc must be reignited in order to continue welding. In a standard 60-Hz ac transformer, the arc is extinguished 120 times per second, resulting in an unstable arc. By superimposing a high-frequency current, ranging from 100 kHz to 2 MHz, upon a standard ac machine, the effective starting time of the arc column is cut down and compressed into a much shorter period. The same type of welding machine usually has an open circuit voltage of 60 to 100 volts, which is the primary method of supplying high-frequency, low-voltage current to the TIG torch.

On dc welding machines, especially old machines that have little current-range control, it is advisable to install an in-line resistor to control low-current ranges, a step not necessary with new dc welding machines. The standard, relatively new dc welding machines may be used for welding with a TIG torch without any modification. The most-used power source for welding with a TIG torch is the rectified transformer with built-in gas timers and built-in water flow circuits that are turned off and on with the control switch that is located on the TIG torch (Figure 16-4).

There are two basic types of TIG torches, the air-cooled and the water-cooled (Figure 16-5). The air-cooled torch is operated in the lower amperage range, usually below 200 amps. During the welding process at 200 amps or

(a)

(b)

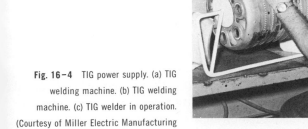

Fig. 16-4 TIG power supply. (a) TIG
welding machine. (b) TIG welding
machine. (c) TIG welder in operation.
(Courtesy of Miller Electric Manufacturing
Company.)

(c)

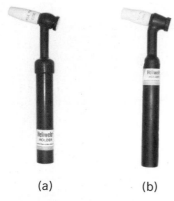

(a)   (b)

Fig. 16–5 TIG torch. (a) Air-cooled torch. (b) Water-cooled torch. (Courtesy of Air Reduction Company, Inc.)

above, a vast amount of heat is liberated. This heat must be dissipated through some medium in order to continue the efficient operation of the TIG welding unit; therefore, special water-cooled torches have been designed to carry away the heat caused by welding. These water-cooled torches can operate successfully up to a maximum of 800 amps. These torches are available in sizes that are based on amperage in approximately 100-amps intervals. There are 200-, 250-, 300-amp torches, and torches that range up to a current-carrying capacity of 800 amps that are water-cooled. The water is controlled by a water control valve solenoid in the welding machine. When the current starts to flow into the electrode, the water solenoid opens the valve. The water that flows into the torch is pure, clean tap water. The water flows out of the torch into a special connection and into a drain (Figure 16–3). The cable that carries the water out of the torch into the drain is also the power cable into the TIG torch.

The electrode clamps or collets that hold the tungsten electrode in the TIG torch vary according to electrode size. There is a specific set of electrode collets for every diameter of tungsten electrode, which insures adequate contact for the clamping of the tungsten electrode. The gas cup or nozzle of the TIG torch is the weakest part of the torch, especially when the gas cups are ceramic. The gas cups can be either ceramic or metal. Metal gas cups are used for welding at low amperages. When metal gas cups are used at higher amperages, the cup melts and clogs the orifice for the inert gas. At all high amperages and elevated temperatures, ceramic nozzles have been found to work satisfactorily. These ceramic nozzles or cups, however, are broken easily because of the intensely hot atmosphere which surrounds them. The cups are also broken by operator ineptness, as in the case of the operator making a small mistake in manipulation and inadvertently dipping the ceramic nozzle into the molten weld puddle, which contaminates the weld bead as well as ruins the ceramic nozzle. Nozzles, both metal and ceramic, are fastened to the torch body by a threaded connection.

Special regulators used in the TIG process are standard single-stage regulators with flow meters attached. The flow meter is a device calibrated to measure how many cubic feet per hour are flowing through a certain point. The

Fig. 16-6 TIG regulator. (By permission of Airco Welding Products Division.)

scale is indexed on a glass tube and the cubic-feet-per-hour rate is indicated by reading to the top of a float ball in the glass tube (Figures 16–6 and 16–7). The ball in the flow meter gage in Figure 16–7 indicates a flow of 40 ft³/hr. Most flow meters have a preset pressure of 50 psi, which is the amount of pressure that is fed into the flow meter. The flow can be adjusted by an orifice control valve. The opening of this valve will increase the flow through the flow meter tube, causing the ball to rise. One requirement in adjusting a flow meter is that the torch be turned on and the gas flowing through the torch tip in order to have an accurate adjustment.

The two major welding accessories used with the TIG welding process are the remote-controlled amperage control, which is identical to the remote amperage control used in shield arc welding, and the gas-saving device (Figure 16–8). A gas-saving device is a means to shut off rapidly the gas supply to the torch. It operates on a straight valve technique with weight placed on a hook. The hook travels downward and then automatically, through a mechanical linkage, shuts off the water line and the gas line to the torch, leaving only the inert gas in the line for postweld waste.

INERT-GAS METAL ARC WELDING

The basic MIG welding system incorporates an inert-gas supply, a regulator, flow meter, appropriate hoses, a welding power supply, a wire-feed unit, and, according to the amperage, a water-cooling unit (Figure 16–9).

The major power supply used in MIG welding is DCRP. The major types of machines used to supply DCRP to the system are the dc generator and the ac rectified transformer. The MIG welding machine must produce either a constant-arc voltage or a rising-arc voltage (RAV). The CAV was the first type of voltage amperage characteristic used in MIG welding machines. The constant voltage amperage characteristic gives an operator a great latitude in the arc length without any appreciable change in voltage. The operator first adjusts the arc voltage to a desired level and then the machine maintains that level over an extremely wide range of amperage settings. A RAV or a CAV machine usually has a voltmeter and an ammeter for voltage and amperage adjustment. The RAV

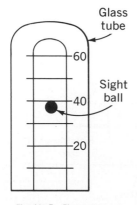

Fig. 16-7 Flow meter gage.

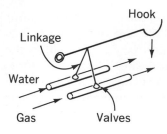

Fig. 16-8 Gas saver.

machines are more applicable to the automated consumable electrodes inert-gas metal arc welding than the CAV machines. Also, the RAV machines are capable of handling larger diameter wire than the CAV machines. The voltage range for the RAV is a greater range than the range of the CAV machines, and as the amperage increases in the RAV machine, the delivered voltage to the arc increases also, a typical characteristic of all RAV machines (Figure 16-10).

The wire-feed units contain all the apparatus that controls the wire feed, the gas flow, and the on-off switch for the voltage and amperage to the MIG gun. Wire-feed units are available that will feed the electrode wire from approximately 200 to 1000 in. of wire per minute, not to be confused with 1000 in. of weld metal deposited per minute (Figure 16-11). Many wire-feed mechanisms that are commercially available have a slow-start nature so that the wire for a part of the first second of welding will feed at a reduced rate to allow the MIG gun operator an easily started weld bead. There are two basic types of wire-feed mechanisms; the type that is used for standard MIG welding and the type for flux-cored welding. A shielding gas is unnecessary in modern flux-cored welding because of the deoxidizers that are placed in the flux inside a tubular electrode. The second type of wire-feed control unit is for magnetic-flux MIG welding. This type of wire-feed

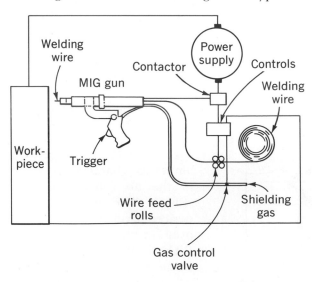

Fig. 16-9 MIG welding system.

(a)

Fig. 16–10 MIG power supply. (a) MIG power supply. (b) MIG welding with a portable control/feeder.

(b)

mechanism has some kind of magnetic-flux reservoir that feeds the flux directly into the MIG gun. Many times that flux container is located on the gun or is piped to the gun.

There are two basic types of MIG guns. They are the wire push gun and the wire pull gun (Figure 16–12). The wire push gun, smaller than the wire pull gun, pushes the wire through the rollers located in the wire-feed mech-

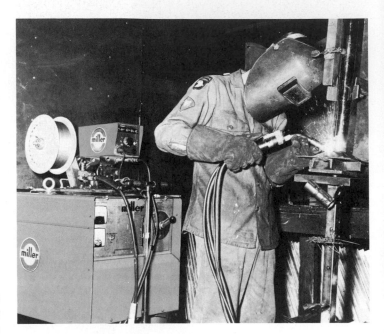

Fig. 16-10 (cont'd) (c) MIG welding
with a standard control/feeder. (Courtesy
of Miller Electric Manufacturing
Company.)

(c)

Fig. 16-11 Wire-feed mechanism.
(Courtesy of Miller Electric Manufacturing
Company.)

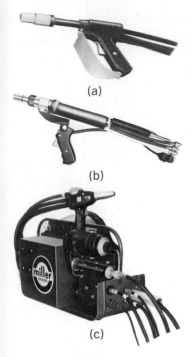

(a)

(b)

(c)

Fig. 16-12 MIG welding guns. (a) Air-cooled push gun. (b) Water-cooled push gun. (c) Air-cooled pull gun. (Courtesy of Miller Electric Manufacturing Company.)

anism. The wire-feed rolls in the wire push gun also act as a power supply to push the wire and to straighten the wire, delivering it as a specific rate to the MIG gun. The pull gun, on the other hand, pulls the wire from the roll. The gun has a set of wire-feed rollers contained within it. In some of the newer manual MIG welding guns, the guns are identified as the push-pull type and can be used either as a push-type gun or as a pull-type gun with but a few simple adjustments. Both types of guns, however, operate in the same manner. There is a simple trigger mechanism in the gun. When this is depressed, the inert-gas control valve is opened, starting the water flowing, starting voltage and amperage, and also starting the wire-feed mechanism. Many of the water-cooled MIG guns have a mechanism that automatically stops the voltage and amperage when water flow ceases, which protects the gun from overheating if something happens to the water supply.

The standard inert-gas regulator flow meters are used in the same way as they are used in the TIG process. The flow meters are calibrated according to the type of gas to be used. An argon regulator flow meter is calibrated differently from a helium flow meter regulator because of the different densities of the gases, a fact true of all the shielding gases.

As in the TIG process and in the shield arc process, the same type of operator protection equipment is used. The standard leather or cloth welding gloves, long-sleeved shirts, leather sleeves and aprons, and standard welding helmets should be used. Many operators prefer a slightly darker lens shading in the welding helmet than when shield arc welding. Because of the lack of contaminants in the air surrounding the arc column, the arc column can be more easily seen, which sometimes leads to eyestrain when a no. 8 or no. 9 lens is used. However, when a no. 10, 11, or 12 lens is used for the MIG welding process, eyestrain is minimized.

QUESTIONS

1. What are the power-supply requirements for the atomic-hydrogen process? The TIG process? The MIG process?

2. Why is it necessary for the electrodes in the atomic-hydrogen process to be tungsten?

3. What are the common electrode sizes and the accompanying current requirements of the atomic-hydrogen process?

4. What are the types of TIG torches? How are they used?

5. What are the types of MIG torches? How are they used?

6. How are flow meters regulated?

7. What are the components of the complete TIG welding outfit? The MIG welding outfit?

8. How does the wire-feed unit in the MIG welding outfit control the current settings and the stickout of the wire electrode?

9. Why are CAV and RAV welding machines preferred over DAV welding machines?

10. Why do some wire-feed units have a slow-start mechanism?

PROBLEMS AND DEMONSTRATIONS

1. Demonstrate the atomic-hydrogen torch and how the arc fan is controlled by the trigger and the current setting.

2. Demonstrate how to set up the TIG welding outfit.

3. Weld a bead with the TIG torch and a small-sized ceramic nozzle. Change the nozzle size without changing either the current settings or the gas flow. TIG weld another bead with the incorrect nozzle and compare the effects of the shielding gas on the weld bead.

4. Weld a bead with the MIG gun with the proper current and wire-feed settings. Change only the wire-feed setting and weld another bead. Notice the effect that wire feed has upon the current requirement.

SUGGESTED FURTHER READINGS

Giachino, J. W., et al.: *Welding Skills and Practices,* American Technical Society, Chicago, 1965.

Griffin, Ivan, and Edward M. Roden: "Basic TIG Welding," Delmar Publishers, Inc., New York, 1962.

Jefferson, T. B., and Gorham Woods: *Metals and How To Weld Them,* The James F. Lincoln Arc Welding Foundation, Cleveland, 1966.

Kaiser Aluminum Welder's Training Manual, Inert Gas Process, Kaiser Aluminum and Chemical Sales, Inc., Chicago, 1962.

17 TIG OPERATION

The TIG welding process can be applied to almost any type of welding that would normally use the oxyacetylene or oxygen fuel process and the shield arc process. The TIG welding process can produce strong beads over an extremely wide range of metals. The basic manipulation of the TIG process is identical to that used with the oxyacetylene process with the exception that the welding temperature is supplied by an electrical rather than a chemical source. The basic joint designs are interchangeable. A few joint designs that are especially suited to TIG welding are small adaptations of the standard weld designs that have been previously discussed.

The major joint designs that are used in TIG welding are the square butt joint, the single-V bevel, and the double-V bevel (Figure 17–1). Of course, any joint that

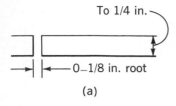

(a)

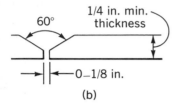

(b)

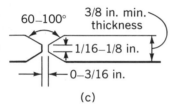

(c)

Fig. 17-1 TIG joint designs. (a) Square butt. (b) Single-V bevel. (c) Double-V bevel.

can be thought up could be applied to the TIG welding process. The joint designs for TIG in Figure 17-1 are merely samples to acquaint the welder with the possibilities and the limitations of the TIG welding process. With nonferrous metals the use of the square butt joint is limited to metal approximately ¼ in. thick. In ferrous metals, the limitation is the same as in oxyacetylene welding. The standard single V with a root distance up to ⅛ in. is used on nonferrous metals that range in thickness from ¼ in. up to approximately ¾ in. Metals that are more than ¾ in. thick require a double-V bevel. These limitations represent a standard practice in the welding industries. Selection of the proper design for a particular weldment depends upon the chemical properties that are desired in the weld; the cost of preparing the joint and performing the weld; the type of metal that is used as the base metal; and the size, shape, and appearance of the part to be welded. Many times, due to the design of the joint, filler metal will not have to be added to obtain the complete physical properties of the weld desired.

Regardless of the joint design incorporated into the weldment, if a good weld is to be made, the base metal must be thoroughly cleaned. This cleaning may be either a mechanical cleaning or a chemical cleaning. If a chemical solvent is used to cleanse the metal, the chemical contents of the solvent must be carefully controlled so that the solvent residue on the base metal will not form a toxic gas when the heat of the arc reacts with it. Mechanical cleaning is generally accomplished with a wire brush.

In most TIG welding applications, the back side of the weld joint must be protected from the atmosphere. On light-gage metal, there are four ways to apply the backing that is to protect the underside of the weld from the atmosphere. This backing protects the weld from the atmosphere, and it also prevents the metal surrounding the weld zone from sagging. The first backup method is the use of support bars, like the carbon block bars that are used in oxyacetylene welding. The second backing method is the introduction of a shielding gas into the atmosphere on the underside of the weld. The third method is the use of support bars coupled with a shielding gas to protect the weld (Figure 17-2). The fourth way to shield the back side of a weld from the contaminants in the atmosphere is to use a flux backing. The flux is generally painted on

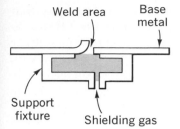

Fig. 17-2   Shielding gas backing.

the underside of the weld and is allowed to dry before welding. When exposed to the extreme heat of welding, this flux is activated and releases an amount of shielding gas, which then protects the back side of the weld. When TIG welds must meet rigid specifications, all atmospheric contaminants, such as oxygen, must be removed from the back side of the weld. This is accomplished by injecting a gas into the back side of the weld (Figure 17-2). Gases used as backing shielding gases include nitrogen, used for stainless steel, and argon, used for aluminum, magnesium, and other metals that oxidize readily. Argon is also used with metals that react with nitrogen at elevated welding temperatures. When an explosion problem does not exist, hydrogen can be used. When a backup shielding support fixture cannot be used, three or four oxygen-hydrogen flames burning continuously can provide shielding on the far side of the weld to keep the atmosphere and its contaminants out of the weld area.

The TIG welding process can be used for the fusion welding of aluminum, magnesium, stainless steel, low-alloy steel, high-alloy steel, mild steel, Monel, Inconel, brass, bronze, tungsten, silver, molybdenum, and a wide range of other metals. Also, this process can be used to weld many dissimilar metals. The TIG process can be used to braze and to supply the heat source for braze welding, and it can also be used as the heat source for the hard surfacing of materials. The major metals that the TIG process is used for are aluminum, magnesium, steel and its various alloys, stainless steel, and the copper-base alloys.

Although the TIG process was originally invented for the fusion welding of magnesium, aluminum and stainless steel have become the major metals welded by this process. For aluminum, as for all metals, the metal thickness determines the settings of the equipment. For example, a piece of aluminum 1/8 in. thick usually requires a tungsten electrode of 3/32 in. (Chart 17-1). The diameter of the

CHART 17-1   TIG-aluminum welding, AC HIGH FREQUENCY (ACHF)

METAL THICKNESS, IN.	CURRENT, AMPS	TUNGSTEN SIZE, IN.	NOZZLE SIZE, IN.	FILLER SIZE, IN.	ARGON, FT³/HR
1/16	60-90	1/16 D	1/4	1/16 D	15
1/8	110-160	3/32 D	3/8	1/8 D	18
1/4	200-350	3/16 D	1/2-3/4	1/8-3/16 D	25
1/2	200-450	1/4 D	3/4	1/4 D	31

tungsten electrode determines the size of the nozzle that shapes the shielding gas being ejected from the TIG torch. Again using the example of aluminum approximately $\frac{1}{8}$ in. thick, the nozzle size would be approximately $\frac{3}{8}$ in. in diameter. With these two things in mind, the electrode size and the nozzle size, the amperage range for ACHF is determined. In the above example, the amperage would range between 110 and 160 amps. The amperage range depends on the particular machine and the expertise of the operator. The novice operator might set the amperage in this case at approximately 110 amps. The filler metal for the $\frac{1}{8}$-in. aluminum should be approximately $\frac{1}{8}$ in. in diameter. The aluminum filler metal can be any of the standard filler metals of the Aluminum Association. The most common filler rods used for TIG welding are the 1100, and 4043, the 5154, the 5183, and the 5356 series of aluminum alloys. All of these aluminum filler rods are available in standard diameters. For TIG welding of aluminum, the arc length ranges between $\frac{1}{8}$ and $\frac{3}{16}$ in. The appropriate amount of argon indicates the estimated amount for helium. Because helium is less dense and dissipates more rapidly in the atmosphere, approximately twice as many cubic feet per hour of helium is needed than is needed to perform a like weld when argon is used as the shielding gas.

The TIG welding process lends itself readily to the welding of copper and its alloys (Chart 17–2). The major alloys of copper that are welded with the TIG process are aluminum bronze, deoxidized copper, silicon bronze, and the brasses. The standard rule of thumb in the welding of copper alloys is that the higher the proportion of the alloying ingredients, or the more ingredients in the copper-base metal, the lower the amperage setting. For example, a pure copper sheet, $\frac{1}{16}$ in. thick, would have a power requirement of approximately 150 amps, because

CHART 17–2 TIG-copper base welding, DCSP

METAL THICKNESS, IN.	CURRENT, AMPS	TUNGSTEN SIZE, IN.	NOZZLE SIZE, IN.	FILLER SIZE, IN.	ARGON, FT³/HR
$\frac{1}{16}$	90–150	$\frac{1}{16}$ D	$\frac{5}{16}$–$\frac{3}{8}$	$\frac{1}{16}$ D	13
$\frac{1}{8}$	120–200	$\frac{3}{32}$ D	$\frac{3}{8}$–$\frac{1}{2}$	$\frac{3}{32}$ D	15
$\frac{1}{4}$	150–375	$\frac{1}{8}$ D	$\frac{1}{2}$	$\frac{1}{8}$ D	15
$\frac{1}{2}$	250–600	$\frac{1}{8}$–$\frac{3}{16}$ D	$\frac{1}{2}$–$\frac{3}{4}$	$\frac{1}{4}$ D	18

of the rapid heat dissipation of copper. However, when an alloy of tin, zinc, or aluminum is added to the copper, as the amount of alloying ingredient is increased, the amperage requirement is decreased. However, the shielding gas remains constant. The extent of the copper-base alloy does not have any affect on the requirements for inert shielding gas. The filler material that is used when welding copper and its alloys is identical to the base metal being welded, which insures a 100 percent strong weld that matches the strength of the base metal.

The TIG process was originally invented for the welding of magnesium, and it is one of the few processes that can successfully and predictably weld magnesium. The amperage settings should be low when welding magnesium and the inert shield gas flow should be higher than when welding some of the less reactive metals. For example, when welding a sheet of magnesium that is $1/8$ in. thick, only 80 to 110 amps of ACHF is needed. Also, the diameter of the tungsten electrode is smaller to meet the lower power requirements. The diameter of the tungsten electrode for welding metal $1/8$ in. thick is $1/16$ in. The inert-gas demands are greater, however, because of the reactive nature of the base metal. The standard filler material, again as in copper-base alloys, is matched to the base metal (Chart 17-3).

CHART 17-3 TIG-magnesium welding, ACHF

METAL THICKNESS, IN.	CURRENT, AMPS	TUNGSTEN SIZE, IN.	NOZZLE SIZE, IN.	FILLER SIZE, IN.	ARGON, FT³/HR
$1/16$	35-60	$1/16$ D	$1/4-3/8$	$3/32$ D	14
$1/8$	80-110	$1/16$ D	$1/4-3/8$	$1/8$ D	18
$1/4$	100-150	$3/32$ D	$3/8$	$5/32$ D	20
$1/2$	200-265	$5/32$ D	$3/8-1/2$	$3/16$ D	25

DCSP is generally used for the TIG welding of mild steel, low-alloy steel, medium-carbon steel, stainless steel, and cast iron. The amperage requirements for all these steels as well as cast iron are constant. The filler metal used in the welding of steel and its alloys is generally one of the commercially available filler materials. DCRP may be used as may ACHF power. ACHF power will not change the amperage, the diameter of the tungsten electrode, or the nozzle size to any great extent; however, when DCRP is used to weld steel and its alloys, the diameter of the tungsten electrode must be increased to compensate for the extra heat that is absorbed into the electrode. The minimum electrode size that is used for DCRP TIG welding is about $3/16$ in. in diameter. Using a tungsten electrode with a larger diameter automatically requires that the size of the metal or ceramic shielding nozzle or cup be increased. As the nozzle size is increased, the amount of shielding gas is also increased. There are certain advantages to welding steel with a shielding gas process. Carbon dioxide, CO_2, can be used as an adequate shielding gas in the TIG process only when welding steel. It may not be used for the welding of cast iron; but for any of the mild, low-alloy steels, CO_2 is an excellent shielding gas. It is also much cheaper than argon or helium (Chart 17-4).

CHART 17-4   TIG mild, low-alloy, and stainless steel welding, DCSP

METAL THICKNESS, IN.	CURRENT, AMPS	TUNGSTEN SIZE, IN.	NOZZLE SIZE, IN.	FILLER SIZE, IN.	ARGON, FT³/HR
$1/16$	80-120	A $1/16$ D	$1/4-3/8$	$1/16$ D	12
$1/8$	100-200	$1/16-3/32$ D	$1/4-3/8$	$3/32$ D	12
$1/4$	200-350	$1/8$ D	$1/2$	$1/8$ D	14
$1/2$	300-450	$1/8-3/16$ D	$1/2$	$1/4$ D	15

DCRP is often used, especially with stainless steel, because of the oxide that is formed on the surface of some metals. This DCRP, as previously mentioned, helps to break up the surface oxides on metal and exposes a clean oxide-free metal in the weld area. Because of this characteristic of DCRP, many welders prefer it for the welding of aluminum, which requires only that larger-diameter electrodes and larger ceramic shielding gas cups be used. There will be no difference in the manipulation of the welding operation.

ELECTRODE PREPARATION   The preparation of the electrode is of prime concern if a strong, clean weld is to be achieved. The asymmetry of the electrode determines the gas flow pattern across the work. For example, in DCSP the electrode is generally ground to a point, or almost to a point. The more concentric this conical shape is, the more even the gas flow will be. The standard dimensions for grinding are so exact

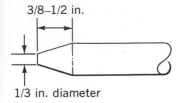

3/8–1/2 in.

1/3 in. diameter

Fig. 17–3 TIG electrode preparation.

that many companies have special grinding machines that prepare the tungsten electrode before welding. If the tungsten electrode were ground with the point off center, the shielding gas would flow unevenly through the tip, causing a portion of the molten base metal or the weld puddle to be unprotected and exposed to the atmosphere, which would result in contamination. DCRP electrodes, as well as DCSP electrodes, are prepared by a precise grinding method. Ac electrodes, especially those used with ACHF machines, do not require a precisely ground point. Instead of a point, the ideal shape for an ac tungsten electrode is a ball. This ball shape makes it easier for electrons to flow off the center of the ball. The ball is formed by striking an arc with the tungsten electrode, using DCRP, and allowing the tip of the tungsten electrode to heat and cool. Generally, the material that is used for the striking of this arc is a piece of copper or a carbon block, which decreases chances of electrode contamination. After the arc has been struck and the tip of the tungsten electrode heats, the ball will automatically form. At this point, the arc is broken and the electrode is either removed or the power source is changed to alternating current. The electrode is then ready to be used (Figure 17–3).

The basic design of the weldment determines how far the tungsten electrode will protrude from the nozzle (Figure 17–4). In flat welding, the electrode may extend 3/16 in. from the cup or nozzle. Fillet welds, because they have more difficult access than flat welding, require that the tungsten electrode extend a little farther out of the cup, up to 1/4 in., which is thought to be the maximum extension for fillet welds. The extension for corner welds ranges from a minimum of 1/16 in. to a maximum of 1/8 in. The minimum distance that the tungsten electrode extends outside the nozzle should be maintained at 1/16 in. Any extension less than 1/16 in. permits too much heat to be absorbed into the nozzle, which causes early nozzle failure.

TORCH TECHNIQUE In ac welding the electrode does not have to touch the workpiece to start the arc. The high-frequency current that is usually used in ac TIG welding enables the arc column to start as the gap between the electrode and the workpiece closes. The arc is started by holding the torch

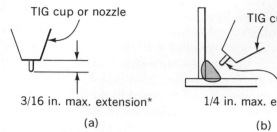

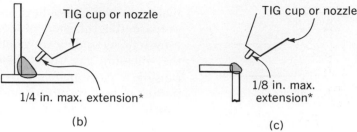

TIG cup or nozzle

3/16 in. max. extension*

(a)

TIG cup or nozzle

1/4 in. max. extension*

(b)

TIG cup or nozzle

1/8 in. max. extension*

(c)

*NOTE: 1/16 in. min. extension

Fig. 17–4 Tungsten electrode extension.
(a) Flat welds. (b) Fillet welds. (c) Corner welds.

in a horizontal position about 2 in. above the workpiece. Then the end of the torch is swung down to approximately ⅛ in. above the workpiece. The downward swing is done quickly to insure the greatest amount of protection to the weld zone. Often an operator will first swing the torch down in a practice run to determine the correct height of the electrode. This practice run can be accomplished without depressing the on-off button, which is located on the TIG torch. Then the torch is swung back up, the power button depressed, and the downward motion begun again to bring the electrode into position to create the arc column.

In dc welding, the same method is used to start the arc column; however, in the dc system, the electrode must touch the workpiece in order to start the arc. As soon as the arc has started, the electrode is then moved outward until it is approximately ⅛ in. from the workpiece. This ⅛-in. distance from the end of the electrode to the workpiece helps to minimize contamination of both the weld puddle and the tungsten electrode. A high-frequency dc current is sometimes superimposed in the beginning to help start the dc arc column. This high-frequency current eliminates the need for the electrode to touch the workpiece and can be, on most dc units with a starting cycle, turned off with a relay after the arc has been initiated. The arc can also be started on some other piece of metal and carried to a starting point. A carbon block, however, must not be used to start the TIG arc because the carbon will contaminate the electrode and cause the arc column to wander. If the torch is taken quickly away from the workpiece, the arc may be broken without any noticeable damage to the work. If the arc is taken away very slowly the work may be marred because of the splatter of the filler material and the wandering, erratic arc.

Arc wandering is the wavering of the arc across the workpiece without apparent reason. There are four causes which may contribute to arc wander. The first cause is low electrode density. At lower current settings, the entire end of the electrode is in a molten state and is completely covered by the arc. It is reacting to the stream of electrons flowing from it or to it. But when the current is too low, only a small portion of the electrode will be in the molten state and a poor unstable arc is the result. A too-high amperage setting will result in excessive melting of the tip and possible electrode contamination, because of metal transfer from the electrode to the base metal. The second cause for arc wander is carbon contamination of the electrode. When the tungsten tip is touched to a carbon block and a current is passed through it, tungsten carbide is formed. Tungsten carbide has a lower melting point than tungsten and will form a large spherical glow on the end of the electrode. This tungsten carbide will cause contamination of the weld metal and will also increase the resistance of the electrode which will lower current density.

A third cause of arc wander could be due to the magnetic effects of the current, known as magnetic arc blow. It is a condition resulting from the magnetic field caused by the current flowing through the workpiece. This magnetic field is always at a 90° angle to the current flow. This arc blow can be prevented by altering either the position of the ground connection on the workpiece or the angle of the TIG torch. Sometimes the arc blow can be dimin-

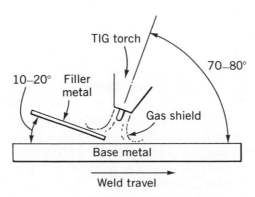

Fig. 17-5 TIG torch and filler-metal angles.

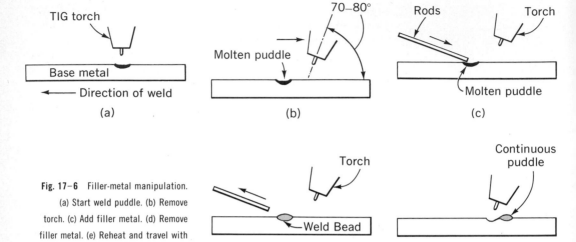

Fig. 17–6 Filler-metal manipulation. (a) Start weld puddle. (b) Remove torch. (c) Add filler metal. (d) Remove filler metal. (e) Reheat and travel with weld puddle.

Fig. 17–7 Results of shielding gas in TIG welding aluminum. (a) Sufficient gas. (b) Insufficient gas. (Courtesy Center for Technology of Kaiser Aluminum and Chemical Corporation.)

ished by tack welding the piece to be welded. Arc blow occurs mostly in dc welding. The last and most minor cause for the wandering of the TIG torch arc column is drafts of air that are directed across the arc column which disrupt the inert-gas shield. This disruption of the inert-gas shield may also be caused by an improper technique of applying filler metal or erratic TIG torch manipulation.

The lower the angle at which the filler metal is introduced into the inert-gas shield, the less disruption of the gas will take place and the cleaner the weld bead will be. The angle of insertion of the filler metal is approximately 10 to 20°, which has been found to disrupt the gas shield the least amount. This 10 to 20° angle is used in all welding positions, the flat, horizontal, vertical, and overhead positions. The filler metal is added in the direction of weld travel and the torch is slanted in the direction of weld travel, as in the oxyacetylene forehand technique (Figure 17–5). There are many exceptions to this rule; however, the beginning inert-gas welder will benefit by using this technique. The basic TIG torch movements are the same as those that are used in oxyacetylene welding, and the frequency of the addition of filler metal is also the same as in the oxyacetylene welding technique.

The only difference in the adding of filler metal is that it should be added only when the TIG torch is not heating the molten pool, which maintains minimum contamination

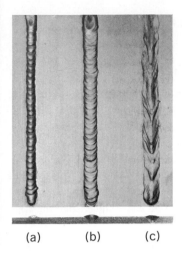

(a)   (b)   (c)

Fig. 17-8 Results of current settings in TIG welding aluminum. (a) Current too low. (b) Current correct. (c) Current too high. (Courtesy Center for Technology of Kaiser Aluminum and Chemical Corporation.)

from the tungsten electrode. The ideal way to add filler metal to the weld bead is to start the weld puddle, which will develop a molten weld puddle, to move the TIG torch away from the weld bead just far enough and to lengthen the arc just long enough to maintain the arc and to maintain the gas shielding of the weld bead, to dip the filler metal rod into the weld puddle for a sufficient length of time to add filler metal, to remove the filler rod, and then to carry the filler metal in the weld torch by manipulation of the torch (Figure 17-6). Many beginning operators of the TIG torch will overcompensate and move the torch too great a distance away from the weld, which will expose the weld bead to the atmosphere (Figure 17-7). Mastering the proper technique in order to maintain sufficient shielding gas, however, is a matter of practice. Another common error in welding with TIG torches is having current settings either too high or too low. When the current setting is too high, the bead will resemble the shield arc welding bead in mild steel when the current setting is too high. If current settings are too low, the bead will be narrow in width and will pile up on the base metal, as in shield arc welding mild steel when the current setting is too low (Figure 17-8). This error can be easily corrected by resetting the amperage on the welding machine.

QUESTIONS

1. What are the major joint designs used in TIG welding?

2. What metals can be welded by the TIG process?

3. Why must the base metal be cleaned prior to TIG welding?

4. Why is shielding gas used on the back side of the weld joint?

5. What are the advantages of the DCSP power supply? The DCRP power supply? The ac or ACHF power supply?

6. How does metal thickness affect the tungsten electrode size? The nozzle size? The shielding gas flow?

7. How is the tungsten electrode prepared for use with an ac power supply?

8. How is filler metal added to the puddle in the TIG welding process?

9. Why should the filler metal rod enter the shielding gas atmosphere at a very low angle?

10. What causes the arc to wander?

11. How can the TIG torch operator tell if there is an insufficient amount of shielding gas?

12. What is the appearance of a TIG welded aluminum bead that has been welded with the current too low? Too high?

PROBLEMS AND DEMONSTRATIONS

1. Select a piece of aluminum and TIG weld a bead with the current and shielding gas settings correctly adjusted. After welding one bead, reset the current 50 amps lower and weld another bead. Reset the amperage control to 30 amps higher than the original setting and weld another bead. Compare the weld beads.

2. Select a piece of metal and adjust the welding machine and shielding gas to their correct settings. Weld a bead and then decrease the shielding gas flow to approximately 5 ft³/hr. TIG weld another bead. Compare the weld beads during the operation and after the weld has been completed.

3. Select a piece of aluminum and set the TIG welding equipment to its proper adjustments. Weld a bead and purposely dip the tungsten electrode into the molten weld pool. Notice how the weld bead and the electrode become contaminated.

4. Insert a $1/16$-in. diameter tungsten electrode into the TIG torch. Choose a piece of metal and set the welding equipment accordingly; however, use DCRP as the power supply. Try to weld a bead. Pay particular attention to the tungsten electrode and how quickly it deteriorates when using a small electrode and DCRP.

SUGGESTED FURTHER READINGS

Bruckner, Walter H.: "Metallurgy of Welding," Sir Isaac Pittman and Sons, Ltd., London, 1954.

Griffin, Ivan H., and Edward M. Roden: "Basic TIG Welding," Delmar Publishers, Inc., New York, 1962.

Linnert, George E.: *Welding Metallurgy*, American Welding Society, New York, 1965.

Littlefield, L. L., and E. A. Cheek: Advancement of a Numerically Controlled Gas Tungsten-Arc Welding Process, *The Welding Journal*, vol. 49, no. 3, p. 183, March 1970.

Saenger, J. F.: Gas Tungsten-Arc Hot Wire Welding, *The Welding Journal*, vol. 49, no. 5, p. 363, May 1970.

18 MIG OPERATION

The gas metal arc welding processes are quickly replacing the shield arc stick electrode process because the MIG processes can easily be applied to an extremely wide range of metals, both ferrous and nonferrous. The MIG welding processes are also popular because they can deposit a large quantity of weld metal in a relatively short period of time. The MIG processes lend themselves to both semi-automatic and fully automatic operation. The skill of the operator when using semiautomatic MIG equipment is not as critical as in the operation of the TIG welding process. Modern-day welding equipment is available in the spray arc, the short arc, or the pulsed arc welding systems, and can be used whenever almost any of the other welding systems are used. Another advantage of the MIG welding process is that the same equipment can be used to weld a large variety of metals. The only changes required in the MIG system in order to weld various metals are in the electrode wire and the shielding gas. The major metals that are welded by the metal arc welding process are aluminum; magnesium; stainless steel; low-, medium-, and high-alloy steels; Monel; lead; brass; tungsten; molybdenum; silicon steel; copper; pure silver; titanium; all of the copper alloys; and all of the nickel alloys. Because of this wide range of application, the small amount of operator skill involved, and the high rate of metal deposited, the MIG process is becoming increasingly popular in the welding industries.

The major types of power supplies used for welding in the MIG process are either the dc generator or the ac transformer that has been rectified. The major type of power supply is the CAV power supply, where voltage is maintained over a wide amperage range (Figure 18-1). This CAV power supply permits a wide range of physical arc lengths to be used while maintaining an adequate voltage amperage supply to the arc column. This type of power supply also is helpful in preventing the wire electrode from stubbing, from burning back into the wire contact tips of the MIG gun. The major reason for the use of the CAV power supply is that, as the arc length shortens a small amount, there is a large increase in welding current that automatically increases the burn-off rate of the electrode. The amperage requirements decrease until the standard arc length is reached. If for some reason the operator pulls back or keeps the MIG gun away from the work so that the arc length becomes longer, the amperage

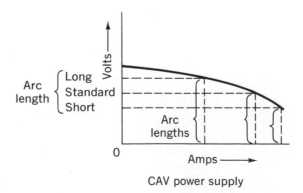

Fig. 18-1 Burn-off balance.

CAV power supply

of the CAV power supply would demand that the welding current be lessened. Then the operator would have to compensate by lowering the MIG gun. Therefore, the CAV power supply determines and compensates for the arc length, relieving the operator of a large amount of skilled responsibility. Because of this effect of arc length upon the current requirements for the MIG arc column, the amperage is adjusted by the amount of wire feed. Many of the power supplies for the MIG welding process simply have a voltage adjustment on the machine, and the amperage adjustment is maintained by the feed control of the wire feed.

The MIG welding processes are slightly more complex than the TIG welding process because consumable electrodes and a wider range of shielding gases are used. Three variables of the MIG process must be considered before the application of the MIG welding process: the operator-manipulation-controlled variables, the machine-controlled variables, and the base-metal-controlled variables. These three variables are characteristic of all welding processes but are not as far-reaching in their importance in other processes as in the MIG processes because the MIG process has such a wide range of application.

The operator-manipulation-controlled variables are those to which it is difficult to assign any precise measurement, such as the stick-out of the electrode, the nozzle angle, the angle of the MIG gun, or the wire-feed speed. These variables are controlled in the semiautomatic welding operation by the welder; however, they cause some adjustment of the machine-controlled variables.

The amount of stick-out of the electrode is preset on most wire-feed control mechanisms. The wire stick-out determines the amount of penetration of the arc column. The stick-out length controls the current density of the electrode. The longer the stick-out of the electrode, the more resistance the electrode has because the electrode that is in the stick-out position heats. As the temperature of the consumable electrode rises, its resistance also rises, which places higher current requirements on the electrode. These current requirements determine the amount of digging action of the electrode (Figure 18–2). This stick-out rate, then, is controlled by both the current settings and by the wire feed in inches per minute; but

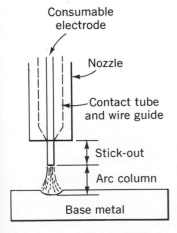

Consumable electrode

Nozzle

Contact tube and wire guide

Stick-out

Arc column

Base metal

Fig. 18–2 Stick-out.

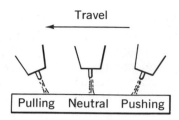

Fig. 18-3 Nozzle angle.

mainly the stick-out is controlled by the distance the operator holds the MIG torch from the work, or the tip-to-work distance.

Another operator variable is the angle at which the MIG gun is held to form the weld bead (Figure 18-3). The three possible angles of a MIG gun are the pulling or trailing angle, which would be comparable to the oxyacetylene method of backhand welding; the neutral position, with the MIG gun held perpendicular to the workpiece; or a pushing position, which resembles the oxyacetylene welding position of the forehand welding technique. As in oxyacetylene welding, these three techniques determine the penetration of the arc column. However, if the penetration characteristics need to be changed significantly, changing the angle of the MIG welding gun would not be the correct manner in which to manipulate the penetration. The machine variable of amperage or voltage control should be used, or the primary selection of wire size should be reevaluated in order to change the penetration of the arc column.

The last major type of operator control that can be accomplished with the MIG torch is the manipulation patterns that are used by the welder. The most standard manipulation pattern is known as the drag pattern, in which the torch is simply moved in a straight line with no oscillation. However, the nozzle should not be allowed to contact the work (Figure 18-4). Often when materials are welded in other than the flat position, the drag method would not be the best manipulation pattern to use. The choice of pattern, however, is often based on the preference of the welder. The whip pattern is often used for out-of-position work, as are the C pattern and the U pattern. These three patterns are especially suitable to weld-puddle manipulation in the horizontal, vertical, and overhead positions. Many times the welder is called upon to perform a cover pass, such as in pipe work. A lazy-8 manipulation pattern can be used to make a cover pass, for it is wide and will generally cover approximately 3 to 7 times the width of the normal bead (Figure 18-4).

The machine-controlled variables include the arc voltage, the welding current, and the travel speed of the weld bead. These three variables are controlled by the machine; however, they are interrelated with the base-metal-

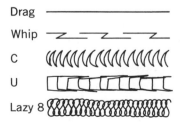

Fig. 18-4 Torch manipulation patterns.

controlled variables of the MIG process. The machine variables of arc voltage and welding-current control will determine whether the spray arc welding system, the short arc welding system, or the pulsed arc welding system is used. These two controls, the voltage and amperage in combination, determine the optimum travel speed of the weld bead. The amperage used to perform a weld in the MIG processes determines whether or not the metal transfer system is the short arc or the spray arc. The maximum amperage at which the short-circuit effect can take place in the MIG welding system is approximately 250 amps. Above this ampere setting, the short-circuiting effect will not take place. Because of the increased emf, the metal will be removed by a spray effect above approximately 150 amps. The point at which this effect occurs is determined by the electrode-wire size plus the voltage-amperage settings on the machine (Chart 18–1).

CHART 18–1   Pulse or DCRP minimum current requirements

ELECTRODE-WIRE SIZE, IN.	CURRENT	
	SPRAY ARC, AMPS	SHORT ARC, AMPS
0.030	150	50
0.035 ($^1/_{32}$)	175	75
0.045 ($^3/_{64}$)	200	100
0.0625 ($^1/_{16}$)	275	175
0.09375 ($^3/_{32}$)	350	*

*Note: short circuiting is limited to below 250 amperes.

The variables that are selected before the machine variables are those determined by the composition of the base metal. The base metal determines the amount and the type of shielding gas that must be used. The base metal also determines the basic type of joint design, the electrode-wire type, and the electrode-wire size that should be used. The type of metal determines also the welding position and the mechanical properties required in the finished weld. All these variables are indicated by the base metal to be welded.

The major shielding gases that are used for the MIG processes are argon, helium, carbon dioxide, and oxygen. Argon is the principal inert gas used to weld nonferrous metals. Helium is used for better control of porosity and better arc stability because of its greater density. The major gases used to MIG weld aluminum are argon, in its pure form, or a mixture of argon and helium. This mixture may be a combination as high as 25 percent argon and 75 percent helium or a combination of 5 percent helium and 95 percent argon. In many instances, pure argon is used. The effects of the shielding gas help to remove the surface oxides from the aluminum. When helium is used, it not only helps to remove oxides but it also exhibits a certain amount of porosity control over the weld bead (Chart 18−2). When steel and its alloys are welded, carbon dioxide

CHART 18−2   MIG shielding gas

METAL	GAS	GAS EFFECTS
Aluminum	Argon	Helps to remove surface oxides
	Argon 25%, helium 75%	Oxide removal plus porosity control
Copper	Argon	Reduces sensitivity to surface cracks
	Argon 25%, helium 75%	Counteracts high thermal conductivity
Magnesium	Argon	Helps to remove surface oxides
Nickel	Argon; or argon 50−25% helium 50−75%	Helps to control base-metal fluidity and provides good wetting
Steel, low-alloy	Argon 98%, oxygen 2%	Eliminates undercutting
	Argon 75%, oxygen 25%	Reduces undercutting
Steel, mild	Carbon dioxide	Low splatter
Steel, stainless	Argon 99−95%, oxygen 1−5%	Oxygen adds to arc stability and reduces undercutting
Titanium	Argon	Improves metal transfer

is the major shielding gas used in the MIG welding process because it has an extremely low cost when compared to the other inert gases, such as argon or helium, and it is capable of producing a sound weld. The major portion of welding is done on mild steel because it is the most-used metal in our society. The carbon dioxide gas metal arc process is one that reduces operator skill significantly while insuring proper weld beads. The gas flow in cubic feet per hour is dependent upon the size of the nozzle of the MIG welding torch. Gas flow, however, generally ranges from 8 to 35 ft³/hr. Gas flows that range between 15 and 20 ft³/hr are the most common.

All the variables have a certain effect upon the weld bead, the penetration of the weld bead, and the rate of the deposit of weld material (Chart 18–3). Voltage is usually

CHART 18–3 Cause-effect weld variables

| VARIABLE | BEAD | | | | PENETRATION | | DEPOSIT RATE | |
	SMALLER	LARGER	WIDER	NARROWER	DEEPER	SHALLOWER	FASTER	SLOWER
Voltage			Increase	Decrease				
Amperage	Decrease	Increase			Increase	Decrease		
Travel	Increase	Decrease						
Nozzle angle			Pushing	Pulling	Pulling	Pushing		
Stick-out	Decrease	Increase	Decrease	Increase	Decrease	Increase	Increase	Decrease
Wire size					Decrease	Increase	Increase	Decrease
Wire feed	Decrease	Increase					Increase	Decrease

used to either increase or decrease the width of the weld bead, and amperage is used to increase or decrease the overall size of the weld bead as well as the penetration depth of the weld bead. The size of the weld bead can also be changed by the rate of travel the welder uses to lay down a bead across the work. When the travel speed is decreased, the bead will naturally be larger, as in the case of shield arc welding and oxyacetylene welding. The amount that the consumable electrode sticks out of the welding guide and collet affects all the factors that are concerned with the weld bead, the penetration, and the rate of deposit of the weld metal. The stick-out distance is one of the most important distances that affect the weld.

Two major effects that can be visually checked while welding with the MIG torch are the welding current and the shielding gas effects. The welding-current setting can be either too high, too low, or correct (Figure 18–5). Welding current that has too low a setting results in the metal piling up on the work instead of penetrating into the work, causing a high weld bead. Conversely, welding current that is set too high heats the base metal to such an

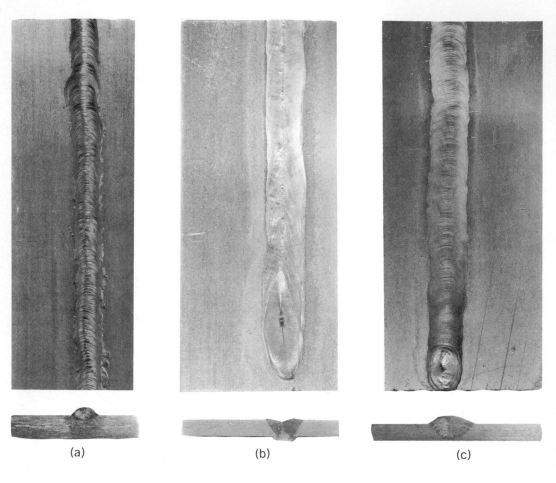

(a)   (b)   (c)

Fig. 18-5 Visual check of welding current. (a) Current too low. (b) Current too high. (c) Current correct.

extent that the weld bead sags (Figure 18-5*b*). Welding current that is set correctly in the MIG welding process yields a bead that is similar to the bead resulting from the use of the iron-powder electrode in the shield arc welding process after the bead has been cleaned of slag.

The effects that shielding gases have upon the base metal are generally noticeable by the welder, especially when welding nonferrous metals. When sufficient shielding gas is being used and is protecting the weld sufficiently, the weld bead has a clean appearance and is splatter free (Figure 18-6). When insufficient shielding gas is used, however, there is a tendency for the weld bead to have surface porosity and the ripples are disrupted.

APPLICATION   The application of the MIG welding processes depends on the type of base metal to be welded. After the type of metal to be welded has been identified, it is determined whether it is a metal that can be welded with an inert shielding gas or a metal that can be welded with a carbon dioxide shielding gas. The major metals that are welded with inert shielding gases are aluminum and its alloys, stainless steels, nickel and its alloys, copper alloys, carbon

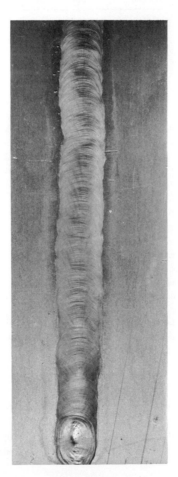

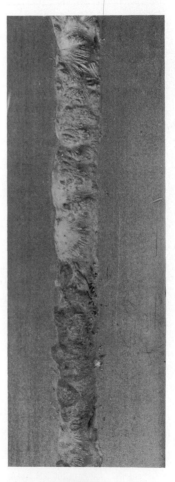

Fig. 18–6   Visual check of shielding gas. (a) Sufficient gas. (b) Insufficient gas. (Courtesy Center for Technology of Kaiser Aluminum and Chemical Corporation.)

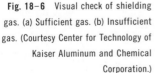

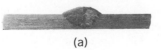

(a)

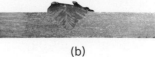

(b)

steels, low-alloy steels, and high-strength steels. The carbon dioxide shielding gas is limited to the welding of steel and its alloys and is not capable of performing a high-quality weld on any of the nonferrous metals (Chart 18–4). The

CHART 18–4 Quick selection guide to MIG welding processes

	INERT GAS		CO₂ GAS		
	SPRAY ARC	PULSED ARC	BURIED ARC	SHORT CIRCUITING	FLUX-CORED WIRE
Metals to be welded	Aluminum and aluminum alloys, stainless steels and PH steels, nickel and nickel alloys copper alloys, titanium, etc., as well as carbon steels	Same as spray arc	Carbon steels	Low- and medium-carbon steels, low-alloy steels, some stainless steels	Low- and medium-carbon steels, low-alloy, high-strength steels
Metal thickness	12 gage (0.109 in.) to $3/8$ in. (no preparation), single pass welding up to 1 in. max. thickness practically unlimited	Light gage	10 Gage (0.140 in.) up to $1/2$ in. (no preparation), practical max. 1 in.	20 Gage (0.038 in.) up to $1/4$ in., above $1/4$ in., not economical	$1/4$ in.–$1/2$ in. (no preparation), max. thickness practically unlimited
Welding positions	All positions	All positions	Flat and horizontal	All positions	Flat and horizontal
Major advantages	Welds most nonferrous metals, no flux required, all positions (with small wire), visible arc, negligible cleanup	Spray transfer in all positions, low spatter, excellent puddle control, porosity-free deposits, wide operating range, good dilution control, easy handling	Low cost, high speed, deep penetration, visible arc	All positions, thin material will bridge gaps, min. cleanup, visible arc	Smooth surface, deep penetration, sound welds, visible arc
Limitations	Thickness practical limited, cost of gases, spatter removal sometimes required	Thickness practical limited, cost of gases	High spatter	Uneconomical in heavy thickness, gas coverage essential	Lower efficiency slag removal
Quality	Good properties, x-ray	Good properties, x-ray	Fair properties	Good properties, x-ray	X-ray
Appearance of weld	Fairly smooth, convex surface	Similar to TIG	Relatively smooth, high spatter	Smooth surface, relatively minor spatter	Smooth surface, some spatter
Travel speeds	Up to 100 IPM (auto), 20–25 (manual)	Vertical: 12–14, horizontal: 20–25, (manual)	Up to 300 IPM (auto)	Max. 30 IMP (manual)	Up to 150 IPM (auto), 20–25 (manual)
Range of wire Sizes, in.	Diameter: 0.035, 0.045, $1/16$, $3/32$, $1/8$	Diameter: 0.035, 0.045, $1/16$, $3/32$	Diameter: 0.045, $1/16$, $5/64$, $3/32$, $1/8$	Diameter: 0.020, 0.030, 0.035, 0.045	Diameter: $3/32$, $7/64$, $1/8$, $5/32$

CHART 18−4 (cont'd)   Quick selection guide to MIG welding processes

	INERT GAS		CO₂ GAS		
	SPRAY ARC	PULSED ARC	BURIED ARC	SHORT CIRCUITING	FLUX-CORED WIRE
Range of welding current	50 min.−600 max., amperes	35 min.−300 max., amperes	75 min.−1000 max., amperes	25 min.−250 max., amperes	200 min.−700 max., amperes
Electrode and shielding cost	Expensive electrode wires, relatively expensive gas	Same as spray arc but can use electrode one size larger	Reasonably priced electrode wires, least expensive gas	Reasonably priced electrode wires, least expensive gas	Relatively expensive wire, least expensive gas
Overall welding costs	Least expensive for nonferrous metals, medium to heavy thicknesses	Least expensive for nonferrous metals, thin work and/or out-of-position welding	Least expensive for medium thickness	Least expensive for thin material	Least expensive on low-alloy steels

Courtesy of Ted B. Jefferson, "The Welding Encyclopedia," 16 ed., Monticello Books, Morton Grove, Ill., 1968, p. M−57.

two major types of MIG welding processes are the spray arc processes and the pulsed arc processes. The pulsed arc is a representative of a process that employs a short arc type of weld. The spray arc and the pulsed arc, or the short-circuit arc, are capable of welding the same metals. The pulsed arc, as in short arc welding, is limited to light-gage metal while the spray arc has the capability of welding metal up to 1 in. in thickness. With appropriate joint designs, the spray arc is capable of welding metals of unlimited thicknesses. The costs of the two welding systems are practically the same; however, the spray arc systems are more adaptable to automation. Many of the spray arc systems are capable of depositing 20 to 25 in. of weld per minute, as is the pulsed arc in the manual process. The current range for the spray arc process ranges from 50 to 600 amps while the current range for the pulsed arc MIG welding system ranges from 35 to 300 amps maximum.

The four types of carbon dioxide processes are the buried arc, the short-circuiting arc, the flux-cored wire, and the magnetic flux. The costs of the flux-cored-wire and the magnetic-flux processes are practically the same, and the characteristics of the flux-cored-wire and the magnetic-flux processes are comparable.

The buried arc process is a slight adaptation of the standard gas metal arc process. In this process, there is no metal transferred in the arc column; rather the arc of the buried arc process digs a crater in the base metal which fills, producing a molten pool. The weld metal transfers within the weld pool, making this method a third type of metal transfer. The buried arc process causes an extremely high amount of splatter and is usually available only in an automatic system. However, because of this type of metal transfer, weld beads can be produced at approximately 300 in./min. The amperage range because of this type of metal transfer can be exceedingly high. The welding current ranges from 75 to 1000 amps maximum. The buried arc process is generally limited to low- and medium-carbon steels. It is used successfully for the fabrication of railroad cars and in the auto industry where long, high-speed welds of fairly heavy material are used. The welding of truck frames is an example of the application of the buried arc process.

The carbon dioxide MIG welding short arc or short-circuiting process is capable of welding steels from approximately 20 gage to 1/4-in. thick with little or no joint preparation. Short arc welding of metal more than 1/4 in. thick is too slow. Since the amperage settings must increase as the thickness of metal increases, short arc welding is not used to weld thick metal. Because of the limitation on metal thickness, the maximum speed of travel is higher than in the spray arc process. Generally, it is possible to weld as many as 30 in./min in a manual operation. The resultant weld bead is smooth and has little splatter to be cleaned up after welding. The short arc process is limited to a maximum of 250 amps. Above this amperage, the short-circuit pinch effect will no longer take place, and the spray arc effect will be initiated. The higher the amperage, the more the spray effect will be apparent. This amperage limitation also limits the metal thickness. Carbon dioxide short arc welding is many times called microwave welding because of the wire size. The diameter of the wire electrodes used for the short-circuiting carbon dioxide MIG welding process ranges from 0.020 to 0.045 in. Wire diameters greater than 0.045 in. require higher amperage settings, which do not allow the short-circuiting effect to take place.

The flux-cored-wire and the magnetic-flux carbon dioxide MIG welding processes are capable of welding the low-, medium-, and high-carbon steels and the low-alloy, high-strength steels. Metals ranging from 1/4 to 1/2 in. thick can be welded with these two processes with no joint preparation. These processes are capable of welding metals of unlimited thicknesses when joints are used. The flux-cored-wire and the magnetic-flux processes deposit a smooth, deep-penetration weld bead. However, because of the flux involved, there is a slag, as in the shield arc welding process, that must be removed. These processes are available either as automatic or manual welding processes. The automatic welding processes generally have the capability of depositing as high as 150 in. of weld metal per minute, and the manual welding processes are capable of depositing between 20 and 25 in. of weld per minute. The flux-cored-wire and the magnetic-flux MIG welding processes are the least expensive that can be used for depositing x-ray quality weld beads upon low-alloy steel.

QUESTIONS

1. What metals can be welded by the MIG process?

2. What types of power supplies can be used for MIG welding?

3. How does the power supply compensate for differences in arc length?

4. What is stick-out?

5. What are the penetration differences between pulling and pushing the MIG gun?

6. What controls does the MIG gun operator have over the welding variables?

7. Explain the machine-controlled variables of the MIG process.

8. Why is 250 amps the maximum current allowed in the short arc process?

9. What is the pulsed arc?

10. How do voltage and amperage determine the size of the weld bead?

11. Why is oxygen sometimes used with argon as a shielding gas?

12. What are the major variables that control the weld bead?

13. What happens to the weld bead when there is not sufficient shielding gas?

14. What are the basic carbon dioxide MIG welding processes? How do they differ?

PROBLEMS AND DEMONSTRATIONS

1. Select a piece of metal, the proper wire electrode, the correct shielding gas, and a CAV power supply. Weld a bead with the MIG gun and purposely vary the arc column length. Notice how the voltage and amperage change as the arc column length changes. These changes will be noticeable by the change in the sound of the arc column.

2. Weld a bead with the MIG gun using the pulling weld technique. Weld a bead with the MIG gun using the pushing technique. Compare the amounts of absorbed heat in the base metal. Notice especially how the width of the weld bead increases as the entrapped heat increases.

3. Weld a piece of ferrous metal with a steel wire electrode 0.030 in. in diameter, using 50 to 100 amps. After welding one bead, increase the amperage to 150 amps and weld another bead. Compare the sound of the arc and the penetration of the arc column.

SUGGESTED FURTHER READINGS

Adle, J. M., ed.: *The Welder's Training Manual, Inert Gas Processes,* Kaiser Aluminum and Chemical Sales, Inc., Chicago, 1962.

Giachino, J. W., et al.: *Welding Skills and Practices,* American Technical Society, Chicago, 1965.

Jefferson, Ted B., and L. B. MacKenzie: "The Welding Encyclopedia," Monticello Books, Morton Grove, Ill., 1968.

Lockwood, Lloyd F.: Pulse-Arc Welding of Magnesium, *The Welding Journal,* vol. 49, no. 6, p. 454, June 1970.

Micro-Wire Welding, Hobart Bros. Technical Center, Troy, Ohio, 1964, EW 300.

Pender, James A.: "Welding," McGraw-Hill Book Company, New York, 1968.

IV WELDING PROCESSES

19 PLASMA ARC WELDING

The plasma arc torch was first developed in 1881, but it was not until nearly 45 years later, in 1925, that the first plasma test device was developed. The founder of the first plasma test device was a German scientist by the name of Gerdien. His first attempts at harnessing the plasma arc used water. He discovered that temperatures as high as 100,000°F could be obtained by making a small vortex of water squeeze through an arc stream in a small channel. For a number of years, the plasma arc process was considered to be for only the intellectuals. The process was not practical because water could be used in very few applications and because the carbon electrodes that were used to power the system were consumed too quickly. In 1959, the plasma arc flame was used to simulate conditions that nose cones of rockets would encounter when reentering the earth's atmosphere, causing a spurt of interest to create a plasma arc device for use in industry.

More research was not conducted earlier on the plasma arc welding process because it was thought that it was impossible for a high-current electronic arc to be forced through a small hole in a solid nozzle without consuming the nozzle. This problem was conquered through the use of water-cooled metallic nozzles and the use of the plasma gas itself to shield the nozzle from the extreme heat of the arc. The use of plasma arc in cutting, welding, and metal spraying was further developed in 1953, in the Linde Research Laboratories in Tonawonda, New York, by Robert M. Gagu, who was then a developmental engineer for the company. The developments in plasma arc were discovered as the result of a study of another problem, mainly the melting of titanium ingots through the use of an arc. In this problem, an inert gas was used to surround the ingot during the melting process. While engaged in his experiments with inert gases and arc streams, Gagu observed a similarity in the appearance of the ordinary gas flame and the long electric arc. This observation led to the idea that better control of the intense heat and of the velocity of the arc might be accomplished by putting these forces through a small metal nozzle. His theory proved valid and the use of the plasma arc flame process became practical for modern industry.

Any high-current arc is comprised of plasma, which is nothing more than an ionized conducting gas. The plasma gas is forced through the torch, surrounding the cathode. The main function of the plasma gas is shielding the body of the torch from the extreme heat of the cathode. Any gas or mixture of gases that does not attack the tungsten or the copper cathode can be used; argon and argon mixtures are most commonly used.

The plasma arc, or jet, has a controlled composition and can cut any metal since it is primarily a melting process. Plasma jet energy is virtually unlimited. The greater the power used, the greater the temperature for melting the metal. Many plasma jet torches have a temperature capability of approximately 60,000°F. Theoretically, the temperature range for other types of plasma jet torches extends into the thermonuclear range. The arbitrary temperature of 200,000°F separates thermal plasma from thermonuclear plasma. As a comparison, oxyacetylene welding is limited to the maximum temperature of the chemical reaction, or to approximately 6500°F. The ordinary electric arc, because of its diffuseness, can attain a maximum temperature of approximately 20,000°F. The maximum temperature achieved by the ordinary welding electric arc is thought to be approximately 10,000°F. The design of the plasma arc torch constricts the arc force through a small opening and at the same time bombards the arc stream with gas particles, causing multiple collisions of electrons within the particles; consequently, the energy released is dependent upon the amount of electrical energy induced into the system.

Plasma arc consists of an electronic arc, plasma gas, and gases used to shield the jet column. The plasma gas does not provide enough shielding protection because of its low pressure, which must be maintained to prevent turbulence around the area of the cut or the weld. If there were high pressure in the plasma column, it would cause a displacement of the molten metal in the weld bead, or in the kerf, the cut. Therefore, supplementary shielding gases must be supplied for this purpose.

The equipment necessary for plasma arc welding includes a conventional dc power supply with a drooping volt ampere output and with 70 open line volts. This type of power supply is suitable for most applications in which argon or argon mixtures are used.

When the amount of hydrogen exceeds 5 percent of the amount of argon, two power supplies must be connected in series to obtain the necessary open voltage for igniting the arc, or the jet stream. Both rectifiers and motor generators can be used for the power source, although rectifiers are preferred because they produce better arc stability as the machine warms up to an operating temperature.

The two main types of torches for welding and cutting with plasma arc are the transferred arc and the nontransferred arc. The transferred arc plasma jet torch is similar to the TIG torch, except that it has a water-cooled nozzle between the electrode and the work. This nozzle constricts the arc, increasing its pressure. The plasma, caused by the collision of gas molecules with high-energy electrons, is then swept out through the nozzle, forming the main current path between the electrode and the work-

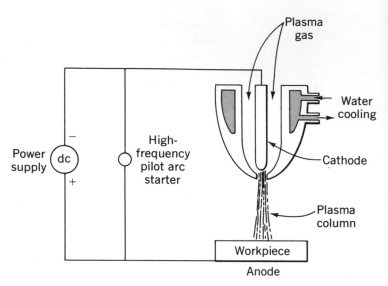

Fig. 19-1 Transferred arc.

piece (Figure 19-1). The plasma arc and the transferred arc are generated between the tungsten electrode, or cathode, and the workpiece, or anode. The second type of plasma jet torch is called the nontransferred arc torch (Figure 19-2). This torch extends the arc from the electrode, or the cathode, to the end of the nozzle. The nozzle acts as the anode. This type of plasma jet is completely independent of the workpiece, with the power supply contained within the equipment.

The arc force in the transfer arc torch is directed away from the plasma torch and into the workpiece, which means that the arc is capable of heating the workpiece to a higher temperature than with the nontransferred arc. In the nontransferred arc, the arc force generated from the electrode is absorbed in the water-cooled nozzle, making the nontransferred arc more adaptable to metal spraying and welding. The transferred arc is more adaptable to melting or cutting metal. The gas tungsten arc cutting process or the plasma jet process is done simply by striking an arc between the tungsten electrode in the torch and the surface to be cut. A pilot arc circuit that connects the torch nozzle to the ground by a current-limiting resistor makes the arc easier to start. This starting unit disconnects itself once the arc is started. The electrodes in the metallic nozzle of the torch are usually cooled with water. The electrodes and the arc-constricting nozzles are responsible for about 5 percent of the input energy being used as heat and for the remainder of the energy entering the plasma column. The plasma gases used for the cutting of nonferrous metals and stainless steels are mixtures of argon and hydrogen, with 100 percent argon being used to start the cut and approximately 25 percent hydrogen added to the argon after the cut has been started. The plasma arc cutting process has been applied with tremendous results to the cutting of carbon steels, aluminum, stainless steels, Inconel, Monel, many of the hard-to-cut steels, and other metals. When carbon steel is cut with the plasma arc, dross-free cuts with smooth surfaces, sharp edges, and almost square faces can result. When aluminum is cut with the arc, a mixture of 65 percent argon and 35 percent hydrogen is used. The plasma arc process for cutting aluminum has proven a successful application of this process. High-quality, dross-free cuts can be produced in aluminum that is up to 5 in. thick. The amount of material that can be cut in a designated period of time by the plasma arc exceeds that of any other method for cutting aluminum. Because of its successful application to the cutting of aluminum, new concepts in both manual and automatic cutting of aluminum have come into use. The plasma jet has been adapted for cutting risers on castings, cutting manholes in tank shells, and cutting heavy bulky items that are impossible to cut by means of shears or a saw. Its ability to cut intricate contours aids in reducing operating costs in the metalworking industry.

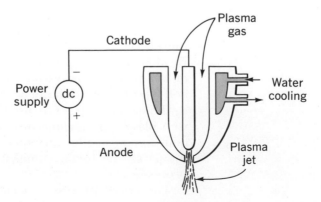

Fig. 19-2   Nontransferred arc.

In 1958, a major breakthrough was made in plasma jet cutting. The plasma jet was then developed so that it could cut stainless steel economically, which is its most common use today.

In the current applications of the plasma jet cutting process, the plasma causes virtually no distortion; whereas, distortion may exist in the chemical flame cutting process. The corrosion resistance of stainless steel is not affected by the plasma jet except for a microscopically thin layer of metal on the surface of the cut. In shape cutting, cuts can be started at any point on the workpiece because of the ability of the torch to cut stainless steel up to 2 in. thick. The newest of the plasma arc processes is the welding process. The welding process uses a nontransferred arc that has electrical circuits similar to those used in the plasma arc cutting process. The process uses a power supply with a high-frequency pilot arc starter. The electrode used for stainless steel welding and most other metals is a straight-polarity tungsten electrode. When aluminum is welded, reverse polarity is used with a water-cooled copper electrode. The plasma arc welding process is used in aerospace applications, and in the welding of reactive metals and thin materials. It is capable of welding stainless steel, titanium, maraging steel, and high-carbon steel. The plasma jet torch that has been designed for welding has one extra passage within the nozzle of the plasma torch, providing a passageway for the shielding gas during the welding operation. The plasma gas itself does not sufficiently protect the metal of the weld bead. The gases used for this shield are the same as those used to weld in the TIG welding process, argon, helium, and CO_2, or mixtures of these gases. Plasma gas, shielding gas, and water to cool the nozzle are piped to the lower portion of the torch. By changing the tip of the nozzle of the torch, the shape of the weld can be varied. The penetration of the weld and type of metal that can be welded are greatly influenced by the size and type of tip used. Practically all welding done with plasma arc is done mechanically, for the process requires great stability and speed. The high temperature of the jet stream further limits a manual application of this process.

The actual process of welding with the plasma jet is

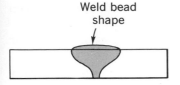

Fig. 19-3 Wineglass effect.

done with what is called the keyhole method. The jet column burns a small hole through the material that is to be welded. As the torch progresses along the material, the hole progresses also; however, it is filled up by the molten metal as the torch passes. Welding with this method automatically insures 100 percent penetration.

Because the plasma jet strikes the surface of the metal, a larger area is melted at the surface, resulting in a unique cross-sectional weld bead design called the wineglass design (Figure 19-3). The nontransferred arc plasma jet torch is also used for the spraying of metals, especially those metals that have a melting point above 600°F. The metals to be sprayed are injected into the plasma jet stream either approximately where the shielding gas is injected into it or just outside the nozzle (Figure 19-4). The wire or powder can be injected into the plasma stream, which is hot enough to melt any of the metals and blast them out into the jet stream at sonic velocities, forcing the now-molten metal onto the part to be sprayed. The filler materials that can be plasma arc jet sprayed are ceramics; some of the nonferrous, and all of the ferrous, metals including tungsten and tungsten carbides; and metals that range in hardness up through vanadium and the vanadium carbides. The feed rate for metal spraying depends on the fusion point of the filler material used. Aluminum can be sprayed from approximately $1/10$ to $7\frac{1}{2}$ lb/hr, while tungsten can be sprayed from $1\frac{1}{2}$ to 25 lb/hr. Generally, when spraying, the plasma gun is held from 2 to 6 in. from the workpiece. Most of the filler materials have a deposit rate of around 60 to 75 percent. With aluminum oxide, this rate is approximately 85 percent.

There are several dangers connected with use of a plasma jet. More electrical equipment is used, which raises the electrical hazards considerably. Also, ultraviolet and infrared radiation is present, requiring the use of welding tinted lenses from no. 9 to no. 12. Also common to this process is a high-pitched noise, which requires that the operator wear ear plugs. The noise ranges around 100 dB (decibels) if it is unrestricted. This noise level must be reduced to 80 dB in order to prevent damage to the inner ear. The ultraviolet and infrared radiation sometimes

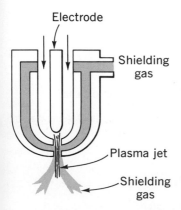

Fig. 19-4 Plasma jet shielding.

is so intense that it can cause violent sunburning, even through clothing.

Manual and mechanized equipment is available for plasma arc cutting. The manual equipment sells for approximately $1100 for the torch body and approximately $3000 for the power supply. For plasma arc welding, only mechanized equipment is available and the cost ranges from $3000 to $10,000 per unit. These initial costs might be assumed to prohibit the use of this method; however, over an extended period of time, the plasma jet method is considerably more economical than the oxygen-fuel gas method because the higher rate of deposits reduces the amount of finishing required on the weld deposit and on the base metal around the weld deposit as well as reduces the amount of deposit material used during the operation.

Because of the versatility of the plasma jet as well as the economical features of the process, it has become an important asset in the metal-cutting industry. The quality of the finished product is superior to, and the man-hours required are considerably less than, any of the other plain cutting processes.

QUESTIONS

1. What is the keyhole method of welding?

2. Why is the plasma jet sometimes referred to as the fourth state of matter?

3. What is the difference between a plasma gas and a shielding gas?

4. What are the two basic types of plasma torches? What are their characteristics and applications?

5. What is the difference between thermal plasma and thermonuclear plasma?

6. How does the heat generated by the plasma jet compare with the heat generated by the electric arc? The oxyacetylene flame?

7. Why doesn't the nozzle of the plasma torch melt?

8. What is the most common use of the plasma torch?

9. What metals are used for the cathodes in the plasma torch?

10. What are the operational hazards in using the plasma jet torch?

SUGGESTED FURTHER READINGS

Dato, J. E.: Trends in Arc Welding, *Machinery*, vol. 47, p. 74, January 1968.

Filipski, Stanley P.: *Plasma-Arc Welding*, Union Carbide Corporation, Linde Division, Electric Welding Library, New York, 1967.

Gorman, E. F.: New Developments and Applications in Manual Plasma Arc Welding, *Welding Journal*, vol. 47, p. 547, July 1968.

Obrien, Robert L.: *Plasma-Arc Welding Applications: A Report to Metals/Materials Congress*, American Society of Metals, Philadelphia, 1964.

Phillips, Arthur L.: *Current Welding Processes*, American Welding Society, New York, 1965.

20 RESISTANCE WELDING

Electrical resistance can best be explained as opposition to electric current as the current flows through a wire. This resistance has been given the name ohm. Resistance has been used as a source of heat generation for welding for a long time (Figure 20-1). In resistance welding, two factors perform the welding application: the resistance heating of the two pieces to be joined and the forging pressure exerted joining the two pieces of metal. The first factor involves resistance heating, which can be justified by the formula $H = I^2RT$. H is the heat generated indicated in joules, I is the current in root-mean-square amperes, R is the resistance in ohms, and T is the time of the current flow in seconds. Basically, this formula states that the amount of heat is directly proportional to the square of the amperes, times the resistance, times the current flow

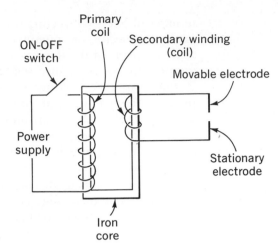

ON-OFF
switch

Primary
coil

Secondary winding
(coil)

Movable electrode

Power
supply

Stationary
electrode

Iron
core

Fig. 20-1   Typical resistance welder.

in seconds. The formula also indicates that in simple
resistance welding, a high-ampere current is necessary for
an adequate weld. Also, the amperes are the major controlling element for the generation of heat in the metal.
Heating metal to the temperature necessary for welding,
therefore, is indicated by a direct application of Joule law
to resistance welding. Joule law, applied to a law of conduction, is that a poor conductor heats up to a higher
degree and to a greater extent than a good conductor with
the same amount of amperes passing through it. Heating
takes place on the surfaces of the two pieces of metal to be
spot welded because of the poor conduction between the
two. The spot-welding machine passes a high current at
low voltage through the electrodes to the spot where
welding is to take place. As the machine cycles, exerting
pressure on the metal, a joint is formed and coalescence
takes place.

The second factor in the welding process, the forging
factor, is the next step after the metal has been heated to a
plastic state. Pressure must be exerted on the pieces being
joined in order for the forging process to take place. The
pressure exerted comes from the electrodes extending
from the arms of the welding machine. These electrodes
also carry the current which provides the heat to the
pieces. The electrodes provide good contact between the
pieces to be joined and insure against shrinkage pockets.
The mechanical force that brings about the necessary
forging pressure exerted by the electrodes employs any

number of energy-directing sources. Springs, levers, cams, or hydraulically or pneumatically controlled electrode arms are some of the devices that have been used for forging pressure by resistance welding manufacturers.

The resistance welding machine is cycled in order to produce the needed heat timed to coincide with the pressure exerted by the electrodes upon the surfaces of the metal pieces to be joined (Figure 20–2). The approach or the squeezing time is the first stage in the common four-period resistance welding cycle. This squeeze stage takes place when the electrodes are clamped onto the pieces to be welded. The next stage is the heating stage or the weld time, which occurs when an area in the two pieces of metal are brought up to the welding temperature, indicated by S in Figure 20–2.

The welding time can be adjusted to a rheostat in the primary winding. After the metal has reached the fusion point, the current is shut off and additional pressure is applied by the electrodes. This pressure is held until the metal is cooled; then it automatically shuts itself off. This pause begins a new cycle. Welding time is usually controlled by the machine, but the operator can set the machine to a desired cycle time by means of a rheostat. The

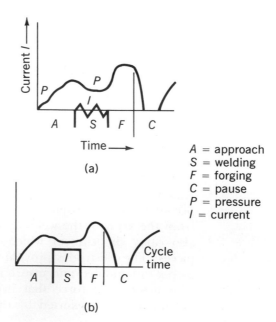

A = approach
S = welding
F = forging
C = pause
P = pressure
I = current

Fig. 20–2 Four-period resistance welding. (a) Single-phase ac (b) Dc.

welding time commonly runs from 3 to 120 Hz with a
60-Hz current or standard current; 120 Hz will yield
approximately a 2-sec welding timing cycle. The four-
period cycle is used because hardenable alloys are prone
to cracking when one surge of current and uncontrolled
cooling is employed for welding. However, with a controlled approach to the welding, forging, and cooling
cycle in the machine, hardenable alloys can be spot welded
with a high degree of success. Chart 20−1 lists various

CHART 20−1   Resistance weldability

METAL	RESISTANCE WELD (SPOT, SEAM, OR PROJECTION)
Low-carbon steel	
Sheet	Common
Plate	Occasional
Medium-carbon steel	
Sheet	Occasional
Plate	Rare
High-carbon steel	
Sheet	Rare
Plate	Not used
Low-alloy steel	
Sheet	Occasional
Plate	Rare
Stainless steel	
Sheet	Common
Plate	Occasional
Aluminum	
Sheet	Common
Plate	Occasional

types of steels and the type of resistance welding that can
be used to join them.

Spot welding and butt welding are the two major methods of applying heat energy and mechanical force in
resistance welding.

SPOT WELDING   One of the most common pieces of welding equipment
found in industry and in school shops, especially when
relatively light-gage metal is used, is the spot welder. It
is characterized by low cost, speed, and dependability,
making it a common electrical resistance welding process.

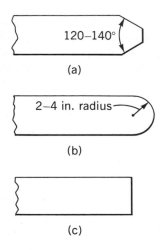

Fig. 20-3 Tip design. (a) Pointed electrode. (b) Domed electrode. (c) Flat electrode.

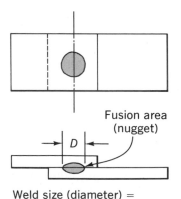

Weld size (diameter) = 0.10 in. + 2 × thickness of thinnest member

Fig. 20-4 Spot-weld fusion.

The electric resistance spot welders most used are of two different types. One is a stationary welder, which is available in different sizes according to its use. The other is made like a stationary welder, in that it has a stationary transformer, but the electrodes are in a gun form. The spot welding gun is powered and actuated through flexible leads and hoses. These two different types of spot welders make spot welding a useful tool to industry because of its versatility.

The electrodes on a spot welder are made of low-resistance, hard-copper alloy. The electrodes are either air-cooled or butt-cooled by water circulating through the rifled drillings in the electrode. The electrodes are capable of transmitting the heat that is generated in the contact points, away from the work to the horns or arms, which hold the electrodes. The electrode points should be kept clean at all times. If the points are dirty or scaly, there will be excessive heat in the points, which will cause imbedding, burning, or splitting. The electrode points on a spot welder should be approximately the same size and should always meet on top of each other in order to obtain the best welds. There are three basic electrode point designs, which are used according to the weld desired (Figure 20-3): the pointed, domed, and flat electrodes. Pointed electrodes are the most widely used and are used for ferrous materials. Domed electrodes are designed to withstand heavier current loads and more mechanical force, which make them applicable to nonferrous metals. Flat electrodes are used when spot-welding deformation is not wanted or when the spot weld should be inconspicuous. One flat electrode is usually used with one domed electrode.

Process The diameter of the electrode determines the diameter of the fusion zone of the weld spot (Figure 20-4). The diameter of the fusion zone is 0.01 in. plus 2 times the thickness of the thinnest member being spot welded; however, this preset weld fusion size is determined by the application; that is, a sheet-metal shop would determine the optimum sizes for their spot-welding needs, and an automotive body shop might require different electrode sizes. The amount of lap that is required for a good spot weld depends upon the size of the fusion weld. Usually

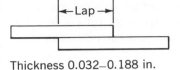

Thickness 0.032–0.188 in.
Lap = 2 × spot size + 1/8 in.

Fig. 20–5 Lap-joint design.

spot welding is limited to metals that range in thickness from 0.03 to 0.188 in. The lap joint is 2 times the diameter of the spot weld, plus ⅛ in. for alignment. If the two pieces to be spot welded are held in a fixture, the ⅛ in. for spot alignment can be eliminated (Figure 20–5). If the spot weld is made too close to the edge of the metal, the metal will become overheated, and cracking will result around the weld. Spot welding too close to the edge of a lap joint causes metal to explode from the surface of the metal being welded.

As the current passes through one electrode and the work to the other electrode, a small area is heated. The temperature of this weld zone is approximately 1500 to 1700°F. Because of this flow of current from one electrode to the other, the heat mark shows a tendency to pull toward the center. After the weld is completed, it is allowed to cool, producing a fine strong weld. During this complete weld cycle, the welding time can be as slow as 0.01 sec, or it can be as long as 5 sec. The amount of discoloration and burning or weakening of the base material is related to the length of the weld cycle.

During the welding process, the electrodes are put through harsh treatment. They must be able to withstand a terrific amount of punishment. They must have a considerable degree of hardness and the ability to maintain their shape at high temperatures. If the resistance of the electrode itself is too high, there will be a heat loss in the electrode, resulting in a loss of welding energy. No matter how high the quality of the electrode, there is at least one drawback that must be considered: the presence of the zinc coating or galvanized steel on the electrode with its low melting point of 787°F. This zinc coating has a tendency to erode and pit the electrode surface which does not improve the wear performance of the electrode. The more the electrode deforms, the greater the area of contact with workpieces, and the less concentrated is the weld current. A decrease in weld quality results. When this deformation occurs, the tips must be refiled in order to restore them to their original shapes. Flat tips and pointed tips are usually filed by a single-cut mill file. The domed tips are generally refinished by machine because of the radius, which is too difficult to re-form by hand. An advantage of spot welding is that it has a small heat-affected zone. Large heat-affected zones waste energy. The spot-welding process also has a fast welding rate and quick set-up time, plus a low unit cost per weld, which makes this process economical for production.

SEAM WELDING

Seam welding is an adaptation of the basic spot-welding process; the only difference is that the joint is being continuously welded instead of spot welded. The equipment used for seam welding utilizes the same principles as any of the resistance welding processes. Seam welding uses a two-step transformer with a primary and secondary winding. The secondary winding is hooked to the horns or arms that support the electrodes. The horns are movable and can apply pressure to the work, as in spot welding. However, the electrode is in the shape of wheels and can be rotated. These copper wheels carry the current to the pieces to be welded. As they rotate, ignitron tubes and thyratron tubes are used to switch the current on and off through the rollers. This switching of the current produces a continuous or overlapped spot weld, which is called a seam weld (Figure 20–6).

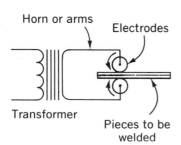

Fig. 20–6   Seam-welder schematic.

When seam welding was first developed, there were many problems, such as the heating of the electrodes and also the maintaining of a fairly constant heat range on the revolving electrodes. Overheating of the electrode resulted from the current continuously passing through it to produce the overlap spot weld. A first attempt to cool the electrode was to inject an external spray of water, but this external spray was messy and difficult to control. Modern machines are cooled by refrigerant fluids that flow inside the copper wheels. Cooling is necessary if adequate control of the overlapped spot welds is to be maintained. If cooling is not used or is insufficient, distortion takes place in the work because of the hot electrodes, and the copper alloy used in the electrodes breaks down because of the excessive heat in welding. Seam welding machines are available in the same types of machines as spot-welding machines, the portable gun type or the stationary machines. Naturally, the stationary machines are the larger. Seam welding is restricted to the welding of thin materials, with the metal ranging from 0.10 to 0.187 in. thick. Seam welding is also further restricted to metals that have a low

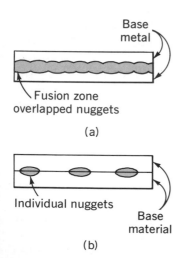

Fig. 20–7   Seam-welding fusion areas.
(a) Stitch. (b) Roll.

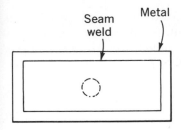

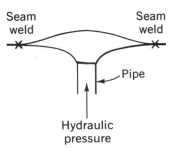

Fig. 20-8 Pillow testing.

hardenability rating, such as hot-rolled grades of low-alloy steel. When seam welding light linings to heavier or thicker pieces of metal, higher current and greater welding pressure must be used. Many times the result of this greater pressure and higher current density settings is a deeper surface indentation of the seam weld.

Basically, two types of weld can be formed in seam welding with the individual nuggets: the stitch weld and the roll weld (Figure 20-7). The stitch weld is made by turning the current on the rolls off and on quickly enough so that a continuous fusion zone is maintained. The fusion zone will not be parallel but will be in the shape of each overlapped bead. The roll weld occurs when the current to the copper-rolled electrodes is turned off and on intermittently which causes individual nuggets to be formed (Figure 20-7b). Seam welding or stitch welding is used more for joints for use with liquid or gas, while roll welding is used for simple joining of two pieces of metal. The roll weld will not be water tight, liquid tight, or gas tight.

Testing The pillow test is the most common test for determining the strength of seam welds (Figure 20-8). It involves seam welding two pieces of metal to enclose a cavity. An appropriate pipe fitting is either put on with a nipple or welded on to the two pieces that were seam welded together. Hydraulic or air pressure is pumped through the fitting, expanding the cavity into a pillow shape. The pressure at which the pillow bursts is recorded and compared to the fracture strength of the base metal. Failure should always occur in the base metal and not in the welded seam. If the weld seam fractures, then the weld will not support that particular metal.

PROJECTION WELDING Projection welding uses the same equipment as spot welding. The only difference is that the electrodes used are flat on the ends and slightly larger in diameter than the flat electrodes used in spot welding. Successful projection welding depends greatly on the surface preparation of the pieces to be welded. Projections, small deformations that will touch the surface of the material to be welded, are made on the weld areas. One of the main advantages of these projected points is that welding areas can be located

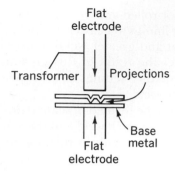

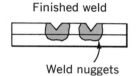

Fig. 20-9 Projection welding.

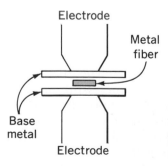

Fig. 20-10 Metal-fiber welding.

easily, which makes projection welding a high-production welding technique (Figure 20-9). As the current flows through the two parts to be welded, the projected points are the main contact area. Wherever a point or a projection touches the metal, a weld nugget will start. These weld points soon reach the plastic state, and the force applied by the electrodes finishes the weld nugget. The cycle time is the same as the spot-weld time. As the points reach their plastic state, the metal is compressed so that the finished weld is similar to the spot weld with the exception of the small indentations created by the projections. Projection welding, as spot welding, requires no protective atmosphere in order to produce successful results. It is also useful in securing small circular components to larger pieces of metal. For projection welding, a large, flat-surfaced electrode is used, which covers the areas of the projections. The areas of projection, one or more, and the spots where they touch on the other piece of metal will be the only places where fusion takes place, even with a large electrode. Projection welding has been used by auto manufacturers for many years. Auto bodies have many areas that are projection welded. Another area that uses projection welding is stud welding, which has also been utilized for car bodies.

Projection welding reduces the amount of current and pressure needed in order to form a good bond between two surfaces. With this reduction there is less chance of shrinkage and distortion in the areas surrounding the weld zone, one reason why it has been incorporated into many manufacturing processes.

Resistance welding is used widely in the assembly areas of manufacturing because of its good joining properties, its inexpensiveness, and its great speed. With little previous experience, a person can learn to operate a resistance welder in a comparatively short time, usually from 10 to 15 min.

An adaptation of the projection welding method is metal-fiber welding, which uses a metal fiber rather than a projection point (Figure 20-10). This metal fiber may be composed of various metals such as any brazing metal or any metal that can be used for a filler material. It is generally a felt material. Tiny elements of the filler metal are used to produce a thin sheet of felt metal cloth, which is placed between two pieces of metal and is projection welded in the usual manner. The metal fiber used in the filler material makes it possible to weld dissimilar metals such as copper to stainless steel, stainless steel to a ferrous material, and copper to brass. The metal-fiber welding process is more expensive than the projection welding technique.

BUTT WELDING Welding two pieces of metal together with a butt weld can be done by many resistance welding processes. The butt weld consists of joining two pieces of metal together either on face or on edge. Some of the major types of resistance butt welding are flash welding, upset welding, study welding, foil butt welding, high-frequency welding, and percussion welding (Figure 20-11).

FLASH WELDING Flash welding is the resistance welding process that received its name from the arc or flash that is created between the two pieces of metal to be welded. Metal to be welded is clamped into specially designed electrodes that have one stationary platen and two movable platens. As the current is turned on, the two pieces of clamped metal are brought into close proximity. As the metals move closer, their near or slight contact causes an arcing. This arcing immediately raises the temperature of the metal particles to a welding temperature. The arc stream then projects incandescent particles or metal. At this point, the current is turned off and the movable platen presses the molten areas together to form a fusion weld (Figure 20-12).

Flash welding is used to assemble rods, bars, tubing, sheets, and most ferrous metals that have a low hardenability rating. Flash welding generally is not recommended for zinc and its alloys, cast iron, lead and its alloys, or copper and its alloys. The special, water-cooled electrodes that serve as clamping devices for the metal to be welded must be performed to exacting shapes that will fit the shapes of the base materials being welded. The durability of the clamping dies has been greatly improved by the use of water-cooled dies with alloyed steel inserts. A major advantage of flash butt welding is that many dissimilar metals with different melting temperatures can be flash

| Metal | Metal |

Fig. 20-11 Butt joint.

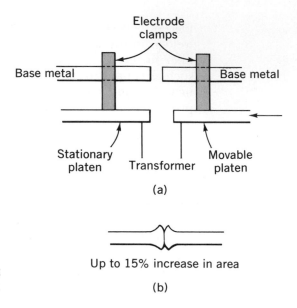

Fig. 20-12 Flash welding. (a) Welding system. (b) Finished weld joint.

welded. The major restrictions of flash welding are that the shapes of the metal to be flash welded should be similar and that the weld zone will increase in size because of the pressure that forces the two soft ends together (Figure 20-12b). The most commercially acceptable increase in size is a 15 percent increase in the weld zone area.

RESISTANCE BUTT WELDING

Upset welding uses the same type of equipment as flash welding. The basic difference between flash welding and upset or resistance butt welding is that an arc is not allowed to occur between the two pieces of metal to be welded. The two pieces of the base metal to be welded are brought together to a single interface. A heavy current is then passed from one piece of base metal to the other, which causes an electrical resistance to be set up between the two materials, heating the metals. After the base metal has reached the welding temperature of 1600 to 1700°F, the two pieces of base metal are pressed together more firmly. This pressing together is called upsetting. Upsetting takes place while the current is flowing and continues until after the current is shut off. The upsetting action mixes the two metals homogeneously and pushes out many of the impurities of the atmosphere. It also reduces the heat-affected zone to a minimum. There is a slight increase in the size of the heat-affected zone. The resistant butt welding process is a slower process than the flash welding process because more time is needed for the current to create resistance enough to bring the base metals up to a melting temperature, but it is an easy process to fully automate, which makes it an excellent commercial fabrication process.

STUD WELDING Stud welding is a process similar to flash welding as well as to metal arc welding. Stud welding incorporates a method of drawing an arc between the stud, a rod, and the surface of the metal. Then the molten surfaces are brought into contact with each other under pressure. This process was discovered by H. Martin around 1918, and was used extensively by the British navy. It was not until about 20 years later that Ted Nelson perfected a similar process in the United States.

Stud welding eliminates the need for drilling or punching holes in the main structure and saves time by mechanically fastening objects, such as bolts, screws, or rivets, to the main structure. The stud welding apparatus most widely used during the war was the stud gun, which was used to weld studs rapidly and economically in flight deck planking, wire ways, plumbing, pipe hangers, and other installations.

The same methods are used to operate both the portable and the stationary welding equipment. The equipment required for stud welding consists of a stud welding gun; a device to control the time of the current flow; a source of dc welding current, usually 300-, 400-, or 600-amp dc current; and studs and ferrules. The Nelson and the Phillips stud guns are most commonly used in the United States. Both stud guns weigh approximately 5 lb and resemble a pistol with an oversized barrel. The frame is made from fabric-reinforced Bakelite, with the welding leads and control circuits entering the gun through the pistol grip. A push-button switch or trigger is located in the handle for starting the welding cycle. The mechanism of the gun has a copper coil that is cast integrally with the frame of the gun. Inside this heavy copper coil is an armature, connected through linkages to the stud chuck or holding device. When the trigger is pulled, a slight lateral shift of the armature sets up the necessary arc gap, establishing the flow of welding current. After a predetermined, preset elapsed time, the current is interrupted. Then a spring in the gun barrel moves the stud into the weld pool to complete the weld.

The control unit used consists of a high-speed relay and timer. The timer is adjusted to the desired number of cycles to be used, with the cycles based upon 60 Hz. The time interval will vary from 3 to 45 Hz, depending upon the size of the stud to be welded.

There are two different types of stud welding equipment commonly used in this country. The cyc-arc equipment operates by drawing an arc between the studs and the base metal to which it is being welded. The current is set at a predetermined rate and the stud is lowered into a molten pool formed in the base metal by the arc stream. The gun is removed and the fusion weld takes place between the stud and the base metal. The Phillips stud welding process uses a timing device that is activated by a cartridge placed on the end of the stud. The cartridge, which is a semiconductor, starts the arc between the stud and base metal. While the welding is taking place, that part of the cartridge connected to the stud fuses away and releases the stud, which is then pressed into the molten weld puddle. The size of the generator used depends on the size of the stud to be welded, and the circumstances under which the welding is done. Direct-current generators from 300 to 600 amps that recover their peak voltage from 1 to 2 Hz are the most desirable for good stud welding performance. The ac generator is not widely used because its use is restricted to the flat welding position, which limits considerably the application of the studs.

Ferrules used are either ceramic or porcelain. They perform several functions. The ferrule serves to concentrate the heat of the arc in the weld zone, to act as a dam by confining the molted metal to an area, to protect the operator from the harmful effect of the rays of the arc, to prevent overheating of the base metal, and to protect the weld puddle from contamination. Any stud may be used, provided it can be welded by the generator available. Most commercially available studs come with a ferrule. There

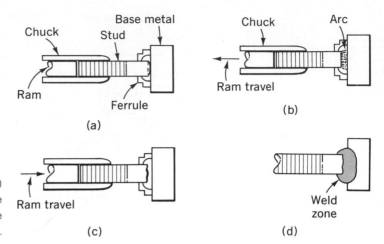

Chuck Base metal Stud Ram Ferrule

(a)

Chuck Arc Ram travel

(b)

Ram travel

(c)

Weld zone

(d)

Fig. 20-13 Stud welding sequence. (a) Contact stage. (b) Arc stage. (c) Forge stage. (d) Finished stud (with ferrule removed).

are many types of studs, such as straight, female, bent, and threaded.

The stud welding cycle involves several steps (Figure 20-13). The stud is placed in the chuck of the gun and the ferrule is slipped into the position over the stud. Then the gun is placed in the proper position against the surface of the base metal to be welded. The trigger is pushed which causes the stud to be retracted automatically from contact with the base metal by a solenoid coil inside the gun, causing an electric arc to be established between the stud and the base metal. The arc melts the base metal and part of the stud. The arc is then automatically shut off by the timer, deenergizing the solenoid coil, which releases the main spring. The spring drives the stud into the molten pool of the base metal, causing a fusion weld to take place between the stud and the base metal (Figure 20-14).

A large number of ferrous and nonferrous metals can be stud welded successfully. The ferrous materials include carbon steel, stainless steel, and low-alloy, high-strength steels. The nonferrous metals include lead-free brass, bronze, aluminum, and chrome-plated metals. Studs can be fastened to thin sheets of metal, but the strength of the assembly is not too sturdy because the base metal lacks toughness. Some of the applications of stud welding can be found in the installations of conduit, pipe hangers, planking, insulation, and corrugated roofing. Stud welding is used widely in shipbuilding and in railroad, automotive, and construction industries.

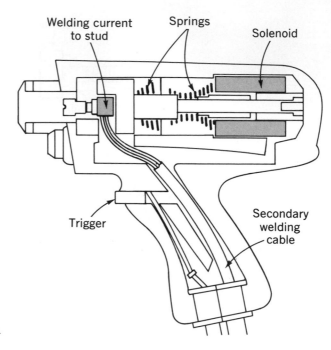

Welding current to stud · Springs · Solenoid · Trigger · Secondary welding cable

Fig. 20-14   Stud-gun cross section.

HIGH-FREQUENCY RESISTANCE WELDING

The phenomenon of high-frequency current which causes it to flow at or near the surface of a conductor and not through the entire thickness of a conductor makes it possible to resistant weld extremely thin pieces of material, as thin as 0.004 in. Another phenomenon of high-frequency current is the proximity effect, the current following a path of low conductance rather than low resistance, which means that the effective resistance of the metal being welded is much higher with high-frequency current than standard 60-Hz current. Therefore, the amperage requirement for a given amount of calorie release or heat release is but a fraction of that needed for standard resistance welding. This characteristic, coupled with the fact that the low-inductance path is the one that is closest to the conductor of the circuit, determines the design of the high-frequency resistance welding machine. Supplied to the high-frequency contacts placed on the base metal are 450,000 Hz of ac power. Because of the extremely fast cycling, the conductor of the current assumes the shape of a V between the conductors (Figure 20-15). This V path acts as a

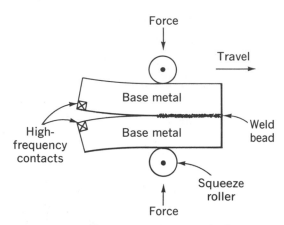

Fig. 20–15   High-frequency welding.

return conductor for the low inductance, which causes the surfaces of the two pieces of base metal to be heated. At the point of the V, there are two rollers which force the metals together, which slightly upset the base metal, causing the weld to take place. Materials can be joined that range from 0.004 to 0.012 in. in thickness at welding

Fig. 20–16   High-frequency welding. (a)
High-frequency welder.

(a)

speeds from approximately 200 to 1000 ft/min. The high-
frequency resistance welding process can be used to join
copper and its alloys, nickel and its alloys, aluminum, and
many types of steels. The process is used mostly for the
welding of tubing, especially with a butt weld or a butt
seam weld (Figure 20–16b and c).

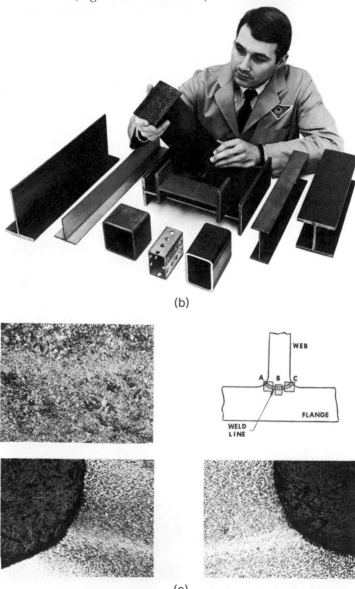

(b)

(c)

Fig. 20–16 (cont'd)   (b) Possible
weldments. (c) Welded I-beam. (Courtesy
of American Machine and Foundry Company—AMF Thermatool, Inc.)

QUESTIONS

1. What is resistance welding?

2. What are the two cycles in resistance welding?

3. How is seam welding an application of spot welding?

4. What are the advantages of a pointed electrode?

5. What is a nugget?

6. How is the size of a resistance spot weld determined?

7. What are the advantages of a stitch weld over a roll resistance weld?

8. Why is projection spot welding a higher production output welding method than a standard spot welding?

9. What are the advantages of metal-fiber spot welding over projection spot welding?

10. How do flash welding and upset (resistance butt) welding compare?

11. What are the four stages in the stud welding cycle?

12. What is the proximity effect?

13. What are the limitations of high-frequency resistance welding?

SUGGESTED FURTHER READINGS

American Welding Society: "Resistance Welding Theory and Use," Reinhold Publishing Corporation, New York, 1956.

Freytag, Norman A.: A Comprehensive Study of Spot Welding Galvanized Steel, *Welding Journal,* vol. 44, p. 145, April 1965.

Houlderoft, Peter T.: "Welding Process," Cambridge University Press, Cambridge (Eng.), 1967.

King, James: Which Way to Weld, *Product Engineering,* Oct. 12, 1964, p. 77.

Oates, J. A.: *Modern Arc Welding Practice,* George Newnes, Ltd., London, 1961.

Resistance Welding Manual, Resistance Welder Manufacturers' Association, Philadelphia, 1948.

Schaller, Gilbert S.: "Engineering Manufacturing Processes," McGraw-Hill Book Company, New York, 1959.

21 ELECTROSLAG AND ELECTROGAS WELDING

The electroslag and electrogas welding methods are used to fuse two sections of thick metal, forming a seam in a single pass. Elimination of the need for making multiple passes and special joint preparations make these methods commonly used welding processes when joining heavy ferrous metal.

The basic idea for electric welding was conceived about 1900, but it was not until approximately 1950 that reliable equipment demonstrated its economic feasibility and proved its metallurgical features. The Paton Institute in Russia performed the basic research to perfect electroslag welding as a reliable, workable tool. The Bratislave Institute in Czechoslovakia experimented in parallel research so that the United States received many of the patterns through the Czechoslovakian designs. The electroslag/gas welding processes were developed because a process for welding joints of thick sections of metal was needed. With this process, plates of 1 in. or more can be joined without costly multiple passes and without special joint preparations, such as bevel joints, V joints, and U or J grooves being performed. These processes have reduced costly time in the fabrication of large vessels and tanks. There is theoretically no limit to the thickness of the weld bead. Weld beads ranging in thickness from 1½ to 15 in. have been performed with presently existing equipment.

ELECTROSLAG The process of electroslag welding is a vertical, uphill process. Two copper shoes, dams, or molds must be placed on either side of the joint that is to be welded in order to keep the molten metal in the joint area. One or more electrodes may be used to weld a joint, depending upon the thickness of the metal. The electrodes are fed into the weld joint almost vertically from special wire guides. Electrodes need not be of a special deoxidized nature but they may contain a flux, if it is needed. A mechanism for raising the equipment as the weld is completed and an ac power source that has approximately 1000 amperes output and a 100 percent duty cycle are needed (Figure 21–1).

Electroslag welding depends upon the generation of

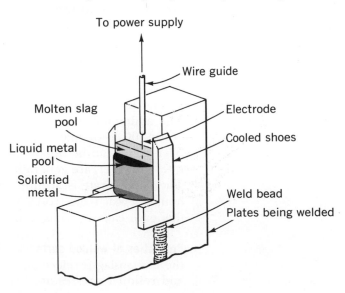

Fig. 21–1   Electroslag welding.

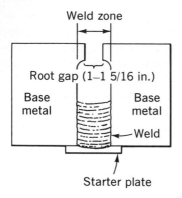

Fig. 21–2   Weld zone.

heat that is produced by passing an electrical current through molten slag. The space formed between the root faces of the weld joint and the copper, water-cooled molding shoes contains the molten slag pool into which the electrodes are immersed. The current passing through the electrode and the base metal heats the metal to a high temperature, increasing its electrical conductivity. The temperature of the molten slag pool must exceed the melting point of the base metal and filler metal in order to produce a weld bead. Consequently, the weld puddle melts the surrounding surfaces of the base metal (Figure 21–2). The molten base metal and the filler metal collect at the bottom of the slag pool, forming the weld pool. When the weld pool solidifies, the weld bead is formed, joining the faces of the base metal. As the welding action continues, the flux flows to the top of the weld and cleans impurities from the molten metal. A powdered flux is fed into the weld pool in order to reduce oxidation and also to continue the welding process after the start of the weld. The flux forms a molten slag after the initial arc, and this slag becomes an electrical conductor which circulates vigorously, melting the parent metal and filler metal. After the initial starting arc, there is no arc because the conductivity of the flux creates the necessary heat required to melt the metal. There is also no need for a continuous flow of flux into the weld area because once a supply is acquired it does not vaporize or disintegrate rapidly. Only small additions of powdered flux are required periodically. As the weld metal begins to fill up the lower limits of the sections that are to be welded, a mechanical device elevates the shoes and the wire-feed mechanism so that the weld continues upward until it is completed. The additional flux is added at a rate of 5 lb of flux for every 100 lb of filler wire. As much as 105 to 135 lb of filler metal can be deposited in 1 hr. These high deposit rates, however, are generally accomplished with electrodes. The deposit rate is around 35 to 45 lb of weld metal per hour for each electrode. One requirement, however, is that each electrode has its own power source.

Properties of welded surfaces   In welds produced by either the electroslag or electrogas welding processes, distortion and residual stresses are kept to a minimum. The tensile strength of steel plate welds ranges from 54,000 to 58,000 psi. Because the temperature attained in the immediate weld area is approximately 3500°F, the grain structure is large, which produces a brittle portion to the finished product. Usually it is desirable to normalize the product by heating the base metal to approximately 100° above the lower transformation level of the metal and allowing it to cool slowly. Low-carbon steels produce excellent welding properties with these processes.

Joint design   With electroslag welding, it is necessary to have only a square butt joint or a square edge on the plates that are to be joined. A plate is placed at the bottom of the joint for initial starting. This starter plate is removed after the weld is completed. The allowable root gap for this process is from 1 to $1^5/_{16}$ in. During the weld, there may be approximately 2 in. of molten metal that is in a constant swirling state, and the slag puddle may be from $1^1/_4$ to $1^1/_2$ in. deep. The length of the weld depends only on the material to be welded. The plate thickness, however, is limited to a minimum of 1 or $1^1/_2$ in. for economic operation. There is theoretically no maximum plate thickness.

Welding procedure   A starting tab or plate must be used in order to build up the proper depth of conductive slag before the molten pool comes into contact with the work. The electroslag process is initiated, much like the conventional submerged arc process, by starting an electric arc beneath a layer of granular welding flux. When a sufficiently thick layer of hot flux or molten slag is formed, all arc action stops and electrical current passes from the electrode to the workpiece on through the conductive slag pool. At this point, the process is truly electroslag welding. In order to ensure that the filler material fuses properly, the base metal is initially heated to a high temperature, approximately to its melting point. The heat generated by resistance to the flow of current through the molten slag is sufficient to melt the edges of the workpiece and also to melt the welding electrode. During the process of electroslag welding, since no arc exists, no splattering or intense arc flashing occurs. The slag bath temperature is around 3500°F. The liquid metal coming from the filler wire and the heated base metal collects in a pool beneath the slag bath and slowly solidifies in the weld. This slow solidification causes the weld area to develop a radial crystalline pattern, which helps to increase the resistance of the weld to hot cracking. Electroslag welding in which the weld is molded differs from both the automatic arc welding processes in a number of ways. These differences must be taken into account before this technique is employed. When an electric current passes through molten slag, the evolution of gas and the resulting slag splatter are not as intense as in arc welding. Once the electroslag process has become stabilized, no slag splatter of any kind takes place.

A small amount of slag is consumed when this process is used. The slag consumption averages 5 percent of the weight of the deposited metal. This 5 percent includes losses by dispersion, or slag that has been absorbed into the molten weld pool. This 5 percent consumption is approximately 20 times less than that of the submerged arc welding process. The electroslag process requires less electrical power per kilogram of deposited metal than either the submerged arc welding processes or the shield arc welding processes. An even more important practical advantage is the possibility of single-pass welding with metals up to 3 in. thick when using a single electrode and with metals of unlimited thickness when using a multiple electrode. Electrowelding has other important assets. The low flux consumption and the small addition of flux required for the molten metal to pool result in the chemical composition of the welding metal being more consistent than in arc welding. The uniform heating of the weld area keeps distortion and residual stresses to minimal amounts.

Safety   As in the other electric welding processes, the operator of the electroslag or electrogas welding machine should take the precaution of wearing safety goggles, especially in electrogas welding. Protective clothing and safety shoes should also be worn to protect the feet from the liquid slag that might fall to the floor while the weld is in operation.

ELECTROGAS WELDING   Electrogas welding operates on the same general principles as electroslag welding, with the addition of some of the principles of submerged arc welding. The major dif-

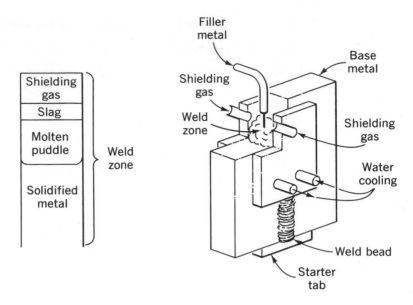

Fig. 21-3 Electrogas welding.

ference between electroslag welding and electrogas weld-ing is that an inert gas, such as carbon dioxide, is used to shield the weld from oxidation, and there is a continuous arc, such as in submerged arc welding, to heat the weld pool. The joints and the use of flux to cleanse the weld are the same as in the electroslag process. The shoes that are used to form the weld, as in the electroslag process, are also used in the electrogas process to control the weld zone through water cooling. However, the flux, instead of being issued to the weld zone through a hopper mechanism, is incorporated within the electrode itself in the form of a cored wire (Figure 21-3).

QUESTIONS

1. What is the difference between electroslag and electrogas welding?

2. What is the function of the molten slag pool?

3. How is flux added to the molten pool?

4. What is the deposit rate per electrode?

5. Why is distortion kept at a minimum in the electrogas process?

6. What are the power source requirements for electrogas/slag?

7. Where did the electroslag process originate?

8. What is the thickness limit that electroslag can successfully weld?

SUGGESTED FURTHER READINGS

Irving, R. R.: The Turning Point is Reached for Electroslag Welding, *The Iron Age,* vol. 198, no. 10, p. 117, March 10, 1966.

Paton, B. C.: "Electroslag Welding," Van Nostrand Reinhold Company, New York, 1962.

Pierre, Edward K.: *Gas Metal Arc Welding,* Miller Electric Manufacturing Company, Appleton, Wis., 1965.

Thomas, B. R.: Electroslag Welding, *Welding Journal,* vol. 39, p. 111, February 1960.

22 SOLID-STATE BONDING

Solid-state bonding is the placing of two extremely clean metal surfaces in such intimate contact that a cohesive force between the atoms of the two surfaces holds or welds them together. Examples of solid-state bonding are the precision Johansen gage blocks, which are one of the standards of measurement in modern industry. They require a force of approximately 250 psi to pull them apart once they are brought together. However, they are not an example of a perfect bond because of the imperfections in their surface and the lack of laboratory cleanliness in their bonding area. Theoretically, if metal could be made of a single crystal and if the single crystal could be clean and smooth and could possess only one plane, two such crystals could be brought together and no force would be required of a single crystal but have a polycrystalline structure and present many surfaces. The major types of bonding mechanisms or bonding processes that are currently approaching the possibility of solid-state bonding are explosive welding, ultrasonic welding, friction welding, and diffusion bond welding.

EXPLOSIVE WELDING The fastest growing branch of explosive metal working is the welding, joining, and cladding of metals. Strong metallurgical bonds can now be produced between metal combinations which cannot be welded by other methods or processes. For example, tantalum can be explosively welded to steel although the welding point of tantalum is higher than the vaporization temperature of steel. In critical space and nuclear application, explosive welding permits fabrication of structures that cannot be made by any other means; and, in some commercial applications, explosive joining is the least costly method.

The major advantage of explosive welding include the simplicity of the process; the extremely large surface that can be bonded; welds that can be produced on heat-treated metals without affecting the heat-treated processes, also, explosion welded bonds do not have heat-affected zones, incompatible materials can be bonded, and thin foils can be bonded to heavier plating.

The development of explosive welding In the past, explosive welding was considered a form of cold pressure welding, but this view has been substantially altered in recent years. Although cold pressure welding was used as early as 3000 B.C. for welding sponge iron, the first scientific demonstration took place in 1724, when two lead balls were pressure welded, forming a bond as strong as the parent metal. The production of clad material and composites formed by passing plate through the rolling mill became an important industrial process around 1930.

A major limitation of pressure welding is that a severe deformation is needed to obtain a bond. Welding by directly applying pressure demands that the metal surfaces be deformed in order to expose clean, oxide-free surfaces. A minimum range of 40 to 60 percent of deformation of the parent metal is necessary for pressure welding. However, interfacial shielding during welding breaks up the film and considerably reduces the extent of deformation required before the surfaces weld together.

In explosive welding, the plastic deformation is restricted to the interface or the area between the two faces of metal, making unnecessary the area reduction required for cold pressure welding. The explosive impulse is used to provide both extremely high normal pressure and a slight, relatively sheer or sliding pressure between the two surfaces, or the interface. Explosive welding and cladding are carried out by bringing together properly paired metal surfaces with high relative velocity at a high pressure and with a proper orientation to each other so that a large amount of plastic interaction occurs between the surfaces. While a variety of procedures in the geometric arrangements of the explosive charge and the parts to be welded have been successfully employed, the major techniques of explosive welding and cladding can be divided into contact techniques and impact techniques.

Contact welding techniques achieve sufficient plastic interaction at the weld interface by positioning the explosive charge so that the shock wave is delivered at an oblique angle to the parts being welded.

The impact techniques have largely supplanted contact methods of explosive welding. The usual procedure is to project explosively the pieces to be welded together toward each other so that they impact at a high velocity. This velocity may range from 500 to 1000 ft/sec. Achieving these high velocities means that the detonation velocity caused by the explosives usually averages approximately 21,000 ft/sec in the detonation front. The pressures produced by the velocity at the interface of the two materials to be welded together range from 100,000 to 1,000,000 psi. If the initial parts are driven together at an angle so that the point of contact between the colliding pieces sweeps across the weld area, plastic deformation is accomplished with predictable control. When the velocity of impact and the angle of collapse are properly selected for the material being welded, intense plastic flow at the surface, called surface jetting, will produce a high-strength weld (Figure 22–1).

Surface jetting has been studied by numerous investigators. As in pressure welding, the requirements for a strong weld between two metal surfaces are that the sur-

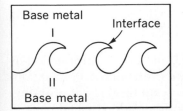

Fig. 22–1   Interface jets.

faces be free of contaminants and that sufficient pressure
be applied to bring surfaces within interatomic distances
of each other. Studies of high-velocity metal impacts
show that at certain ranges of speed and angles of impact
a high-velocity metal jet is formed from the bottom surface
of the flyer plate and the top surface of the base plate. The
surface jet removes surface contamination and oxides,
leaving a clean surface on both metals. As a consequence,
two surfaces are exposed and brought together under
pressure as the jet passes. To achieve maximum weld
strength, the speed, the angle of incidence, and the speed
of the detonation front must be correctly chosen so that a
wave formation is produced during jetting. An angle of
incidence from 1 to 10° has been experimented with for
metals. This rippled surface enlarges the welding area,
increasing the strength of the joint. Under conditions of
properly controlled jetting, metallurgical bonds as strong
as the weaker of the two components can be achieved
without the deformation resulting from the conventional
pressure of cold pressure welding, a substantial advantage
in the cladding of exotic metals in sheet form, such as
are used in the aerospace industries.

Explosive welds Properly formed explosive welds exist
at 100 percent efficiency so that the strength of the joint
equals the strength of the weaker parent metal. Various
types of tests are used to determine the strength of the
weld, including the shear test, tensile test, and band test.
In most instances, failure takes place in the parent metal
rather than at the weld zone. In general, the microhard-
ness of the metal near the weld interface is slightly increased, probably because of the plastic deformation and
strain hardening effects of the explosion and the jetting
action. This hardness zone is apparently only a few thousandths of an inch in thickness. Otherwise there seems to
be no variation in hardness across the section of explosively
welded metals.

Metal cladding A commercially successful introduction by
Du Pont of explosive-clad plate is one of the most signifi-
cant recent developments in explosive metal working.
Du Pont first offered the explosion-cladded metal plate in

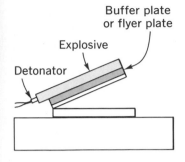

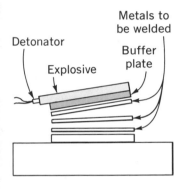

Fig. 22-2 Impact explosion welding.

1963. Today, explosion-clad plate products meet most existing specifications, for clad metals are available in commercial sizes and quantities. Metal plates as large as 7 by 30 ft, weighing up to 25 tons, have been successfully clad. Explosively solid-state bonded titanium clad has been the most commercially attractive of the various clad metals. Since there are production problems associated with the cladding of titanium sheets by either direct mill rolling or brazing, the explosively clad titanium sheets are in demand. In addition to titanium clad, other explosion-bonded products such as nickel, stainless steel, Hastelloy, tantalum, and carbon steel are available. Some major applications for these clad products are heat exchanger tube sheets and pressure vessels. The explosion bonding process for stainless steel clads is expected to be more competitive in thicknesses over 2 in. The major advantages of metal cladding with explosives include the wide range of thicknesses that can be explosively clad together. The size of the metal is not important; either large or small objects can be clad. Chemical properties and physical properties of the parent metal will not be altered significantly, and there is no practical limit to the backing plate thickness; the thinness of the plate is the limiting factor.

The limitations to explosive cladding are concerned with the brittleness of the alloys. Metals to be bonded by explosive cladding must possess some ductility and some impact resistance. Brittle metals or brittle backer plates cannot be used because they fracture during cladding. Figure 22-2 illustrates the two most common setups in which the welding conditions are controlled by inclining the flyer plate at a carefully preselected angle to the base and adjusting the explosive charge to drive the plates together at the optimum impact velocity. In the standoff technique shown in Figure 22-3, the weld conditions are controlled by the standoff distance, the charge density,

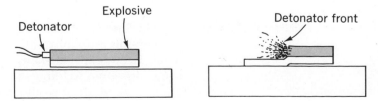

Fig. 22-3 Contact explosion welding.

the detonation velocity, and the deformation character-istics of the flyer plate. A welding arrangement used for special part shapes and specific applications is shown in Figure 22–4.

SPOT WELDING Spot welding by a small explosive charge offers great advantages in joining difficult-to-weld metals and may prove indispensable for space applications, such as emergency repairs to spacecraft or the erection of devices in space.

A compact, hand-held explosive spot welder that weighs approximately 10 lb and produces ³/₈-in. diameter spot welds is now available. The unit contains its electric power supply for ignition and is equipped with multiple safety interlocks. Various weights of commercial PTN explosives (pentaerythritetranitrate explosives) are employed with a standard cap. The explosive force is directly coupled to the metal sheet being welded. On some metals, a thin protective sheet of plastic may be needed to avoid burning the surface. While the welding machine is equipped to vary the explosive standoff distance, the general practice is to use as small an explosive charge as possible to weld the metal surface.

The spot welder has given excellent welds in approximately 35 different metal combinations. More are added to the list continually. Particularly good results have been obtained with austenitic stainless steels with cobalt-based alloys, which are used for high-temperature applications, and with nickel-base alloys such as Inconel and nickel. Good results are also obtained in spot welding aluminum alloys. However, all aluminum alloys must be chemically cleaned to remove the oxide before welding. The welding should be completed within 4 hr to avoid buildup of the oxide.

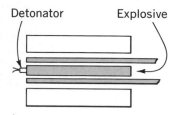
Detonator   Explosive

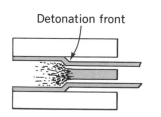
Detonation front

Fig. 22–4 Explosion welding application.

ULTRASONIC WELDING In 1938, Ludwig Bergmann and some colleagues were experimenting with ultrasonic waves and their effects on metal. He found that many metals could be combined by using ultrasonic welding that could not be joined by any other method. It was also found that any other metal could be strengthened by subjecting the metal in its molten state to ultrasonic vibrations. The ultrasonic effect upon the molten metal generates a smaller grain size, giving the metal more strength.

One of the most easily understood ultrasonic devices is the so-called silent dog whistle. The sound radiated by the whistle is beyond the range of human hearing but well within the range of the hearing of a dog. This ultrasonic whistle operates above 20,000 Hz. The upper range of ultrasonics used in welding is approximately 60,000 Hz. Ultrasonic welding combines pressure and high-frequency vibration motions to form a solid-state bond. This cool, strong weld is capable of joining such combinations as aluminum to steel, aluminum to tungsten, aluminum to molybdenum, and nickel to brass. Ultrasonic welding can even join a large metallic object to a piece of foil, a feat made possible by the ability of ultrasonic vibrations to create the proper heat in each object being joined. Ultrasonic welding has also made it possible to join metals with vastly different melting temperatures, making strong rigid joints. Thus, many applications previously considered unweldable can now be reevaluated.

The cost involved in ultrasonics is usually the limiting factor in its applications. If another method of welding is possible, then it would probably be more economical to use it rather than ultrasonic welding. In many cases, this high cost factor can be compensated for by the fact that ultrasonics can be used to join metals once thought to be unjoinable. In short, ultrasonic welding is far superior in its almost unlimited application to any other form of welding. Ultrasonics welding usually begins where the other processes stop.

In the ultrasonic process, as in resistance welding, pressure is applied on two sides of the work while a hydraulic piston forces the welding pieces against a solenoid that vibrates them at about 20 kHz/sec. An ultrasonic weld is completed in about as much time as it takes to strike an arc in arc welding.

Ultrasonic welding differs from resistance welding, which requires a generation of heat by electrical resistance at a strategic point of the weld, in that it does not depend upon the similar melting points of two metal workpieces. Although ultrasonic welding requires two metals of similar hardness, vastly different metals in a variety of thicknesses can be joined with a minimum of heat. As an example, ultrasonic welding is used to seal containers filled with reactive chemicals, such as nitroglycerin, and produces practically no heat; whereas, resistance welding would involve more than enough heat to cause combustion.

In comparison with fusion welding, which requires electric arc or gas to heat metals to their proper melting points, ultrasonic welding does not create any great quantity of heat to distort metals being joined or to cause a heat-affected zone near the fusion area. The gas of ultrasonics does not require any protective shielding equipment for either the metal being joined or for the welder's eyes, as is the case in fusion welding.

Electron-beam welding, which is presently the most sophisticated form of fusion welding, is still far from surpassing ultrasonic welding. Many metals that react to the vacuum which electron-beam welding requires for the generation of intense heat cannot be welded properly. Some metals that do react to a vacuum are magnesium, zinc, cadmium, and chromium. These metals can be welded with ultrasonic sound; however, in some instances, electron-beam welding can be used in welding many of the same metals joined by ultrasonics. Because electron-beam welding is a slower process, it is used more when speed is not a major factor in production. It can be used with metals ranging in thickness from light gage to heavy sections.

Cold pressure welding is similar to ultrasonic welding in that it, too, requires a small amount of heat for fusion. It does require a great amount of pressure, however, which in most cases can result in the deformation of the metals being joined. In the case of ultrasonics, not enough pressure is required for welding to deform even foils that are being welded. Also, cold pressure welding is restricted to spot welding only and cannot be used for seam welding, as ultrasonics can be.

There are still fundamentals in ultrasonic welding that have not been solved. The bonding mechanism is not fully understood and the welding tip is subject to cracking and sticking. Definitive metallurgical theory has yet to be established to explain adequately the formation of this quality bond. One theory is that the bonding takes place because of an interfacial chemical reaction. Other theories suggest that it is because of an interfacial atomic bonding, a mechanical interlock bonding, or an interface fusion. There has been the suggestion that ultrasonic bonding is a result of a combination of all these reactions.

The ultrasonic welding unit is composed of two smaller units, a power source, and a transducer (Figure 22–5). The power source is no more than a frequency converter which converts 60-Hz line power into a high-frequency electric power. The transducer, then, converts the high-frequency power to ultrasonic revibratory power. The ultrasonic vibrations are produced by an oscillator that sets the number of vibrations per second or the frequency. An amplifier then steps up the low power of the oscillator as high as 25,000 watts, and then the transducer converts the electrical vibrations into mechanical vibrations.

The transducers are of two different types. The piezoelectric crystals vibrate at a natural resonant frequency when a current flows through them. The magnetostrictive transducer is composed of laminations of nickel, or a nickel alloy, which are surrounded by a coil of wire. The laminations expand and contract in step with the vibrations from the amplifier because of the deformation of the nickel upon heating. This transducer is used exclusively because it can operate at high temperatures, near 1300°F, and because it can handle high power. Quartz is a typical piezoelectric crystal. Crystals can also be made of Rochelle salt, ammonium dihyrosphaten, and barium titanate ceramic. The remaining parts of the welder are arranged according to the type of job to be done. Setups differ depending on manufacturers and the applications. Generally, there is a coupler or horn to funnel the vibrations to a tip which is placed directly over the weldment. The weldment is placed on the anvil, which is generally in the part of the machine where access is permitted, and pressure is applied to the tip by means of a clamping device.

Ultrasonics has many uses in the fields of industry. A transducer immersed in a cleaning solution will rupture the water, forming small almost invisible bubbles which are under great pressure, up to 75,000 psi. When the bubbles collapse, they release a tremendous amount of energy which blasts away any foreign materials that are clinging to an object submerged in the solution. Ultrasonics can machine materials as hard as diamonds. The transducer equipped with a small tip will erode away mate-

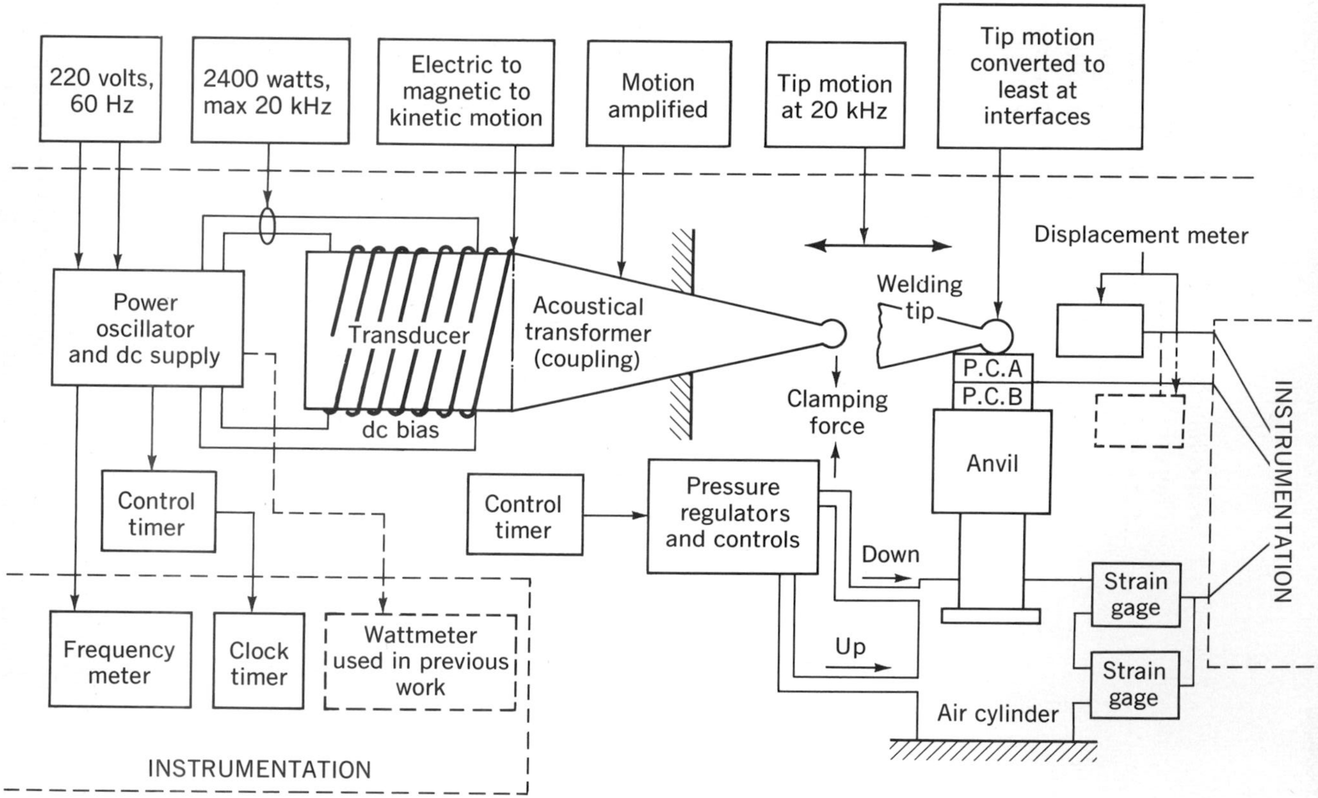

Fig. 22-5 Ultrasonic welding system and energy flow.

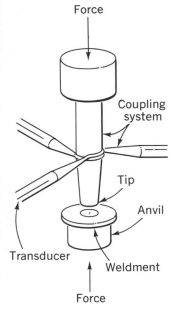

Fig. 22-6 Ultrasonic weld tip-spot welder.

rial in front of it with great precision. Nondestructive testing is another use of ultrasonics. As the vibrations travel through the metals, they are partially reflected back if they hit a crack or imperfection. With sensitive instruments, trouble can be located and the part can be either repaired or discarded. Ultrasonics is also used in welding in which it literally beats the metal together.

The ultrasonic welding process consists of clamping the material to be welded between the transducer and the back-up piece (Figure 22-6). The ultrasonic vibrations are induced into the materials. At the point of contact, the materials are heated by internal friction to nearly 35 percent of their melting temperature. Some authorities believe it to be closer to 50 percent of their melting temperature. The metal is plastic at the point of contact and the vibrations drive the material together. The frequency and the pressure exerted on the material depend on the type and size of the materials, and the make of the welding machine.

There are many advantages to ultrasonic welding. Because of the low temperatures that are involved, the characteristics of the materials are not altered and are continued through the weld zone. With the exception of pure aluminum, the ultrasonic weld is stronger than a weld resulting from resistance welding, a spot welder, or a seam welder. The strength of the ultrasonic weld is from 65 to more than 100 percent as strong as the base material. The weld can be stronger than the base material because the ultrasonic vibrations decrease the grain size, thereby increasing the strength of the materials. In some applications, welding by ultrasonic vibrations is the only means available to weld aluminum to stainless steel. To weld glass is impossible by any other means. Ultrasonics can weld aluminum and glass to similar materials because the vibratory energy disperses the moisture and the oxides and brings the two surfaces into intimate contact, allowing the weld to take place. Other processes involve the melting of the materials. When the materials solidify at different temperatures, they separate. The ultrasonic process is fast; up to 400 in./min can be welded with even penetration. The vibrations break up and disperse the surface oxides, giving the base materials a clean surface for bonding.

A disadvantage of ultrasonic welding is that its use is

restricted to aluminum. Also, the thickness of one piece can be no thicker than ⅛ in. although the other material to be joined can be of any thickness. This method can also join material as thin as 0.0002 in.

The future of ultrasonics is open; new ways are being developed to use this sound power. More research is needed in this field by both private industry and government, and, undoubtedly, this process will be used more in the future (Chart 22-1).

CHART 22-1 Welding comparisons

WELDING PROCESS	DISSIMILAR METALS	USUAL THICKNESS RANGE, IN.	SPECIAL CLEANING OF ATMOSPHERE	EXTENT OF HEAT-AFFECTED ZONE	DISTORTION	RELATIVE SPEED	OPERATOR SKILL
Fusion	No	¹/₃₂–4 (multiple passes)	Usually required	Most	Most	Comparatively slow	High
Electron beam	Some	0.001–2.5 steel to 6 aluminum	Yes	Little	Little	Fast welding overall cycle; time long	High
Resistance	No	0.010–1 steel; 0.010–½ aluminum	Normal cleaning	Little	Little	Fast	Low
Cold pressure	Some	0.005–⅛	Cleaning and wire brushing	None	Little	Fast	Low
Ultrasonic	Wide range	0.0002–⅛ aluminum to 0.04 steel	Normal cleaning	None	None	Fast	Low

FRICTION WELDING The knowledge that metals can be joined by the heat resulting from friction is not new. Machinists had used metal on cutting tools and on insufficiently lubricated tail stocks for years. Recently, however, the utilization of this apparent process by the Russians has brought about a quick and relatively inexpensive method of welding. Friction welding is the fusion of two metals by the creation of resistance between the surfaces to be joined (Figure 22-7). A compressive force in the plane of contact is used during their motion.

The parts to be friction welded are axially aligned so that one part can be rotated against a stationary part. The frictional heat is regulated by the speed of rotation and the axial pressure of the nonrotating piece. As the temperature at the interface of the two pieces increases, the pieces come up to welding temperature. At this point, the forging

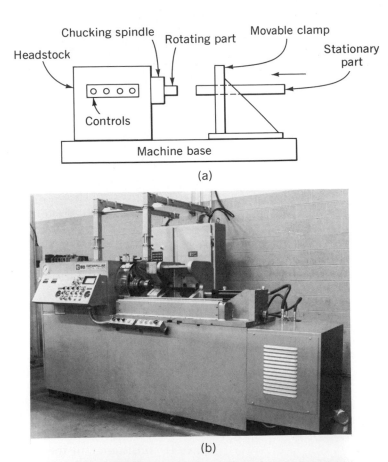

(a)

(b)

Fig. 22-7 Friction (inertia) welding machine. (a) Fundamentals. (b) Production inertia welder. (c) Dual-station inertia welder. (Photographs courtesy of Caterpillar Tractor Company.)

(c)

phase takes place. The rotation is stopped and pressure is increased until the weld is completed. Welding time usually lasts between 2 to 30 sec depending upon the material to be welded. A ½-in. diameter low-carbon steel rod with the welding temperature of 1650°F can be joined with a contact pressure in the range of 5000 to 10,000 psi, while rotating at approximately 3000 rpm for about 5 sec. Medium- and high-alloy steels are friction welded at the same rotation speeds as low-alloy steels but require heating pressures ranging from 10,000 to 30,000 psi and forging pressures between 15,000 and 60,000 psi (Figure 22–8).

There is a fundamental difference in opinion between Russian and American investigations concerning friction welding. The Russians seem to favor a rotational speed of about 1500 rpm with a high axial force, while the Americans favor a rotational speed of 3000 rpm with a lower axial force. However, it has been found that the rpm and the force applied are dependent upon the material to be friction welded. The harder the material, the higher the rpm and the axial force. Forging pressures can be as high as 70,000 psi. Metals that can be welded by the friction welding process include carbon steel, stainless steel, copper, aluminum, and titanium. Research is currently being done on friction welding other types of metals.

Generally, it may be expected that friction welding is done on an engine lathe. This is true up to a certain degree. If consistency in the quality of the weld is desired, it has been found that the use of the lathe is objectionable

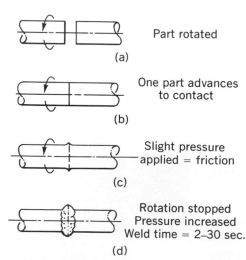

Part rotated	(a)
One part advances to contact	(b)
Slight pressure applied = friction	(c)
Rotation stopped Pressure increased Weld time = 2–30 sec.	(d)

Fig. 22 –8 Friction welding procedure. (a) Rotation phase. (b) Contact phase. (c) Heat phase. (d) Forge phase.

because of the high axial forces at high speeds which soon disable the lathes. The engine lathe is not designed for the endurance needed for friction welding, nor is the lathe designed for the special disengagement devices that must disengage the spindle instantly. Because of these reasons, the use of the engine lathe in friction welding does not allow consistency in the quality of the weld.

It is important to have a machine that will accurately control three variables: the rotational speed, the axial pressure, and the time of the contact. A machine must control all three variables accurately, otherwise there will be inconsistency in the quality of the weld. The metals to be welded can be of any shape as long as they share a common axis. The process can be used only when one part rotates about an axis of symmetry. The length of the materials is limited to the size of the machine. The advantages of this process include its ability to join a large range of materials with high quality and consistency by using simple and compact machinery that requires economical use of power. Also, the heat produced in the friction weld is not enough to melt the base metals, causing little or no warping. There is a burr surrounding the weld area, however, which can be machined off later.

The type of weld joints are limited with friction welding. There is actually only one joint which can be welded on a friction welder, and this is a butt joint. It is possible to weld two pieces of round stock, either pipe or rod, together, or to weld a piece of round stock to a plate (Figure 22–9). Even with these limits for the use of a friction welder, it has the three advantages of speed, accuracy, and economy. Even though the friction process of welding requires one of the parts to be rotated, this does not seriously affect the usefulness of friction welding as a production process.

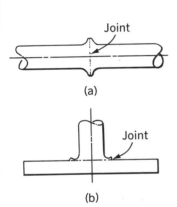

Fig. 22–9 Possible joint designs.
(a) Round to round.
(b) Round to plate.

DIFFUSION BONDING Diffusion bonding is a process that does not necessarily need heat to produce a fusion weld. Rather, it needs two kinds of surfaces that can come into intimate contact under pressure. This pressure is applied for a period of hours. Because of the intimate contact of the two pieces of metal, the pressure ranges from 5,000 to 10,000 psi. In this process, although heating is not essential, if the temperature is raised, the diffusion rate will be cut suf-

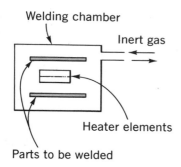

Welding chamber

Inert gas

Heater elements

Parts to be welded

Fig. 22–10 Diffusion bond welding.

ficiently. It might take many hours to perform a certain bonding, but with heat the time element can be cut to a matter of hours or minutes. The one basic requirement for this process is that the material be surfaced accurately so that an intimate bond can take place, as in all solid-state bonding. The oxides that hinder diffusion dissipate over a period of time. Because of the pressure, the dissipation of these oxides will be greatly increased with the addition of heat. Therefore, moderate heating temperatures are used, usually below 2000°F. This process makes it possible to join metal to metal, metal to ceramics, and metal to metal with intermediate bonding materials. The diffusion bonding of steel requires an intermediate fusion material. Temperatures that approach approximately 1600°F are used. This extreme temperature limits diffusion bonding of steel (Figure 22–10).

Diffusion bonding incorporates three basic techniques: gas-pressure bonding, vacuum fusion bonding, and eutectic fusion bonding. Gas-pressure bonding uses an inert gas for the pressure applied on the parts and also uses a high temperature to bond the parts. Gas-pressure bonding uses a heating system that resembles closely an autoclave. The parts to be joined are placed together in intimate contact and then heated to around 1500°F. During the heating cycle, the high-pressure gas is built up to provide pressure over all the surfaces of the parts to be joined. This technique is usually used for bonding nonferrous metals only because of the high temperatures and pressures required for steel and its alloys although it can be used with steel for limited production. Vacuum fusion bonding is diffusion bonding done in a vacuum chamber. The parts are pressed together by a mechanical force rather than by gas pressure. Heating is done the same way as in gas-pressure bonding. Vacuum fusion bonding lends itself readily to the diffusion bonding of steel and its alloys because of the high pressures that are easily accomplished with mechanical means. For example, a temperature of approximately 2100°F and a pressure of 10,000 psi are needed for the diffusion bonding of steel.

Eutectic fusion bonding employs the lowest temperature possible in the diffusion bonding techniques. Eutectic bondings use a thin piece of material between the pieces to be joined. This piece of material forms a eutectic compound with the parent metals. If the parent metals are held at an elevated temperature long enough, the filler material completely diffuses and disappears into the eutectic of the alloy. Many times the intermediate material will completely disappear. The material used for the intermediate material is usually a dissimilar metal made of foil or an electroplated thickness of approximately 0.001 to 0.0002 in.

QUESTIONS

1. What is solid-state bonding?

2. What are the major types of solid-state bonding?

3. What is surface jetting?

4. What is the impact explosive technique? Contact technique?

5. How do high-frequency sound waves weld?

6. What metals are capable of being joined by ultrasonic welding?

7. What are the components of the ultrasonic welder?

8. How does ultrasonic welding compare to cold-pressure welding? Resistance welding? Electron-beam welding? Fusion welding?

9. What shapes are adaptable to friction welding?

10. What are the four phases of friction welding?

11. Explain diffusion bonding.

12. What is intimate contact?

SUGGESTED FURTHER READINGS

Bergmann, Ludwig: "Ultrasonics," John Wiley & Sons, Inc., New York, 1938.

Bergman, O. R.: *Explosive Bonding of Metals,* American Society of Metals, Metals Park, Ohio, 1965.

Jones, J. B., and E. Nippes: New Technique for Metal Joining in the Space Age by Ultrasonics, *Journal of Metals,* vol. 16, pp. 244–245, March 1964.

Sadwin, Lippe D.: Explosive Welding with Nitroguanadine, *Science,* vol. 43, p. 1164, March 13, 1964.

Vill, V. I.: *Friction Welding of Metals,* American Welding Society, New York, 1962.

23 ELECTRON-BEAM WELDING

The use of the electron beam in industry is relatively new, but the development of the beam has evolved over the past 300 years or so. Physicists working on separate problems concerning the theories of the nature of the earth and light development studies, which now make up the physical sciences, were working on theories that eventually led to the development of the electron beam. Some of these theories concerned the electromagnetic wave nature of light, the nature of electrons and electrical current such as the cathode effect, also called electron beam. All the findings from these related studies, coupled with much of today's technology, resulted in a device that will emit, focus, and magnify the beam of electrons.

The need for electron-beam welding developed after World War II. At this time, it became feasible to use costly metals such as titanium, molybdenum, and tungsten as structural components. Also, the manufacturing of high-capacity vacuum units took place at approximately this time. The demand for these highly reactive metals introduced the need for a welding process that would function adequately. Many of these metals became contaminated when they came in contact with certain elements in the air. Even the shield processes using argon, helium, and carbon dioxide could not be used despite their ability to produce atmospheric purity as high as 99.99 percent. The electron-beam process is now used to weld these metals, as well as stainless steel, aluminum, and many of the more refractory metals.

Electron-beam welding fusion joins metal by bombarding a specific comfined area of the base metal with high-velocity electrons. The operation is performed in a vacuum to prevent the reduction of electron velocity. If a vacuum were not used, the electrons would strike the small particles in the atmosphere, reducing their velocity and decreasing their heating ability. The electron-beam welding process allows fusion welds of great depth with a minimum width because the beam can be focused and magnified (Figure 23–1). The depth of the weld bead can exceed the width of the weld bead by as much as 15 times. The process joins separate pieces of base metal by the fusing of molten metals. The melting is achieved by a concentrated bombardment of a dense stream of electrons, which are accelerated at high velocities, sometimes as high as half the speed of light. Under most circumstances, the entire process is done inside a vacuum chamber. Most chambers house not only the workpiece but also the cathode, the focusing device, and the remainder of the gun, preventing contamination of the weldment and the electron-beam gun itself (Figure 23–2).

The electron beam is characterized by its intense localized heating. In some instances, temperatures generated are so great that the metal instantly vaporizes. If the beam is properly focused, it will penetrate completely through the base metal, creating a small hole. The walls of the hole are molten. The beam is then moved along the joint, melting the material as it comes in contact with it. The molten metal flows to the back of the hole where it fuses to make a perfect weld for the entire depth of the penetration. This

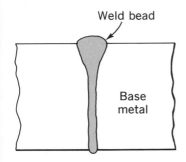

Weld bead

Base metal

Depth to width ratio
25:1 max.

Fig. 23–1   Weld bead.

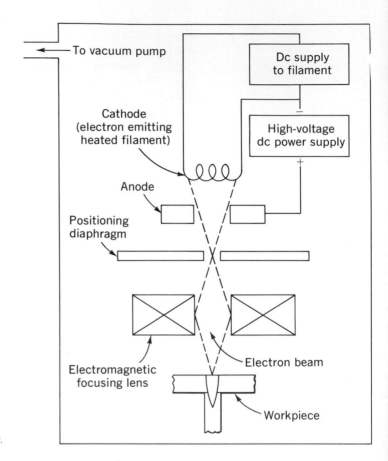

Fig. 23-2 Electron-beam welder.

process is much like the keyhole effect of the plasma arc process. In the conventional arc and gas welding processes, only the surfaces of the metals are melted and fused.

In the electron-beam process, electrons are emitted from the heated filament called the cathode, and then focused into a beam which is directed at the welding point. This electron beam is accelerated by a high difference of potential voltage between the workpiece and the cathode filament. The cathode within the electron-beam gun releases the electrons, which travel through a series of focusing lenses, causing the electrons to converge on one another to form the beam. The electrons are not free flowing but are instead greatly accelerated by the tremendous difference of potential voltage between the cathode and anode. The speeds of these accelerated particles range from 30,000 to 120,000 miles/sec, depending upon the voltage of the unit.

When the beam strikes the welding point, the kinetic energy of the high-speed electrons is converted into heat. The heat is great enough to melt and fuse the metal. The greater the kinetic energy, the greater the amount of heat released. Kinetic energy is determined by the formula $\frac{1}{2} MV^2$. Since the velocity is determined by the difference of the potential between the cathode and the work, the speed of the electrons can be controlled by the amount of voltage applied; thus controlling the amount of heat released. The speed of the electrons is directly proportional to the amount of energy released. Even though the mass is very minute, the velocity is great enough to have a tremendous amount of energy. The density of the electrons, the distance between the emitter, or the cathode, and the work determines the amount of heat at the welding point and the time it takes to perform a weld. Because of the impact of the electrons on the workpiece, the metal is melted and can even boil. The local temperature is so great that even ceramic materials can be cut or welded. First, the stream of electrons penetrates a thick layer of metal to a depth of about 0.001 in. As the stream of electrons travels deeper into the metal, the electrons are scattered, slowed down, and stopped by collisions with the atoms of the crystalline structure, resulting in the heating of a pear-shaped volume of metal. Some of this heated metal is vaporized. The vapor generated helps remove the molten metal, producing a minute hole in the metal. This process is repeated in the second and following layers until complete penetration exists in the base metal. The penetration is accomplished in a very short period of time, usually in fractions of a second, depending upon the energy released and on the thickness and type of material. If the electron beam is allowed to transverse the metal, a seam sequence of penetration is repeated along the weld seam. The metal that is melted is redeposited in the previously melted hole. This keyholeing technique makes it possible for electron-beam welding to be used also as a cutting device or a soldering device simply by increasing the voltage. Increasing the voltage prevents the solidification of the melted metal behind the electron beam.

The amount of heat generated at the welding point is

determined by the power density, which is measured in watts per square inch. The power density of the electron beam is increased by increasing the voltage and decreasing the diameter of the hole. With higher voltages, there are fewer electrons for the same amount of power and less scattering of the bombarding electrons, so that a smaller beam diameter results. Electron-beam voltages vary from 10 to 150 kV, and spot sizes vary from 0.001 to 0.250 in. These factors can produce power densities up to 25,000,000 watts/in.², power levels that are more adequate in order to vaporize metal. Manufacturers of electron guns are constantly working to improve the focusing devices for electron-beam welders so that a more concentrated beam can be achieved at lower voltages. Lower voltages are desirable in order to achieve a thin beam but the penetration power of the beam is then decreased. Penetration is proportional to the amount of voltage and amperage (Figure 23–3). Penetration is also inversely

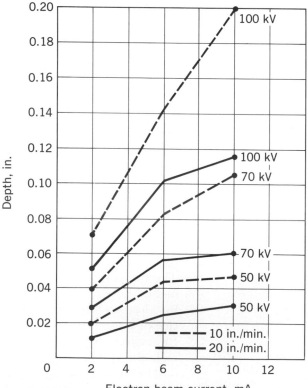

Fig. 23–3 Electron-beam penetration.

proportional to the welding speed. The greater the welding speed, the less the penetration accomplished by the electron beam. Penetration of weldable materials is also affected by thermal conductivity. In most instances, weld penetration increases as the thermal conductivity of the metal decreases.

Most units obtain their power from ordinary line voltage of either 220 or 440 volts, 3-phase, 60-Hz alternating current. The power package is self-contained within the vacuum chamber, but all the controls for the electron beam are outside the chamber at an operator's station.

The intensity of the electron beam is controlled by several variables. First is the speed of the beam across the workpiece. The diameter of the beam itself, the velocity of the electrons, and the beam current are other factors in controlling the intensity of the electron beam. Varying any of these factors provides the appropriate density and size of beam needed for the type of materials that are to be fused. The electron-beam process can be applied to nearly every area in which TIG welding is used. The beam is limited to edge, butt, fillet, or spot welds.

There are two different types of electron-beam guns, the work-accelerated gun and the self-accelerated gun. The work-accelerated gun produces a high-energy electron beam utilizing the difference of potential between the emitter and the work. In the self-accelerated gun, the cathode and the anode are enclosed within the gun itself. The self-accelerated gun is similar to the work-accelerated gun except that there is a positively charged element, called the anode, incorporated in the gun apparatus which attracts the negatively charged electrons. The anode performs the same function as the positively charged workpiece when the work-accelerated gun is used. Each of these acceleration devices is available in equipment using either high voltage or low voltage. The acceleration potential of high-voltage equipment ranges from 70,000 to 150,000 volts; with the low-voltage equipment the range is from 15,000 to 300,000 volts.

Low-voltage units operate at 600 mA (milliamperes), usually using work-accelerated guns. The gun is mounted inside the vacuum chamber and is capable of revolving around the workpiece. As the various metals, or materials, melt, a certain amount of the metal vaporizes, which leaves a film over the gun, making it necessary to clean the low-voltage units frequently. The current of the low-voltage units ranges from 250 to 1000 mA. The current, which is varied by changing the input voltage, determines the intensity of the beam. The voltage power of the beam and the current of the beam at the surface of the work can also be varied by changing the distance between the cathode and the anode. Even though low-voltage equipment has half the penetration power of the high-voltage equipment, it has the advantages of a lower price, easier maintenance, simpler design, and easier operation.

The main components of the electron-beam gun include the electron optical column, the optics system, beaming and focusing devices, and the vacuum chamber (Figure 23-2). The source of the electron beam within any gun is the cathode. Of the many parts within the electron-emitting device, the cathode actually is the source of the electrons. It is made of coiled tungsten or tantalum which is heated to approximately 2600°K/by electrical resistance. At this temperature, the filament discharges about 6×10^{18} electrons per second for each square centimeter of filament area. This filament can also be heated by an outside source of radiation. An ac supply of approximately 3 volts can be tied across the filament to produce the electricity needed for heating. The anode is a perforated plate with a high positive potential applied, which will attract the negatively charged electrons, producing a current flow through the perforations of the anode.

The emissions of electrons by metal can best be explained by the photoelectric effect, accurately explained for the first time by Albert Einstein. Einstein's explanation of the photoelectric effect, besides winning for him the Nobel prize, also ended the age of classical physics by introducing the theory of quantum mechanics.

A metal gives up electrons when an ample amount of energy is applied to liberate an electron from the outermost orbit of an atom in its atomic structure. The energy is supplied by a radiation source or photons, which strike the surface of the metal. If these photons strike with enough velocity, electrons from the metal will be drawn away with great velocity. If the intensity of the radiation source increases, the number of electrons liberated increases, a relationship which is the prime controlling factor of the intensity of the electron beam. The ease with which a metal will give up its electrons is called the thermionic function of the metal. The intensity of the electronic beam is called the emission current.

Most cathode materials are those metals which have a fairly low thermionic function and are reasonably workable. The metal must also be able to function within a vacuum. If the cathode is to be subjected to pressure greater than 1×10^{-5} mm (millimeters) of mercury only refractory metals can be used. Under that amount of pressure, a dispenser cathode should be used, which generates an excess of metal at its surface. This excess is responsible for emitting the electrons. The most common metal used for the cathode is barium.

The refractory cathode is one used when the welding process creates high ampere reading temperatures. Because of the high temperatures, metal with a high thermionic function must be used to avoid an excess number of electrons. Tungsten and tantalum are most suited for refractory cathodes because of their high melting points and high thermionic functions.

Estimating the life of the cathode is difficult, but it is usually estimated as the time required to reduce the diameter of the filament by 10 percent. Oxidation is the usual reason the cathode is reduced. In some cases, evaporation from continued use will reduce the life of the cathode.

In order for the beam to produce the maximum energy required for welding, it must be focused to a fine point of concentration. Focusing can be accomplished by electromagnetic distortion or by electrostatic distortion of the beam. The electromagnetic lens includes a coil encased in iron with a gap in the bore. The electron path, upon entering the magnetic field, is rotated and refracted into a convergent beam. The focal length of the beam can be controlled by a dc current, and the beam can be positioned into axis directions by using deflection plates. The movement of the beam is dependent upon the dc voltage. The beam is either retracted or repelled on the deflection plates by the dc voltage. Electrostatic focusing is accomplished by placing a lens disc close to the filament. The electrons passing through the small hole are affected by the electrical lines of force that surround the hole. Thus the electron beam is focused as it passes through these lines of force. The diameter of the beam can be controlled by moving this lens disc closer to or farther away from the emitter.

The vacuum chamber performs the function of reducing the atmosphere to a vacuum of about 5×10^{-5} mm of mercury. Normal atmosphere is measured at 760 mm of mercury. The props and system required to reach this vacuum are both expensive and complex. This vacuum requirement is approximately 0.0000001 of standard atmosphere. When this vacuum has been obtained in the electron-beam chamber, there is less than 1 ppm (parts per million) in the atmosphere within the chamber. To achieve this extreme vacuum, two pumps are required, a roughing pump and a diffusion pump. The roughing pump reduces the atmosphere to approximately 0.1 mm of mercury, and then the diffusion pump reduces it still further to the desired vacuum. The gas that may be left in the system is entrapped by an oil vaporizing process. The oil is heated in the diffusion pump until it evaporates. Then it is ejected at high speed through a venturi system, which entraps the gases. The gas is then discharged by a mechanical pump into the external atmosphere. Finally, the vaporized oils are condensed and channeled back to the diffusion pump. This piping system is constructed to prevent even the smallest amount of leakage. Rubber or plastic gaskets are inadequate because other elements may boil off residual gases at this vacuum level. All the joints are butt welded.

Since the electron beam is concentrated to a fine point, automatic work-holding fixtures are necessary to make sure the welding, piercing, or cutting is taking place at the correct location on the work. The operation can be done in a straight line around rectangles or circles or in zigzag paths or in any combination of these movements. The standard worktable is actuated on two axes, X and Y. The turntables are usually controlled by either a photocell or by numerical control. Often the fixture can be set up with a pantograph, or tracer control, depending upon the size of the work, the deflection of the beam, and the worktable.

Radiation, a form of x-ray, infrared light, and ultraviolet light are given off in electron-beam welding. The welder must be protected. A radiation-safe optical system is often incorporated into the electron-beam welder that will afford this protection. This system consists of a lens and mirror system that focuses the light rays for visual monitoring and for alignment of the worktable to the beam.

Even though the electron-beam method is an ideal

welding process, it has serious limitations. Equipment is extremely expensive, and portable equipment is rare, but new developments are being made by manufacturers. Some metals, such as zinc, lead, aluminum, and magnesium, cannot be welded easily by this method. These metals boil out of the weld zone and contaminate the welding apparatus. Many of these volatile metals, such as aluminum and magnesium, can be welded with the electron beam if a pulsing procedure is used.

Such limitations are offset by the many advantages of electron-beam welding. The weld zone and heat-affected zone are relatively small in electron-beam welding. Distortion affects only a small area, about 0.001 in. on each side of the weld bead. Penetration depth can also cause distortion. Electron-beam welding is extremely successful in achieving deep penetration with little distortion. The depth-to-width ratio can go as high as 25:1 with electron-beam welding. On stainless steel that is as thick as 4 in., a weld can be made in one pass with full penetration, with the pass performed at a speed of approximately 7 in./min.

Input power is small when it is compared to the power requirements of other electrical welding devices. The total power used is determined by a fine point of concentrated energy, which uses energy more efficiently than the other welding processes. The current usage is low since current only in the milliampere range is required, while with other electrical welding systems, such as in shield arc welding, many amperes are required.

The greatest advantage of electron-beam welding is that it eliminates contamination of both the weld zone and the weld bead because of the vacuum in which the weld is done and because of the electrons doing the heating. The electrons are nonreactive to the metals, which causes no contamination, and because the weld is performed in a vacuum chamber, foreign particles and gases are carried away by the evacuating system. If foreign particles are present, many times these particles are ionized into a positive charge, attracted to the gun, and swept away in the electron beam.

Use of the electron beam as a welding device developed when the demand for beryllium increased for industrial and aerospace components. With other welding processes, beryllium would vaporize rather than melt. Various tests with the electron beam reduced the problem, and now sheets up to $\frac{1}{16}$ in. thick can be welded with a single pass.

Several other metals previously unweldable are now being used in industry because of the electron-beam process. Aluminum has been welded with high voltage and various butt positions. In all cases, porosity was found in the weld, but the strength of the base metal was unaffected. Other metals that are now being welded are the high-strength steels. Unlike arc welding, the electron-beam process does not alter the strength of the steel. When two sections of 300,000 psi ultimate strength steel are fused, a fusion zone is produced with strength equal to or greater than the strength of the base metal.

Electron-beam welding devices range in price from $20,000 to $200,000. For industry to justify purchasing such a unit, the electron-beam welder must be able to in-

Fig. 23–4 (a) Nonvacuum electron-beam welder. (b) Electron-beam application. (Courtesy of Westinghouse Electric Corporation.)

(a)

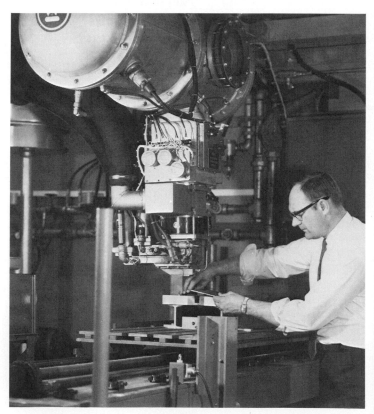

(b)

crease production, improve quality, or cut cost elsewhere. Even though initial costs are high, operating costs are low due to the low power usage. Many of the more costly fabrication methods could be replaced by the electron-beam process.

The narrow beam reduces distortion of the workpiece, making the replacement of costly jigs and fixtures less necessary than when using other types of welding processes. Also, the need for the finishing processes, such as painting and enameling, are reduced because of the close tolerances of the original dimensions that the workpieces must have for electron-beam welding.

A new concept in the electron-beam process has recently been developed, which allows the electron beam to perform its function in standard atmospheric conditions. Although the nonvacuum electron-beam welder has limited applications, it has opened new possibilities for electron-beam welding. This type of electron beam can be used for high-production welding. The maximum stand-off distance for welding is approximately ³⁄₄ in. If a beam length greater than ³⁄₄ in. is used, the beam will be too widely dispersed to operate. The maximum penetration depth of the beam is ¹⁄₂ in. of steel (Figure 23–4). The electrons are freed in a vacuum as in conventional electron-beam welding, but the electrons in the nonvacuum welder go through three vacuum pumping stages that gradually lower the vacuum within the beam-transfer column to standard atmospheric pressure (Figure 23–5).

The electron-beam process is not a dual process. It has limitations but it was designed to perform only specific welding jobs in which atmospheric contamination presents prime problems. Today, few applications of the process exist, but research on its capabilities continues in the welding industry. Perhaps the electron beam will be one method to join some of the high-temperature metals, such as columbium, in the proposed space shuttle.

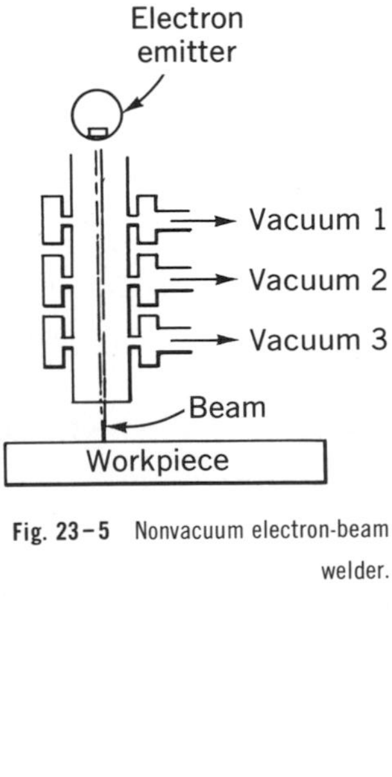

Fig. 23–5   Nonvacuum electron-beam welder.

QUESTIONS

1. Why is a vacuum chamber used with the electron beam?

2. How are electrons accelerated?

3. Why is the depth-to-width ratio so great in electron-beam welding?

4. What is keyholeing?

5. Why is it necessary for the worktable of the electron beam to be automatically driven when welding?

6. What is kinetic energy? How is it applied?

7. What are the power-source requirements for electron-beam welding and cutting?

8. What are the variables that control the electron-beam process?

9. What are the two types of electron-beam guns? Their individual advantages?

10. How are electron beams focused?

SUGGESTED FURTHER READINGS

Bakish, Robert: "Introduction to Electron Beam Technology," John Wiley & Sons, Inc., New York, 1962.

Jefferson, T. B.: "The Welding Encyclopedia," Monticello Books, Morton Grove, Ill., 1968.

Linnert, G. E.: *Welding Metallurgy,* American Welding Society, New York, 1965.

Marton, Ladislaus: "Advances in Electronics and Electron Physics," Academic Press, New York, 1948.

Patton, W. J.: "The Science and Practice of Welding," Prentice-Hall, Inc., Englewood Cliffs, N. J., 1967.

24 LASER WELDING

The word *laser* is an acronym for *light amplification by stimulated emission of radiation*. The laser welding process is the focusing of a monochromatic light into extremely concentrated beams. All light consists of waves, which are similar to radio waves or microwaves. In common light, these waves are incoherent and of varied lengths and frequencies that shoot off in numerous directions. The laser light is a coherent light; that is, the waves are identical and parallel and are capable of traveling over great distances without much reduction in their intensity. The laser welding process, then, employs a carefully focused beam of light that concentrates tremendous amounts of energy on a small area to produce fusion.

PARALLELISM AND INTENSITY

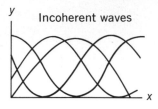

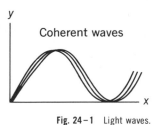

Fig. 24-1 Light waves.

A parallel beam of radiation from a source, such as a flash tube, is produced by placing a radiator in the focal plane of an optical system. The beam produced is not perfectly parallel; it has an angular divergence equal in size to the angular divergence size of the source. Only a small portion of the radiated source is used to improve the sharpness of the beam. Energy from this source is also reduced because the aperture of the optical system acts as a stop, eliminating a large part of radiation. Therefore, a nearly parallel beam of light can be produced by using only a minute fraction of the energy from an ordinary light source. The higher the degree of parallelism, the smaller this fraction of radiant energy becomes. Ordinary light cannot be used as a laser light because radiant energy from an ordinary light source is distributed over a wide spectral range, thus monochromatic single-color sources do not exist. Also, radiated energy from ordinary light is poorly collimated. Collimation, the adjustment of light waves so that they are parallel, cannot be improved without sacrificing intensity. The radiant-energy image from an extended source cannot be increased. The possibility of utilizing ordinary white light for the laser is eliminated because the light is incoherent (Figure 24-1).

FOCUSING

Laser light not only is intense but also can be readily focused without loss of intensity. Ordinary light waves emerge from different points of its source, causing the light, which travels in straight lines, to enter the lens in a variety of different angles. As a result, the light is spread, limiting the minimum size of the area into which it can be focused. Ordinary light also has a great number of different wave lengths. These wave lengths, upon entering the lens, travel from less dense to a more dense medium, causing the bright rays to bend. The angle of refraction changes for each wave length and the ray, upon emerging from the lens, focuses at different points, making radiant energy from ordinary light sources inadequate for a heat source (Figure 24-2).

In order for light energy to be focused properly so that it can perform an operation such as the welding or piercing of metal, it must enter a lens along parallel lines and must be of a single-frequency wave length, or have coherent

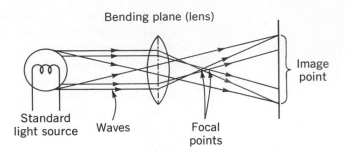

Fig. 24-2   Standard light focusing.

waves, a condition in which the waves are in line (Figure 24-3). Laser light fulfills these requirements and theoretically can be focused into a spot no wider than a wave length of light, or approximately 0.0001 cm (centimeter).

QUANTUM THEORY   There are many forms of radiations; some are radio waves, x-rays, gamma rays, and light waves. All these types of radiation are associated with electron movement and magnetic force fields; or, in other words, they are all forms of electromagnetic radiation. Light is just one of these forms, and it is characterized by having both particle and wave motion.

In the case of visible light, differences in colors are determined by the individual wave length, which is the distance from one wave crest to the next wave crest (Figure 24-1). This distance is measured in angstroms (Å). An angstrom is a unit equal to 0.000001 cm.

Although light sometimes behaves as though it were composed of waves, it also acts as if it were composed of particles of energy, referred to as photons. The basis of all matter is the atom, which has, basically, a nucleus and orbiting electrons. The nucleus contains both protons, positively charged particles, and neutrons, neutrally

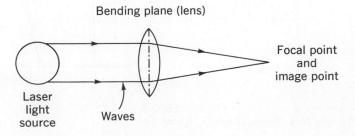

Fig. 24-3   Laser light focusing.

charged particles. The rotating electrons can carry a negative charge and are equal in number to the protons of the nucleus. Hence, the atom is normally in a neutral stage when its orbiting electrons are in their normal orbits. These orbits are called shells. The neutral state is referred to as the ground level or rest state. An atom at its rest state is of no value in a laser device.

In the case of the ruby, if its chromium atom is forced or pumped from its ground level to a higher level by using an external force, such as an xenon flash tube, the atom then absorbs energy. When this energy is absorbed, the electrons increase their spin and expand their orbit. This excited state is short-lived and the atom immediately drops back to an intermediate level or a metastable state. In this dropping back, the atom loses its heat energy but retains its photon energy. After a short time, the atom falls spontaneously and randomly back to the ground state, releasing the photon energy, or quantum, in the form of light (Figure 24-4). This automatic dropping back without being stimulated to do so is referred to as spontaneous emission, and it is the reason why some gems seem to glow or radiate colors when they are exposed to ordinary light, such as the sun.

This random falling of the atoms creates an incoherent light. In order for laser action to be produced, the atoms at the metastable state must be synchronized so that they will emit the photons in phase. The synchronization, which leads to laser action, is achieved by stimulated emission.

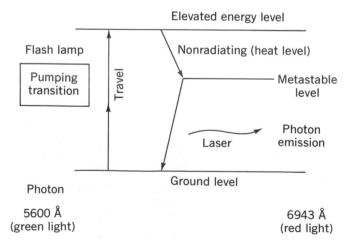

Fig. 24-4 Pumping of atoms.

More simply stated, an excited chromium atom will emit its photon energy if it "sees a similar photon pass by." When an atom in the laser system is induced to make a jump from a lower to a higher energy level, one photon is lost from the radiation field and absorption occurs. When a transition is from a higher to a lower energy level, stimulated emission results and the field gains one photon. If this jump occurs without the presence of a radiation field, it is called spontaneous emission. Of critical importance in laser operation is the fact that stimulated emissions are in phase with the field that causes them. This consistent phase relationship is called coherence. For the laser action to take place, the stimulated emission process must occur more often than the competing process of absorption. Quantum theory predicts that the relative likelihood of these two processes occurring depends only on the relative population of the two energy levels involved. Laser emission is obtained when the upper level is populated at the expense of the lower one, a situation called population inversion. The method of obtaining this population inversion is called pumping.

Laser action occurs in the ruby rod equipped with parallel mirrors if well over half the chromium atoms in the ruby have been pumped to the metastable state. Population inversion at this point is in effect. Action begins when an excited chromium atom decays spontaneously and emits photons parallel to the axis of the ruby. As the photon is reflected, or folded back and forth within the resonant cavity, within an instant, it stimulates the emission of light waves in phase with itself, and a wave front begins to form. This wave front produces an intense monochromatic coherent light beam.

Laser resonant cavities are designed to reflect the waves of light back and forth between two reflective mediums that face each other at different ends of the laser cavity. In this way the light wave is reflected many times from one mirror to the other, providing a light path of almost infinite distance through which the light wave may travel, which enables the wave photons to trigger and release more energy from the excited atoms that are at the metastable state. As the light waves are bounced back and forth between the mirrors, they gather energy until they reach sufficient magnitude to burst through the partially reflective mirrors in the form of a laser beam. This bouncing of the light wave is referred to as folding the light beam. In order to control the direction of the beam, the cavity is constructed with mirrors of different reflectance, one mirror being 100 percent reflective and the other mirror approximately 98 percent reflective. In this way, the beam exits from the 98 percent end of the resonant cavity, allowing the direction to be predetermined. The mirrored surfaces result from two types of reflective coatings. One coating is produced by depositing a thin layer of metal such as aluminum, silver, or gold. However, these metallic reflective coatings can become burnt and lose their reflective quality. A higher-performance reflective coating results when the ends of the lasing device are coated with several nonconductive films, producing a dielectric mirror. The dielectric mirror depends upon the interference among the light waves, which is reflected by the multi-layered films, composed mostly of sulfides and fluorides.

There are three basic types of lasers: the solid laser, the gas laser, and the semiconductor. The type of laser depends upon the lasing source. The solid types are ones that rely upon some type of crystal such as the ruby, the sapphire, and some artificially doped crystals for its lasing ability.

The gas laser operates differently from the solid-crystal laser. The active material of a gas laser consists of a gas, or a mixture of gases, contained in a glass or quartz tube with highly polished mirrors at each end. The first gas laser contained a gas mixture of about 90 percent helium and about 10 percent neon. The light waves were generated by neon atoms. Because of the short angstrom length of neon atoms, the laser-beam wave length was in the infrared spectrum. Later, a gas laser using mercury gas was created that had a visible range in the light spectrum. The mercury laser did not prove to be effective, however, because of its low efficiency; but it was a step toward a better laser that used ions as its energy source. One of the most widely used gas lasers is the carbon dioxide laser, which has an efficiency rating of about 15 percent, a good rating when it is compared with those of other lasers, which range between 1 and 5 percent. The radiant-energy density of the carbon dioxide laser is greater than that of the sun. The carbon dioxide laser can have either a sealed tube of gas or flowing gas within the tube. The flowing-gas principle increases the power about 3 times that of the sealed-tube type of laser material. The wave length of the carbon dioxide laser is 106,000 Å.

Both the gas laser and the solid-state laser devices require a capacitor storage to store energy and then inject the stored energy into the flash tube. The semiconductor injection type of laser does not require the storage of energy or the pumping components, as the other two types do. Instead, electrical power is fed directly into the semiconductor device and is converted directly into laser light. Currently, lasers of this type are of very low power intensities; however, research is being conducted for improving the power intensities.

The laser welding system is composed of an electrical storage unit, a capacitor bank, a triggering device, a flash tube that is wrapped with a wire, the lasing material, a focusing lens mechanism, and the worktable, which is operatable in three axes, X, Y, and Z. The capacitor bank, when triggered, injects energy into the wire that surrounds the flash tube. This wire establishes an imbalance in the material inside the flash tube. Thick xenon often is used in the material for the flash tube, producing high power levels for a very short period of time. The energy released has a spectral distribution comparable to that of daylight, with a pulse lasting from ½ msec (millisecond). The flash tubes or lamps are designed for operation at a rate of thousands of flashes per second. An intense single flash source can have an output ranging up to tens of millions peak candlepower, and a short arc light source can have a flash duration of $1\,\mu$sec (microsecond). By operating in this manner, the lamp becomes an efficient device for converting electrical energy into light energy, the process of pumping the laser. The laser is then activated. The beam is emitted through the coated end of the lasing material. It goes through a focusing device where it is pinpointed on the workpiece. Fusion takes place and the weld is accomplished (Figure 24–5).

APPLICATION In the laser welding device, the process of welding is hindered because of the nature of the light beam. The average duration of the laser weld beam is 0.002 sec.

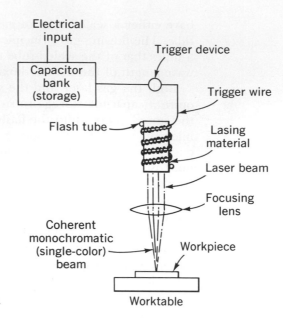

Fig. 24-5   Laser welding system.

Because of this short duration, two basic welding methods have been adapted. In one method, the workpiece is rotated or moved fast enough so that the entire joint is welded with a single burst of light. The other method uses many pulses of the laser to form the weld joint. With the pulse technique, the weld is comprised of round solidified weld puddles each overlapping the previous one by about half the puddle diameter, as in resistance seam welding. An advantage of the laser pulse method is that many lasers pulse approximately 10 times per second so that the workpiece does not even get hot except at one point, which is one of the metallurgical advantages of laser welding. Liquidus is reached only at the point of fusion. Because of this trait, the heat-affected zone is narrow. In laser welds on stainless steel, the heat-affected zone is virtually nonexistent. The major drawback to laser seam welds is the slow welding speeds, resulting from the pulse rates and puddle sizes at the fusion point. Laser welding is appropriately applied to enclosure welding either in a vacuum or not in a vacuum. Also, sensitive materials can be welded and glass-to-metal seals effected, such as in the construction of electrical tubes like klystron tubes. The kystrons are built to generate energy at frequencies as high as 220 GHz (gigahertz = 10^9 Hz). The high temperature of the targets in the small tubes requires that cathodes be made of metals, such as molybdenum, tantalum, and titanium. The high melting points of these metals make them nearly impossible to bond when using resistant welding techniques. A laser has sufficient temperatures to bond such refractory material in small areas solidly and easily. Laser beams also have the capability to penetrate a quartz tube, welding the metal inside without harming the tube itself. Also, because of the small size of the laser welding beam point and its quickness, the weld zone remains uncontaminated.

Three requirements must be fulfilled for a laser beam to perform fusion in a single pulse. First, the beam must be adequate to melt the surface of the metal. Second, the intensity of the beam must be less than the vaporization point of the metal. Finally, the pulse length must be sufficient to allow heat conduction through the metal. The first requirement is governed to a large extent by the reflectivity of the material to be welded. The surface conditions, smooth or rough, black or shiny, play a part only in the initial surface reactions. Once the material is raised to the melting point, these conditions no longer affect the welding. It is obvious that greater intensities are required to weld a smooth, shiny surface than to weld a rough, dull one (Chart 24-1). The second requirement

CHART 24-1   Reflection values of polished surfaces

METAL	LIGHT REFLECTION, %, LUMENS	LIGHT ABSORPTION, %, LUMENS
Nickel	70	30
Copper	85	15
Gold	90	10
Iron	60	40
Tin	55	45
Aluminum	95	5

NOTE: Based on using ruby red laser with a wave length of 6943 Å.

depends on how easily the metal vaporizes. The third requirement is governed by the time needed to transfer heat from the molten pool on the surface into the work, which depends upon the thermal conductivity of the metal, the density and specific heat of the material, and the light beam. It is important to note that heat is developed only

0.020 in. max. thickness

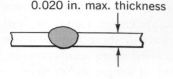

0.002 in. max. thickness

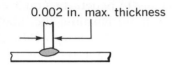

0.001–0.020 in.

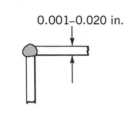

0.015 in. max. thickness

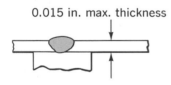

Fig. 24-6 Joint design. (a) Butt. (b) T. (c) Corner. (d) Spike.

at the surface in laser welding so that it must be transferred by conduction throughout the work, contrasting with the resistance welded joint in which the heat is developed at the interface of the materials to be joined. The joints in laser welding are limited to use with thin metals (Figure 24-6). One of the major advantages of laser welding is that the intense heat created by the laser affects an extremely small area, which reduces the energy input demand for producing the weld. Because of this characteristic, the process can be used to weld dissimilar metals with widely varying physical properties. In addition, metals with relatively high electrical resistance and parts of considerably different sizes and mass can be welded. Because the laser is simply a light beam, no electrode is required, so that any part in a particular position can be welded if there is a direct line of sight from the beam to the workpiece. Welds can be made with a high degree of precision and on material that is only a few thousandths of an inch thick. Also the light beam exerts no force on the metal and the heat is instantaneous and highly concentrated. Consequently, laser welding will hold thermal distortion and shrinkage to a minimum by reducing the area of the heat-affected zone in the weld. Its instantaneous heating and cooling cycle limits crystalline growth and shape. Because of the quick heating and cooling cycle, stress-relieving and straightening the weld area are eliminated.

Energy losses with any of the laser systems are high. A relatively efficient system input for the power supply is more than 10,000 joules. The output is only 30 joules. Only a small part of the flash lamp is white light and much of that light passes through the sides of the lasing material. Nevertheless, this energy loss does not restrict the operation of the device, since the focus by the light from the laser is millions of times more intense than the light from the flash lamp.

Costs for systems for laser welding range between $2,000 and $25,000, depending upon the degree of sophistication, capacity, and power supply desired. Laser machines for production welding costs less than electron-beam machines of equivalent sizes. Laser-beam welding has not been adapted to production purposes to the extent that electron-beam welding has been.

QUESTIONS
1. What is coherent light?

2. How is laser light focused?

3. How does the quantum theory of energy apply to the laser?

4. Why does not ordinary light perform in the same way as laser light?

5. What are the basic components of the welding laser?

6. What are the most-used lasing materials?

7. How important is the reflective qualities of base metals when laser welding?

8. Why is monochromatic light coherent light?

9. How is depth penetration accomplished with the laser?

SUGGESTED FURTHER READINGS
Brown, Ronald: "Laser's Tools for Modern Technology," Doubleday & Co., Inc., Garden City, N.Y., 1968.

Dull, Charles E.: "Modern Physics," Holt, Rinehart & Winston, Inc., New York, 1964.

Fishlock, David: "A Guide to the Laser," McDonald Books, London, 1968.

Groner, Warron: Laser, *Electronics World,* vol. 74, no. 2, p. 31, August 1965.

Lancaster, J. F.: "The Metallurgy of Welding, Brazing and Soldering," pp. 106–122, George Allen and Unwin, Ltd., London, 1964.

25 THERMIT WELDING

Thermit welding is the application of a filler material or the joining of two pieces of metal, usually steel, through the use of heat supplied by a chemical reaction. Although not used very much today, in years gone by, a great many repairs were performed by this method. Usually large parts, such as locomotive rails, large gears, and general fractures on large pieces of equipment were repaired by Thermit welding. Hans Goldschmidt is credited with the discovery of the process called Thermit welding. In his laboratory in Essen, Germany, in 1895, he made his initial discovery when he was attempting to reduce chromium and manganese. Instead of discovering a way to reduce either of the two metals, he discovered a safe method of igniting alumina powder so that the heat would change the iron oxide into a liquid steel.

The chemicals used for the heating process are a mixture of aluminum and iron oxide. The Thermit, as the mixture is called, requires an igniter of some type to set off the reaction. The chemical formula for this process is $8Al + 3Fe_3O_4 = 9Fe + 4Al_3O_3 +$ heat. The igniter is usually barium peroxide or powdered manganese. The tremendous heat released, approximately 5000°F, causes the iron remaining in the charge to change to a liquid state.

The Thermit is mixed in a suitable crucible (Figure 25-1), and the igniter is placed on top. It is then lighted with a red-hot metal rod. Wooden matches could be used for this except there is a possibility that hands might be burned. As the igniter burns, temperatures of around 2100°F are reached before another reaction begins. At this point, the metal begins to melt. The reaction then continues to burn throughout the mixture for 25 or 30 sec. As the aluminum in the mixture combines with the oxygen from the iron oxide, it forms alumina oxide, which serves as slag and floats to the top. The heat remaining causes the iron to change from a solid to a liquid. As this steel then becomes molten and reaches the desired fusion temperature, it is channeled into a prepared mold. The fracture, or two pieces to be joined together, have been preheated to the high temperature of the liquid steel, and thus the parts are welded together.

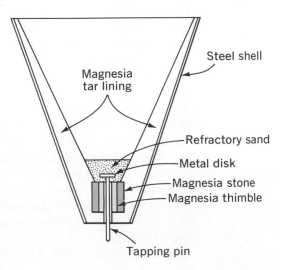

Fig. 25-1 Thermit crucible.

Thermit mixture alone is not hazardous unless the igniter is present. Since the parts to be welded are usually large, several tons of Thermit will be used. Since cooling takes place very slowly, only a minimum of distortion takes place. Although the section that receives the weld is large, it can be machined down to the proper size once cooling has taken place.

The crucible used for receiving Thermit is an important part in this process and must be made of material that can withstand the extreme heat. Not only must the crucible withstand the weight of the Thermit but there are also pressures resulting from the chemical reaction. The outside shell of the crucible is usually constructed of sheet steel and is shaped like a cone, with an opening at one end (Figure 25-1). Magnesia tar is used to line the inside of the crucible. The tar is used only as a binder for the magnesite and is removed when the magnesite is baked on. Located at the bottom of the crucible is a magnesia stone and a thimble through which the tapping pin is suspended. This design allows for fast removal of the liquid steel with no danger of the slag entering the weld because the slag is located on the top, leaving the crucible after the molten metal is channeled off.

There are two types of Thermit welding, plastic, or pressure, Thermit welding, and fusion, or nonpressure, welding. The plastic, or pressure, type is used mainly for butt welding thick vaults, pipe, or rails (Figure 25-2). In the weld, pressure is used instead of fusion to join the metal. The process uses the Thermit reaction only as a heat source. The workpieces are clamped into cast-iron molds and are forced together when the desired temperature is achieved. The Thermit is heated in a crucible above the work. While the Thermit is heating, the lighter aluminum oxide slag rises to the top. When the pouring temperature is reached, the Thermit solution is poured from the top of the crucible into the mold. The system of pouring from the top allows the slag to enter the mold first, putting a protective coating of slag around the joint to be welded, which prevents the iron in the Thermit solution from fusing with the workpieces and makes the removal of the piece from the mold easier. The slag coats the inside of the mold cavity and also the workpieces themselves. When the Thermit mixture has heated the

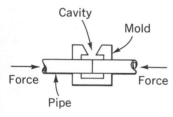

Fig. 25-2 Pressure Thermit.

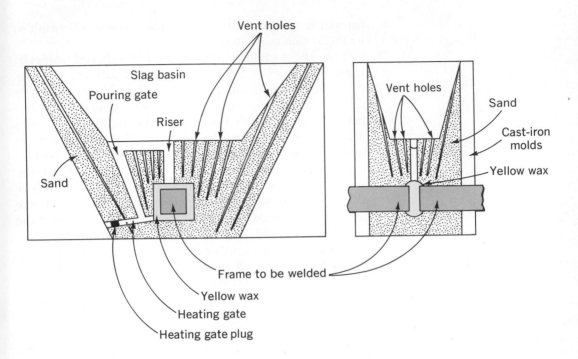

Fig. 25–3 Nonpressure Thermit welding.

work sufficiently, the pieces are forced together, forming a pressure butt weld. The whole process, welding a thick-walled pipe, takes from 45 to 90 sec.

The fusion, or nonpressure, Thermit welding process is completely different (Figure 25–3). In this process, the workpieces are lined up, leaving a space between the ends that are to be welded. Wax is placed between the joint. The whole frame is suspended in a mold, and then the molten metal is poured in.

The first step in preparing a nonpressure Thermit weld is the cleaning of the joint. Cleanliness is important in nonpressure welding because any foreign material leaves a space in the mold when the heat is applied. An oxyacetylene torch, or other suitable means of heating, is used to clean the metal by heating and then quenching it quickly, allowing the expansion and contraction of the metal to pop the foreign material from the steel. An area of 5 or 6 in. back from the fracture to be welded must be thoroughly cleaned. When oxyacetylene is used as a heat source, care must be taken. All loose oxides and scale must be removed. If any grease or other combustible material is present in the mold box, a space will result in the metal that is poured.

When the metal has been cleaned and spaced properly, yellow wax must be prepared to make a pattern. The wax is placed in a container and heated until it reaches its plastic state. The wax is then shaped around the parts that are being welded together. As the forming takes place, a vent must be provided from the heating gate to the riser. A molding box is then placed around the portion to be welded and a molding material is rammed into the box. The size of the box usually depends upon the size of the weld to be produced. Care must be taken as the ramming is started to assure a tight contact between the molding sand and the wax that provides the root filler. The success of the entire operation depends upon whether the sand is packed all around the object to be welded. It is advisable to pack a small amount of sand and use a small ram to make certain all spaces are taken out around the mold.

A preheating opening must be provided and is usually placed at the lowest part of the wax mold. The opening is then extended to the front of the mold (Figure 25-3). This gate is provided to allow a burner or heating device to melt the wax, leaving a mold cavity in the exact shape of the weld. It may be necessary to put the preheat gate farther to one side if a heavy part is to be welded to a light part, which allows all the way to melt at the same rate.

A pouring gate must also be provided to allow the molten metal to run into the mold. The pouring gate should join the preheat gate just outside the wax to prevent excessive cleanup after the mold cools. A riser must be used to allow for the contraction of the metal as it cools. This riser will act as a feeder to the area to be welded taking the metal from the riser and supplying the fusion area, a function that will prevent reductions in the area in the weld. Beginning at the highest point of the wax, this riser should extend upward from the highest point of the wax pattern and feed the largest mass. When there are several high points in the weld zone, a riser at each high point should be provided. The risers also serve as a reservoir to float out loose sand that may wash in during the pouring process. When risers are not provided, the loose sand may prevent the steel from running freely into the mold.

A small area must be provided around the top of each of the gates and risers to act as a basin for the slag and excess liquid steel that may run over as the mold fills up.

After the mold is completed and fully rammed, the sand is vented throughout by forcing an 8- or 10-gage wire down into the sand. Care must be taken not to push the wire into the wax, which would allow the liquid steel to travel up the vent holes. The vent holes allow the gases from the molten steel to escape in the air without holes being blown in the sand. After the gating and risering patterns have been removed, the excess sand should be removed and the edges of the sand smoothed. The mold is now ready to be preheated, after which the preheat hole must be plugged with either a metal or sand plug.

The amount of Thermit used is determined by the amount of wax used to join the pieces to be welded. The Thermit must be mixed before it is put into the crucible. Only a small amount is put in at a time so that the material used for plugging up the bottom of the crucible will not be disturbed. Space must be reserved at the top of the crucible for any excess reaction that might take place. When the ignition powder is lit and the reaction takes place, liquid steel is formed inside the crucible. The plug is then tapped after which the metal is poured. The metal is then allowed to run into the mold, producing a homogenous bond between the Thermit filler material and the preheated parts. The mold is allowed to cool overnight, if possible, or at least for 12 hr.

Thermit welding is used chiefly in the repair or assembly of large components, such as rudder frames, propeller shafts, and steel rolling mill pinions. The amount of Thermit used in a batch depends upon the job. It can range from 25 lb for railroad rails up to 4000 lb for large frame welds. This process is also used to heat steel, and sometimes foundries use it to add alloying materials. The technique of Thermit welding cannot be economically used to weld light, cheap metals or light parts, however.

Another process, called aluminothermic welding, is used for joining lines of cable together (Figure 25–4). This process uses a hand-held device and can be performed by an unskilled operator. A cable or wire is clamped into the two-part mold that is made from cast iron. After the wire is clamped into the mold, a cartridge

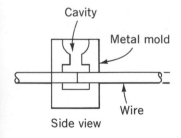

Side view

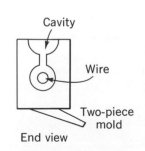

End view

Fig. 25–4 Aluminothermic welding.

that is composed of alumina powder and copper is deposited into the mold and the crucible part of the mold cavity. The mold is then ignited and the copper fuses into the wire. It is allowed to cool; then it is removed. Very little distortion takes place because there are no openings as in the mechanical connecting of wire. One advantage of this process is that the electrical conductivity of the copper is retained throughout the fused portion of the wire. The chemical reaction reduces the copper to approximately 98 percent pure copper, but the small 2 percent impurity in the copper is overcome by the larger cross-sectional area of the weld zone.

QUESTIONS

1. What are the elements in a Thermit mixture?

2. What is the reaction when the Thermit mixture is ignited?

3. Why are the Thermit crucibles designed so that the metals are poured from the bottom?

4. What is the function of the alumina powder in the Thermit mixture?

5. What are the differences in the types of Thermit molding processes?

6. What is the aluminothermic welding process? Its restrictions?

7. What are the major uses for Thermit welding?

SUGGESTED FURTHER READINGS

Houlderroft, P. T.: "Welding Processes," Cambridge University Press, Cambridge (Eng.), 1967.

Jefferson, T. B.: "The Welding Encyclopedia," Monticello Books, Morton Grove, Ill., 1968.

Lindberg, Roy A.: "Processes and Materials of Manufacture," Allyn and Bacon, Inc., Boston, 1965.

Linnert, G. E.: *Welding Metallurgy,* American Welding Society, New York, 1965.

26 METAL FLAME SPRAYING

Flame spraying is a general term used for several different methods of applying a metallic coating to a base metal. The five different methods include the metalizing, thermo-spray, plasma spray, arc spray, and detonation-gun processes.

The original credit for flame spraying must go to M. U. Schoop, a Swiss engineer who first patented the process of flame spraying in 1902. The first commercial recognition of flame spraying came some years later. Originally, the process was developed to build up worn parts that were expensive to replace. Most of the processes of flame spraying create a strong chemical bond between the filler material and the base metal. The filler material is in the form of either a wire or pulverized metal that is used as a powder. Both types of filler material are melted with some

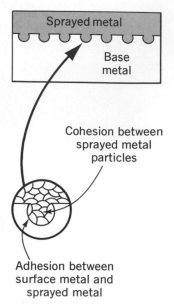

Cohesion between
sprayed metal
particles

Adhesion between
surface metal and
sprayed metal

Fig. 26–1 Surface bonding.

type of fuel or electric arc and are forced into the base metal with tremendous momentum. As the metallic particles are propelled through the air, oxides are picked up from the atmosphere. The minute particles and oxides are splattered on the metal, forming an oxide bond between each two particles. As the particles hit, they spread out and interlock with other particles, forming a tight mechanical bond. In some cases, point-to-point fusion takes place, depending on the process and the filler material used.

The bonding that takes place is essentially mechanical, however. Both adhesion and cohesion take place in order to insure bonding (Figure 26–1). Adhesion of the sprayed metal globules takes place when these atomized particles force themselves into the cracks and tears of a properly prepared surface. Cohesion takes place when the oxide surfaces and even the pure liquid metal particles fuse together into the prepared surfaces.

The compressive strength is low when the coating is compared with the original material, or the base metal. The applied material could be compared to a centered powdered metal because of the grainy appearance and the strength of the bond between the particles.

Several applications of thermal or flame spraying in use today include the coating of metal to increase its resistance to corrosion and oxidation and the hard surfacing of metal to increase its wearability. A hard surface can be applied up to 0.001 in. thick. Any coating thicker than 0.001 in. has a tendency to flake off. When worn parts are built up, as much as 1/4 in. can be successfully applied. Hard surfacing is usually applied to pump shafts; large, slow-moving shafts; and, in some cases, large pistons, such as diesel pistons that can be built back up to a standard size.

Heat derived from an oxygen fuel source is used for metals with melting points under 5000°F. For filler materials with melting points greater than 5000°F, metal spraying equipment with the plasma or electric arc is used. Because little heat is necessary on the workpiece, metal spraying can be used on wood, plastic, and other soft materials. The air stream that helps carry the metal particles from the metal spraying device to the base metal also has a cooling effect on the base metal, which prevents the base metal from melting. With this advantage, metal spraying can be used in making molds for mass production of small parts and molds for foundry work as well as in the plastics industry.

SURFACE PREPARATION When a surface preparation for metalizing is selected, attention must first be given to the shape of the base metal. The choice of surface preparation depends upon whether the surface of the base metal is flat or round, and whether a hard, smooth, wear-resistant surface or a soft, bearing-like surface is required. The seven processes used to increase bonding qualities between sprayed metal and the base metal include rough threading, grit blasting, using self-bonding materials, electric bonding, studding, and sanding or grinding.

Before any of these methods can be utilized, it is necessary to clean the metal to be sprayed of dirt, oil, oxides, and any other impurities that may weaken the bond between the sprayed metal and the base metal. Cleaning is done before the surface preparation so that contamination does not take place when the base metal is being prepared for the bonding metal. Oil-free solvents are usually employed. In the case of porous metals, such as in metal casting, oil may removed by heating the casting to a temperature that will vaporize the oil. Oxide coating may be removed by a mechanical method, provided that the metal surface is sprayed within a reasonably short time after the mechanical preparation has been used.

UNDERCUTTING Undercutting is a method of cutting down the surface of base metal to offer a uniform surface for spraying and to prepare the surface of the metal being sprayed for an even coat. The thickness of the coating should be held to a minimum to make sure that the sprayed metal will provide an adequate service life. Undercutting may be accomplished by grinding, cutting on a lathe, milling, or by a number of various mechanical ways. In order to insure a strong bond at the end of the shaft or wherever the undercutting ends, collaring is often used. This process is used to minimize the separation of sprayed metal and base metal at the end of the shaft or the worn surface (Figure 26–2).

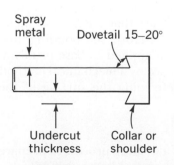

Fig. 26–2 Undercut or dovetail.

ROUGH THREADING

Rough threading is a process of preparing a roughened surface to increase the adhesive qualities of sprayed metal to the base metal. The process involves simply the threading of a shaft, which produces a rough surface of broken or torn threads. A more sophisticated alternative to this method is to knurl the threads that have been rough-cut on the base metal. This knurling will flatten the tops of the grooves that have been formed on the base metal thereby producing a surface that has excellent bond characteristics. Knurling the chased threads will deform the crest of the thread by flattening it, thus forming independent dovetailed grooves (Figure 26–3). After a shaft is prepared in this manner, the sprayed metal should be sprayed at an angle other than 90° from the surface of the metal to insure full surface contact in the dovetailed zones.

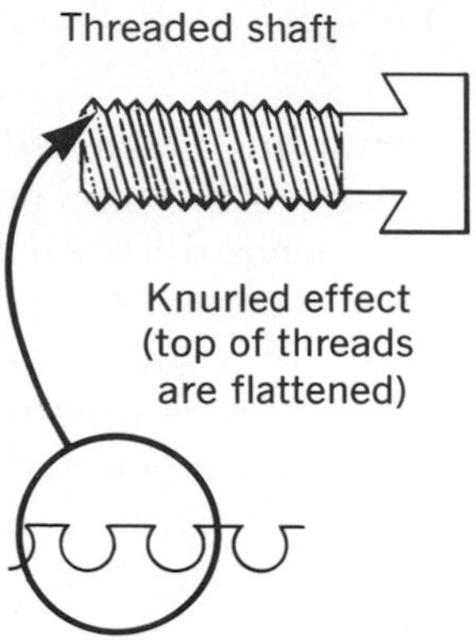

Threaded shaft

Knurled effect
(top of threads
are flattened)

Fig. 26–3 Dovetail grooves.

GRIT BLASTING

Grit blasting is also used to produce a roughened surface. Grit blasting, however, incorporates the use of a special grit, free of contamination, that is used to bombard the surface of the metal being prepared. The effect of peening that is produced by grid blasting produces a compressive strength on the surface of the metal that improves its fatigue resistance. Grit blasting is an inexpensive way to prepare large surface areas of metal if blasting equipment is available.

SELF-BONDING MATERIALS

Molybdenum has been used in the metal-spraying industry as an almost universal base coat between two metals. Almost any other oxide-free metal will adhere to molybdenum. This metal can be used to produce special results on the finished surfaces of a part that has a base metal shaft containing other metals. Before self-bonding materials such as molybdenum are applied to base metals, it is advisable to prepare the surface by a mechanical method such as grit blasting or by deforming the surface of the metal.

ELECTRIC BONDING

Electric bonding incorporates a welding process to prepare the surface of the metal to be sprayed by coating it with a special electrode that has the characteristic of good adhesive qualities and low heat emission. When the electrode comes into contact with the base metal, which must be free of impurities, a small, rough coating of metal is deposited. This method uses special tools and special welding equipment, which are quite expensive.

STUDDING

Studding consists of placing studs on undercut metal surfaces. These studs increase bonding qualities when they are used on sprayed surface areas that are small. A stud can be anything from a machine screw to a rivet. Usually it is necessary to prepare a stud surface by grit blasting first. Studding is an impractical method to use if large surface areas are prepared because it involves too much time to place the studs in the surface.

SANDING OR GRINDING

Grinding is used on hard alloys that cannot effectively be prepared by undercutting or rough threading. It is also used if grit blasting would produce an undesired result of increasing stresses on the surface of the base metal. Grinding uses special abrasives that do not contaminate the surface but produce the desired roughened surface that will bond the metal spray. Rotary hand tools are often used for this method of surface preparation.

WIRE METALIZING

The wire metalizing process was one of the first metal-spraying processes used in the United States, being used as early as the 1920s. The equipment used in a metalizing system is an air compressor with a capacity of at least 30 ft³/min, a standard oxygen cylinder, a standard acetylene cylinder with regulators, a flow meter, an air-pressure control, a wire straightening and feeding device for the gun, and the gun itself (Figure 26–4). The gun feeds metal wire into an oxygen-fuel flame, which is encased in a stream of compressed air that forms an air envelope around the outside of the flame and the liquid metal. The compressed air cools the metal and helps to accelerate it on its path to the base metal. As the wire is melted by the high temperatures it is atomized by compressed air and blown onto the base metal. Among the several fuels used to melt the wire are acetylene, which can achieve temperatures up to 5660°F, and methylacetylene propadiene,

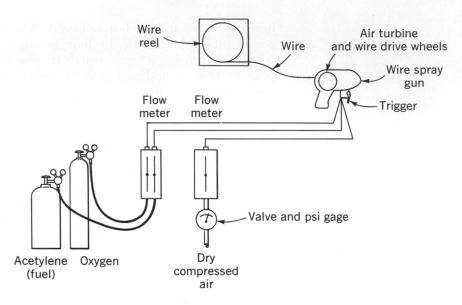

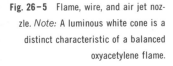

Fig. 26-4 Wire metal spraying system.

Fig. 26-5 Flame, wire, and air jet nozzle. *Note:* A luminous white cone is a distinct characteristic of a balanced oxyacetylene flame.

which attains temperatures of 5300°F; hydrogen propane is also used extensively for the softer metals, such as zinc, lead, aluminum, and tin (Figure 26-5).

The size of the gun varies depending upon the manufacturer, but it usually weighs from 3 to 6 lb for the normally operated gun and up to 10 lb for the machine-mounted gun. A gun is often mounted on a lathe or some other mechanism so that it can travel evenly thus producing a more even surface.

The wires used range in size from a no. 2 Brown and Sharpe up to 3/16-in. diameter wire. For easy use the wires can be purchased in rolls. Wire can also be purchased in

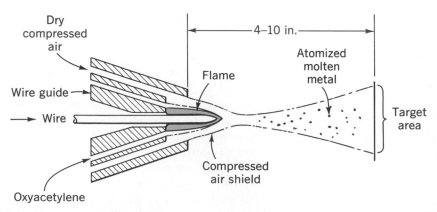

lengths from 3 to 8 ft, which is done if the wire type is too rigid to coil. The amount of filler material deposited on the base metal depends upon the type of wire used and the size of the wire. Depending on its diameter, steel wire can be sprayed at a rate of 17 in./lb and zinc wire at approximately 55 lb/hr. Lead wire can be applied at about 100 lb/hr.

There are two parts to a wire metalizing unit. The first part is a power unit which is used to feed the wire at the desired speed through the nozzle. Most power units use a high-speed air-driven turbine to feed the wire (Figure 26-6d). The turbine operates through a series of gears that reduce the speed of the turbine to a slow rate. This speed is adjusted by increasing or decreasing the speed of the turbine, or by restricting its air intake. The wire is then fed into the oxygen-fuel chamber. Another type of wire feed consists of a variable, dc electric motor, which is bulkier and used primarily for mounted spray units. The second component of the wire metalizing gun is a gas head, which has a wire nozzle, and an air cap, which controls the fuel gas, oxygen, and compressed air mixtures (Figure 26-6a, b, and c). Usually 10 psi of acetylene and 16 psi of oxygen are sufficient to complete the spraying operations. Compressed air must be kept at a constant 35 psi.

Several nozzles are available for a variety of jobs. Extension tubes can be used for spraying metal inside cylinders of small diameters. Broad flat nozzles can be used to produce a fan-shaped spray for large areas. The metal particles that are sprayed are extremely small and range in size from 0.0001 to 0.0015 in. in diameter when sprayed from the optimum distances of a minimum of 4 in. to a maximum of 10 in. When spraying metals from these minimum and maximum distances, approximately 95 percent of the metal sprayed will adhere to the prepared base metal.

The two critical adjustments when using the gun are the adjustments of the wire feed and the neutral flame. A slightly carburizing flame is acceptable; however, an oxidizing flame will cause the wire to oxidize too much and will prevent some of the metal particles from adhering properly to the base metal. When the wire travels too fast, the wire will not atomize properly. When it travels too

Fig. 26-6 Metal powder spray system.
(a) Pressure system. (b) Nonpressure
system. (c) Compressed-air-driven wire
spray gun. (d) Electric-motor-driven wire
spray gun. (Photographs courtesy of
Metallizing Company of America.)

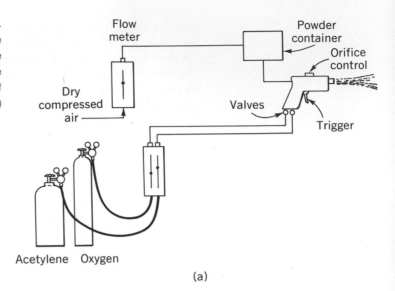

(a)

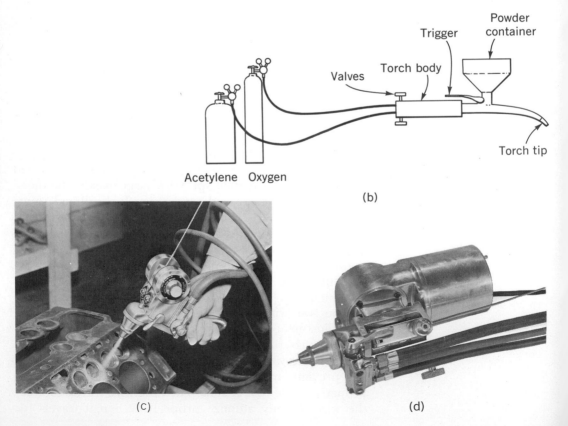

(b)

(c)   (d)

slowly, oxidation takes place or the wire may fuse to the nozzle, causing failure of the gun.

Metal surfaces are metalized in order to increase resistance to corrosion and to increase wearability. The most frequent reason for metalizing is to produce corrosion-resistant surfaces. In most cases, the metal sprayed is applied in thin layers that are less than 0.010 in. thick. Heavy coatings may be applied on some metals in a single pass, but the multiple-pass method is recommended because of its greater reliability.

THERMOSPRAY The thermospray process is similar to the metal metalizing process, except that a powder is used instead of a wire and it can be used for hardfacing. The thermospray process does not involve as much equipment as does the metalizing process. The thermospray system includes an oxygen and a fuel supply with regulators for each, a flow meter, and a thermospray gun (Figure 26–7a).

The thermospray guns that are manufactured in the

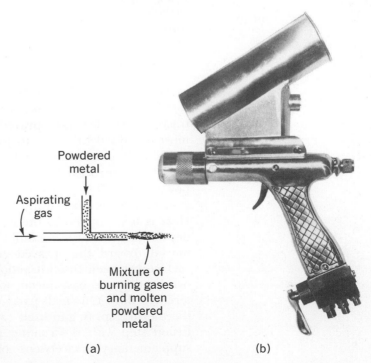

Powdered metal

Aspirating gas

Mixture of burning gases and molten powdered metal

Fig. 26–7 (a) The venturi principle. The aspirating gas may be a mixture of oxygen fuel, oxygen, or compressed air, depending on the spray equipment manufacturer. (b) Application of the powder spray gun. (Photograph courtesy of Metallizing Company of America.)

(a)   (b)

United States operate on the venturi principle; that is, they employ a neutral oxygen-fuel flame that uses the oxygen as an aspirating gas to draw the powder into the flame. The powder is then melted outside the body of the gun and forced onto the metal by the action of the aspirating gas. The spray distance ranges from 4 to 10 in. Approximately 95 percent of the sprayed powder will strike the base metal at these distances (Figure 26-7b). The thermospray powder may be either gravity fed into the aspirating chamber or it may be fed by an electrical vibrating mechanism, which is attached to the chamber. In this way the passageways would be kept full of powder while each chamber has powder in it. The sizes of the powder particles used in the thermospray process fall into three groups, ranging from 100- to 150-mesh size. However, up to 325 mesh is used for special jobs. The first group of particles fuses from 1800 to 2500°F and consists of nickel-silicon-boron alloys or nickel-chromium-silicon-boron alloys. The second group fuses from 1920 to 2080°F and consists of cobalt-chromium-silicon-boron alloys. The last group fuses from 1950 to 2050°F, and are the tungsten-carbide composites. These various materials may be applied to almost all irons, nickels, coppers, and copper alloys. Often, when materials are applied to austenitic stainless steels or other high-temperature alloys, preheating precedes the application of the metals. The base metal must be between 400 and 900°F to compensate for the expansion that takes place. When metals that contract a great deal are sprayed, additional filler material must be added to make up for shrinkage. Sometimes as much as 20 percent filler material must be added.

A fusion process follows the application of the metal powder. This fusion process is usually done with oxyacetylene or oxygen fuel, using a neutral or reducing flame. Heat is first applied to the area adjacent to the sprayed deposit until the area is a dull red. Then the torch is slowly moved toward the sprayed coating until the complete surface has been fused. Caution must be used to prevent overheating the base metal, which causes the fused material to flow. On small parts, the oxyacetylene flame of the thermal spray gun itself can be used to complete the fusion step with the coating process. On larger pieces, supplementary oxyacetylene torches can be used as heating devices. Some parts can be sprayed while they are rotating in an engine lathe or some other revolving device, The correct rotating speed is indicated when the revolving part comes sharply into focus.

The filler material may be supplied to the gun as a liquid from an external crucible. This special adaptation of the gun has a crucible instead of a powder hopper that is insulated. It has some type of heat source, such as an electric element or gas flame to keep the metal constantly in liquid form. This crucible gun is used only for those metals with melting points below 600°F.

PLASMA ARC SPRAY When higher fusion points (above 5000°F) must be reached, the plasma arc spraying process is usually used (Figure 26-8). The plasma torch is an untransferred arc type that is independent of the material to be sprayed because it is a contained heat source. The only difference between a plasma arc torch that is used for cutting and welding and one that is used for spraying is the addition of a device that injects powder into the arc stream. The electric arc travels from a tungsten cathode to a copper anode on the front of the gun. Gas enters the torch and is expanded by the arc, resulting in a plasma jet, which is forced out of the nozzle at velocities that may reach as high as 20,000 ft/sec. Powdered metal is metered into this plasma stream just as it leaves the expansion cavity in the front of the gun. The filler metal is then carried to the base metal in a liquid state, melting actually in the air

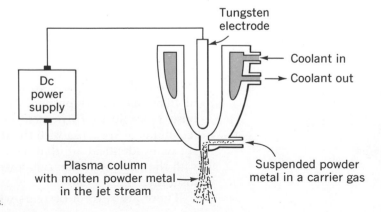

Fig. 26-8 Plasma arc spray process.

and not in the gun because of the high temperatures of
the plasma arc. Because of this high temperature, created
inside the gun, cooling is mandatory and is accomplished
by circulating 50 to 100 lb water pressure through the gun
at 3 to 5 gal of flow per minute. On some models the tungsten electrode is water-cooled. The water must be clean,
pure water in order to prevent corrosion inside the gun.
Many times the plasma arc spray gun maintains temperatures of 30,000°F.

There are several fuels used for plasma arc spraying,
mainly either monotomic or diatomic gases. Argon is a
monotomic gas used for spraying carbides. Helium is
useful as an auxiliary gas. An auxiliary gas is sometimes
injected with the plasma gas to produce certain burning
characteristics. Of the diatomic gases, nitrogen is most
frequently used because it is inexpensive, has a high spraying speed, and has an efficient deposit rate.

The electrical power supply for plasma spraying must
have a 100 percent duty cycle, with ranges of 14 up to
100 kW. The average unit demands 28 to 40 kW with 220
volts, 60-Hz, 3-phase power. All controls are located on
the control console with none on the gun. The various
controls on the consoles include the plasma-gas control,
the power-current control, the stop-start, high-frequency
control, the secondary-gas control, and the powder-gas
control. The current level is adjusted with a rheostat after
the gas and powder speeds have been set.

The filler materials are made of ceramics, some of the
nonferrous metals, all the ferrous metals, and even those
metals that range in hardness through tungsten carbide,
vanadium, and vanadium carbides. The feed rate depends
upon the fusion point of the filler materials used. The
feed rate for aluminum ranges from 0.10 up to 7½ lb/hr,
while tungsten can be deposited from approximately
1½ to 25 lb/hr. The plasma arc gun is held from 2 to 6
in. from the workpiece to insure the largest percentage
of filler material deposited. Most filler materials will deposit from 60 to 75 percent at this distance. The actual
technique of spraying with the thermospray or the powder
spray is identical to that used with oxyacetylene fuel,
with the exception that the powder fuses to the base metal
instantly.

There are several dangers when any plasma jet device

is used. These devices have a high noise level, ranging around 100 dB. This noise level must be lowered to a maximum of 80 dB to prevent damage to the inner ear. There is also intense ultraviolet and infrared radiation, which may cause sunburn, making it mandatory that eyes be protected; usually from a no. 9 to no. 12 eye lens is used.

ELECTRIC ARC SPRAY

The electric arc spraying process was first invented in 1938, but it was not used commercially in the United States until 1964. This process utilizes two electrical guides through which two wires pass forming an arc as they make contact at the center in front of the gun. The arc formed by the two electrodes as they meet produces sufficient heat to melt the wire. Compressed air sprays the metal onto the workpiece in the form of fine particles of metal (Figures 26-9 and 26-10). The ends of the wire make and break contact at the rate of 100 times a second, causing pulsating action that is so fast it seems as if there is one solid stream of filler material. This filler material is then forced by air pressure onto the base metal. The air pressure should be held at a constant 35 psi.

Either a motor generator or a solid-state power source can be used to supply power to the arc spray gun. Both units should be able to supply from 50 to 650 amps. Sometimes the newer solid-state power supply will yield a higher deposition rate. The electric spray gun sprays much faster than other types of spray guns because of the high temperatures created by the arc. Sometimes the travel speed will reach 200 ft/min with deposits of 0.005 to 0.010 in. per pass. The distance from the workpiece to the spray gun should be maintained at approximately 10 to 12 in. This distance will yield optimum metal deposit.

The cost of operating the electric spray gun is small because electrical power is used. At $0.02/kW, the 360-amp unit can be operated for $0.25/hr, and electricity for a 650-amp unit costs only $0.56/hr. The plasma gun in some cases has an operating cost of $25/hr, so the electric spray gun is much cheaper to operate. Another advantage of the electric arc gun is that two different types of filler material can be used, one type for the cathode and one for the anode. The filler materials, meeting in the arc zone, produce a homogeneous mixture that is blown onto

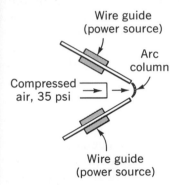

Fig. 26-9 Electric arc metal spray.

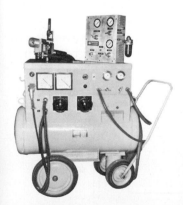

Fig. 26-10 Electric arc metal spray power supply. (Courtesy of Metallizing Company of America.)

the base metal. For example, stainless steel wire could be used on one side in one wire guide, and aluminum bronze could be used in the other guide. These materials would then intermingle and be accelerated onto the base metal, resulting in a long-wearing surface with a desirable bearing surface. The electric arc spray process is not as hazardous to use as the plasma arc process, but it has more hazards than either of the oxyacetylene flame spraying processes. The hazards are created because of the high speed at which the work must be done. This high speed amplifies the fumes, dust, and odors around the work. There is usually a protective shield on the front of the electric arc spray gun, but even with this shield a no. 4 or no. 5 eye-shading lens must be used. Without this built-in protective shield, a no. 10 or no. 12 lens must be used. Noise is not too great a problem with the electric arc spray process.

DETONATION GUN The detonation-gun process is a recently developed surfacing technique. The process is used mainly for depositing hard surfaces of tungsten carbide, cobalt, nickel, or chromium carbide (Figure 26–11). The detonation gun is a long-barrel gun that has four inlets, which act as an ignition device. The four inlets supply powder, nitrogen, oxygen, and acetylene to the ignition chamber. Metered amounts of oxygen and fuel are put into a chamber, along with metal powder particles that are suspended in nitrogen. An electric spark detonates the mixture, and the gas in the ensuing gas stream contains the metal particles. The gas stream leaves the gun barrel at a rate greater than

Fig. 26–11 The detonation gun.

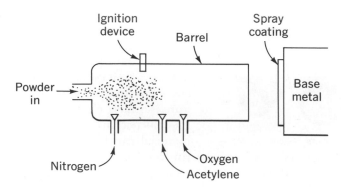

2500 ft/sec. With the oxygen-fuel processes, this rate is 200 to 500 ft/sec; for the plasma arc, it is 1100 ft/sec. With the detonation-gun process, particles actually leave the barrel of the gun at supersonic velocities to strike the workpiece. The initial explosion inside the detonation gun brings the metal particles to a plastic state. The wave front caused by the explosion forces the particles out of the barrel and onto the base metal. The gun itself resembles closely many of the modern-day firearms in its basic operation. Extremely high temperatures are reached inside the gun as high as 6000°F; however, the parts that are being sprayed rarely reach a temperature greater than 300°F. The bonding strength of this process ranges from 8000 to 25,000 psi. The detonation-gun spray process is a completely automated process because of the extreme noise caused by the firing of the gun. The gun is usually located within concrete sound-absorbing walls and is aimed and fired by remote control. The patent for the detonation-gun process is owned by the Linde Air Products Company. The process is available for general use, but the gun itself cannot be purchased.

QUESTIONS

1. What are the surface bonding mechanisms used in metal flame spraying?

2. What are the major types of metal spraying devices?

3. What is the velocity of the particles when the detonation gun is used?

4. What fuel gases are used in the chemical types of metal flame spraying?

5. What are the advantages of using the electric arc spray method over using the plasma arc spray method?

6. Why is surface preparation important?

7. What is a dovetail?

8. What are the major surface preparation methods?

9. What happens to the wire or powder when it enters the flame of metal spraying devices?

10. How is wire used for flame spraying stored?

11. What is the function of compressed air in the wire metalizing process? In the thermo-spray process? In the arc process?

12. Why is a fusion process sometimes applied after the spraying process?

SUGGESTED FURTHER READINGS

Bergman, Ludwig, and O. Simonds: "Finishing Metal Products," McGraw-Hill Book Company, New York, 1946.

Ingham, H. S., and A. P. Shepard: *Metalizing Handbook,* Metalizing Company, Inc., Westbury, N.J., 1954.

Kremith, Roland D.: Plasma Flame Coating, *Machinery,* vol. 76, no. 6, p. 49, February 1970.

Levy, P. S.: Plating with a Metal Mist, *Iron Age,* vol. 194, p. 59, October 1964.

V METALS

27 STRUCTURE OF METALS

The study of metal, or metallurgy, can be approached by two different routes, process metallurgy and physical metallurgy. Process metallurgy is concerned with the extraction and refinement of metal-bearing ore into a pure metal, which in turn is further refined or further processed. These pure metals can be used either in their pure state or combined with other metals to form alloys. Physical metallurgy concerns itself with the structure of metals, the properties of metals, and the influence of fabrication methods on the properties and structure of metals.

Most metals may be defined as having the following characteristics: they are usually opaque and solid at room temperature; they are heat (thermal) and electrical conductors and are reflectors when polished; and they expand when heated and contract or shrink when cooled.

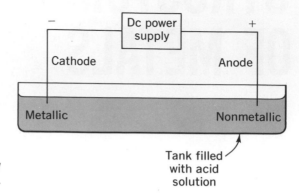

Tank filled
with acid
solution

Fig. 27–1 Element classification by
electrolysis.

These characteristics describe the apparent aspects of metals, but a metal can be identified in a different, more exacting manner. A metal is an element that is, when immersed in a caustic acid solution, attracted to a cathode when a small current is passed through the solution (Figure 27–1). Elements that are not attracted to the cathode but are attracted to the anode are scientifically called non-metallic elements. The only known element that is an exception to this definition is hydrogen. Hydrogen collects at the negative pole, but it is a nonmetallic element. Metalloids also are elements that are attracted to the metallic cathode pole although they too are nonmetallic. A metalloid does not have all the characteristics of a metal. A metalloid has some conductivity, but it has little or no ability to be deformed. Many metalloids, such as carbon, boron, and silicon, can be combined readily with metals.

Metals, as all substances, are composed of separate atoms. All atoms, no matter what the material, have the same basic type of structure (Figure 27–2). An atom consists of a nucleus, with positively charged protons and neutrally charged neutrons. The nucleus composes the major portion of the mass of the atom. Surrounding the nucleus in orbits or shells are electrons. Electrons have relatively small mass and a negative charge. The electrons balance the charge of the protons. The maximum number of electrons that exist in any orbit is represented by the symbols $2N^2$, where N represents the shell or orbit number. Therefore, the first orbit or first shell surrounding the electron may be composed of 2 times 1^2 electrons, or 2 electrons. The second shell would have 2 times 2^2, or 8 electrons, and so on. A characteristic of all metallic ele-

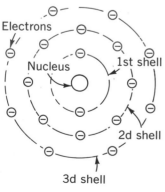

Electrons

Nucleus

1st shell

2d shell

3d shell

Fig. 27–2 Atomic structure.

ments is that there are always free electrons in the outer shell of the atomic structure. This means that the outer shell will not be completely filled with electrons, no matter the shell number. These electrons can be released readily from the atomic structure, and they are used in joining one atom to another or for the precipitation of electrons from the metal, as is the case in the arc column when arc welding. These electrons are known as valence electrons.

Theories explaining how atomic structures are bonded have been hypothesized. One such theory describes an ionic bond, a bond that is typical of many gases or liquids. In this bond, the valence electron in the outer shell is attracted to an element whose outer shell has a minus number of electrons instead of an overabundance of electrons. The valence electron balances the two shells when they are placed in close proximity. Another bond, the covalent bond, on the other hand, is one in which electrons are shared in the outer shell, a sharing that completes the number of electrons in each individual shell. The covalent bond is strong because of the attraction, or the tie-in force, of these shared electrons. The metallic bond is the major way in which the atoms within metals are bonded.

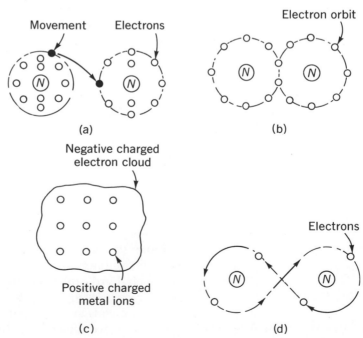

Fig. 27–3 Atomic bonds. (a) Ionic bond. (b) Covalent bond. (c) Metallic bond. (d) Van der Waals force.

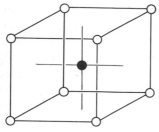

Fig. 27–4 Body-centered cubic space lattice unit cell.

The metallic bond may be thought of as a negatively charged electron cloud that completely surrounds the positively charged metal ions or metal nuclei. This electron cloud is composed entirely of the valence electrons of the metallic element. Another theory postulates the van der Waals force, a weak attraction force that is caused by the separation of the electrons from the nucleus. The electrons often interchange from one nucleus to another. The van der Waals bonding force is characteristic of inert gas (Figure 27–3).

SPACE LATTICES

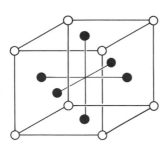

Fig. 27–5 Face-centered cubic.

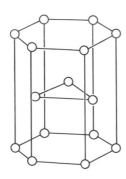

Fig. 27–6 Hexagonal close-packed.

Metals in a liquid state are amorphic, or can be said to have no crystalline structure. When the free-moving atoms within a liquid metal begin to freeze or solidify, they attach themselves to each other in an orderly, well-defined system. This called the space lattice system, the basis for the crystalline form of metal. A metal arranges its crystallinity during solidification the same way every time. The most common space lattice systems in which metal crystals usually arrange themselves are face-centered cubic (fcc), body-centered cubic (bcc), and hexagonal close-packed (hcp) (Figures 27–4, 27–5, and 27–6). Five other crystalline systems, besides the cubic and the hexagonal, are the triclinic; the monoclinic; the orthorhombic; the rhombohedral, which is a triangular design; and the tetragonal. These 7 space lattice systems can produce 14 possible types of crystalline structures.

Iron is the best known metallic element that has a bcc space lattice system. Iron, however, changes its bcc system to an fcc lattice system at elevated temperatures. This change in space lattice systems is known as allotropism. The change from a bcc to an fcc lattice allows carbon to move through the fcc system. Because the atoms are far apart in the bcc system carbon cannot rearrange itself; however, the carbon can and does change position in the fcc system. Common metals that have bcc space lattices at room temperature are tungsten, molybdenum, vanadium, columbium, and iron. The bcc metals are generally extremely hard and have little ductility; however, the bcc metals rate high in strength when compared to fcc metals. Also, bcc metals are not easily cold worked.

Common fcc metals are aluminum, copper, gold, lead,

and nickel. In their pure form, fcc metals are soft and ductile. Generally, fcc metals are easily cold worked and are low in strength.

The hcp system is present in cobalt, cadmium, magnesium, titanium, and zinc. The hcp system is found in metals that cannot be cold worked because cold working will lead to immediate failure of the metallic structure. The strength, however, of these metals varies, some being weak and some being strong.

MILLER INDICES The Miller indices indicate the crystallographic planes in a unit cell. Crystallographic planes are sometimes called atomic planes. The Miller indices describe a set of relationships between certain planes in atomic, or space lattice, structures. Crystallographic planes are expressed by three coordinates, the x coordinate, the y coordinate, and the z coordinate. These coordinates represent the three dimensions in a crystallographic plane. For example, the coordinates 010 indicate that the intersection on the x axis is a plane that stretches indefinitely to infinity; the y axis represents one unit from the origin of the coordinates, and the z axis is a plane that stretches to infinity (Figure 27–7). The infinity sign in the Miller indices is replaced by a 0; therefore, the Miller indices for the plane $CDGH$ in Figure 27–7 are represented by the numbers 010. The major point to remember when using the Miller indices is that 0 indicates a plane infinitely long; and the unit length 1, or a fraction of that unit length, such as $\frac{1}{2}$ or $\frac{1}{3}$, represents a distance from the origin of the coordinates. The Miller indices describe a plane through the atomic structure of a crystal. One set of coordinates describing a plane is the same set used to describe all parallel planes in the structure (Figure 27–8). In the example in Figure 27–8, the Miller indices coordinates are 110, where z is equal to 0, and the x and y coordinates are indicated by the black dots. The set 110 indicates all sets parallel to 110. The distance between each two atomic structures, or crystals, represented by the black dots in Figure 27–8, is measured in angstroms along the $x, y,$ and z coordinates ($1 \text{ Å} = 10^{-8}$ cm). The distance between atoms determines the strength or the softness of the metal. For example, lead, gold, aluminum, copper, and nickel all have fcc

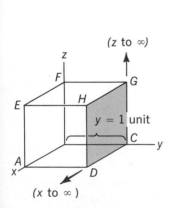

Plane $CDGH$			
Intersection	x	y	z
Numerical	∞	1	∞
Miller indices	0	1	0

Fig. 27–7 Miller indices of a unit cell (crystallographic planes).

space lattice structures. This structure has smaller distances between atoms than bcc structures. Consequently, metals with fcc structures are more ductile because of the closeness of the atoms. They can be bent and twisted without early failure because of their atomic binding. In bcc metals, atoms are farther apart, making these metals, such as iron, molybdenum, vanadium, and tungsten, neither soft nor ductile. In hcp metals, the space lattices present longer distances between atoms in a vertical direction. These metals such as zinc, titanium, magnesium, and cobalt, are not readily cold worked, for they fracture easily when worked.

DENDRITES

Basically, most metals solidify in a set sequence established by the temperature loss, progressing from a liquid state, in which individual atoms of metal exhibit extremely free movement, to a state of complete rigidity. The metal passes through three types of crystalline form, which are based on dendritic growth during this sequence.

A dendrite may be compared in appearance to the trunk and limbs of a tree. Dendritic growth begins with a nucleus which proceeds to grow a skeletal form that fills in with additional metal until the crystal is completely filled. The only interference that stops the crystal from continuing to grow is the lack of space caused by the growth of surrounding crystals (Figure 27–9). As the metal cools slightly, a nucleus, sometimes called a seed crystal, is formed in the metal bath. This nucleus begins the crystalline growth. A primary axis is the first portion to form as the temperature of the bath decreases. The primary axis extends a distance and then the limbs, or the secondary axis, grow. The secondary axis, upon further heat loss, develops a third axis, or a ternary axis. At this point, the skeletal structure is complete, expanding and filling in until a solid dendrite is evident (Figure 27–10b).

In a binary metal, or a more complex metal system, the various possible alloys separate during the solidification process because of the differences in melting temperatures, densities, and other physical qualities of the particular metals in the alloy. This separation can be observed in the full-grown crystal when it is studied under a microscope. The separation of alloys is indicated by a dark area or a

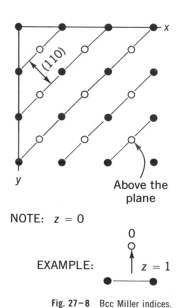

NOTE: $z = 0$

EXAMPLE: $z = 1$

Fig. 27–8 Bcc Miller indices.

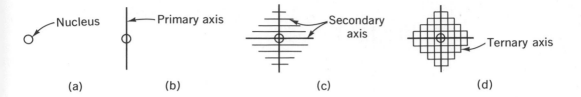

Nucleus   Primary axis   Secondary axis   Ternary axis

(a)   (b)   (c)   (d)

Fig. 27-9 Dendrite growth (crystal).

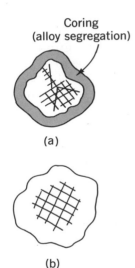

Coring (alloy segregation)

(a)

(b)

Fig. 27-10 Dendrites. (a) Binary alloyed metal. (b) Pure metal.

different colored area around the periphery of the crystal. This area is usually referred to as coring (Figure 27-10a).

The physical solidification process goes through three changes in structural rigidity: liquidus, liquidus-solidus, and solidus. Liquidus is the point at which the first crystalline structure begins to form, or the first grain begins to form. Liquidus-solidus is the point in time and the temperature at which the metal is in a mushy state. The solidus point of a metal is the point at which the grain formation is complete and the metal is a solid even though it is still hot. Graphically, these three characteristics can be represented as a time-temperature function (Figure 27-11). The mushy state, or the liquidus-solidus point, is indicated graphically by a change in the direction of the line.

The start of solidification is that point in time and the temperature at which dendritic growth begins. The transformation of the free atom to an entrapped one within the dendritic structure releases energy in the form of heat, which prolongs the cycle time between the liquid and solid stages of the metal. Many times, the heat evolution of dendritic growth reheats the metal during the growth period. This reheating is called superheating or super-

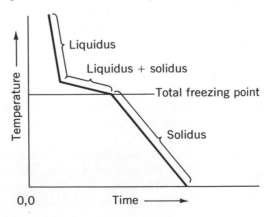

Liquidus

Liquidus + solidus

Total freezing point

Solidus

Temperature →

Fig. 27-11 A typical cooling curve.   0,0   Time →

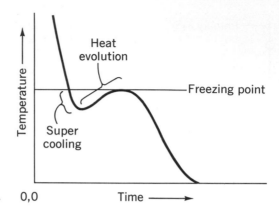

Fig. 27-12   Heat evolution.

cooling (Figure 27-12). The crystalline growth resulting from heat evolution will be short-lived because of the heat dissipation of the particular metal.

Metal shrinkage, or contraction, also occurs during the change in the state of a metal from a liquid to a solid. Contraction is divided into three stages based on temperature ranges: liquid contraction, mushy state contraction, and solid contraction. During each stage of temperature change or loss in a metal, part of its original volume is lost. This volumetric shrinkage begins in the liquid stage as the metal cools to the liquid-solid stage, or when solidification begins. The shrinkage during the liquid stage to the beginning of the nucleation stage is negligible and has not been measured. However, the shrinkage during the mushy stage has been measured. Shrinkage during the superheating stage has been assessed to be between 0.8 and 1 percent of the total volume for each 100° of heat. As the metal proceeds through solidification, an additional 2 to 4 percent loss of volume takes place in the mushy stage. The final and largest volumetric loss due to shrinkage is approximately 6 to 8 percent of the total volume after the metal has completely solidified and is cooling to ambient, or room, temperature. The approximate percentages of volumetric shrinkages can be indicated on a basic time-temperature graph so that they can be compared (Figure 27-13).

The time-temperature studies of metals have led to the plotting of a collection of time-temperature graphs called equilibrium diagrams, which indicate the liquidus point and the solidus point of metals and their alloys

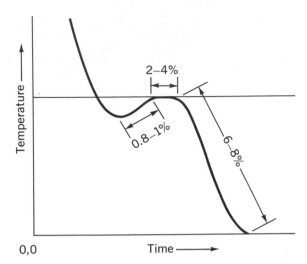

Fig. 27-13 Volumetric shrinkage.

(Figure 27–14). For example, if two metals were mixed, there would be a different time-temperature graph for each of the two metals in the alloy. If an alloy were composed of 60 percent of alloy B and 40 percent of alloy A, the liquidus and solidus points for this mixture would be different from other possible mixtures (Figure 27–15). In equilibrium diagrams, a line connects the liquidus points and the solidus points of the possible alloys, making it unnecessary to use separate time and temperature graphs for each metal in the alloy.

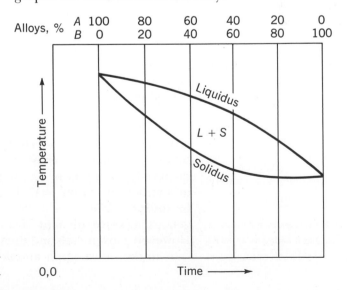

Fig. 27-14 Equilibrium diagram derivation.

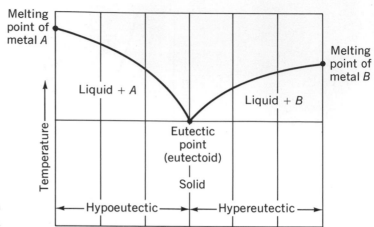

Fig. 27–15 Eutectic equilibrium diagram derivation.

(a)

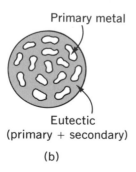

Primary metal

Eutectic
(primary + secondary)

(b)

Fig. 27–16 Solutions of metals. (a) Solid solution (eutectoid metal). (b) Primary metal plus eutectic mixture.

Many times two metals that are alloyed have different melting points. These two metals can be mixed together so that there will be a point at which both are in their liquid state. The point at which the two metals are in balance, the melting point of the combined metals (which can be lower than the melting points of both metals) is called the eutectic point (Figure 27–15). The eutectic point illustrated in Figure 27–15 represents this point of balance for a mixture of 60 percent of alloy A and 40 percent of alloy B. The eutectic point is also helpful in defining various alloys. A hypoeutectic metal, or below-eutectic-point alloy, is a mixture that has more of alloy A plus a eutectic mixture, while a hypereutectic alloy has more of the solid B metal plus a eutectic mixture. Therefore, the characteristics of a hypoeutectic metal are different from those of a hypereutectic alloy. The hypoeutectic alloy would exhibit many of the physical and mechanical characteristics of alloy A, while the hypereutectic alloy would exhibit many of the characteristics of alloy B. An alloy that is represented by the eutectic point is said to be in solid solution. *In solid solution* is a term that describes something dissolved in a liquid or an intermingling of one substance in another so closely that the dissolved substance cannot be distinguished or separated, which would be a homogenous mixture of alloy A and alloy B (Figure 27–16). However, most metals and their alloys represent a different case under most applications. If two metals are mixed

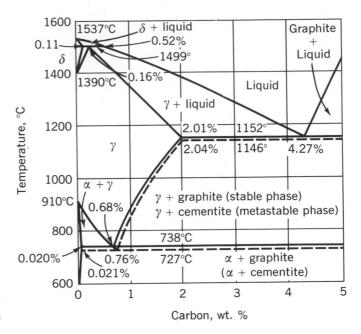

Temperature, °C

1600 — 1537°C — δ + liquid — Graphite + Liquid
0.11 — δ — 0.52% — 1499°
1400 — 1390°C — 0.16%
δ + liquid
γ + liquid — Liquid
1200 — 2.01% — 1152°
γ — 2.04% — 1146° — 4.27%
1000 — α + γ
910°C — 0.68% — γ + graphite (stable phase)
γ + cementite (metastable phase)
800 — 738°C
0.020% — 0.76% — 727°C — α + graphite (α + cementite)
0.021%
600
0   1   2   3   4   5
Carbon, wt. %

Fig. 27–17   Iron–iron carbide equilibrium diagram. (Courtesy of U.S. Steel Corporation.)

together, alloy *A* and alloy *B*, one alloy, supposedly the primary alloy, is the dominant alloy. It will not be able to absorb all of alloy *B*, nor will alloy *B* be able to absorb all of alloy *A*. An alloy can absorb a particular amount of the alloy substance, and then it cannot absorb any more. When a metal has absorbed its maximum amount, it has reached its eutectic point, or its maximum saturation point. The rest of the primary metal will remain in its primary state so that in the crystalline form of the alloy, there will be grains of primary metal surrounded by the eutectic mixture (Figure 29–16). Coring is related to this phenomenon, but refers only to one crystal. The most frequently used equilibrium diagram and one that is an excellent example of the eutectic point is the iron-iron carbide equilibrium diagram (Figure 27–17).

The rate of crystalline growth is proportional to the lapse of time and the temperature loss of the metal. This nucleation rate determines the size and the shape of the crystals; and the size and the shape of the crystals, in turn, determine the strength of the metal. Retarding the nucleation rate causes an excess of new nuclei to develop in the interior of the metal, resulting in extremely large den-

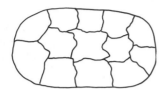

All large crystals (dendrites)

Fig. 27-18 Results of a retarded rate
of nucleation (heat loss).

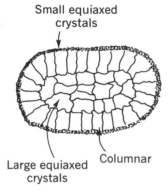

Small equiaxed
crystals

Large equiaxed   Columnar
crystals

Fig. 27-19 Types of crystals.

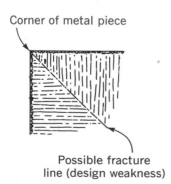

Corner of metal piece

Possible fracture
line (design weakness)

Fig. 27-20 Fracture zones.

dritic structures. These large grains may cause a metal to be physically weaker than if it had solidified at its normal rate (Figure 27-18).

Three shapes of grain structure result during the normal solidification of a metal or an alloyed metal. The outer layer next to the surface of the metal, or the fluid metal, such as in a weld bead, which comes into contact with the slag facing, cools most rapidly. This rapid cooling creates small equiaxed crystals called chill crystals. The chill crystal encircles the periphery of the weld bead. As the metal cools, progressively long, these columnar grains begin to grow at right angles to the bead interface or surface. The growth of the crystals increases the distance between the internal liquid metal and the interface of the weld bead base metal. As a result heat dissipation is lessened, retarding the rate of nucleation and causing the production of large equiaxed crystals in the center of large weld beads (Figure 27-19). Many times, because of the size of the weld bead and because of the metal alloys used for the welding electrode, the only two crystalline shapes represented in a weld bead are the chill crystals and the columnar crystals.

A pure metal usually has only chill and columnar crystals because a pure metal has an extremely wide mushy zone. This solidification zone often extends into the center of the casting or weld bead and allows a more even heat loss, which prolongs the growth time for the columnar crystals. One function of welding flux is to control the solidification rate, which helps to control the growth of columnar crystals. These columnar crystals extend to the center, and at right angles to the surface interface, of a metal. Cooling may be used to control interior crystalline growth. For example, lightly dusted electrodes allow a faster heat loss than semicoated electrodes. While the crystalline shape yielded with both electrodes is generally a columnar configuration, the faster heat loss results in smaller grain size. The smaller the grain size, the stronger the metal. Most weld beads yield a columnar crystal in the interiors of the weld zone and the base metal. While column-shaped crystals are associated with metal strengths, possible planes of weakness in sharp corner designs can result because of their angle of growth. These planes of weakness appear as dark lines when the microstructure of the metal is studied (Figure 27-20).

QUESTIONS
1. What is a metal?

2. What is an atomic shell?

3. What are some of the theories that explain atomic bonding?

4. How are atoms arranged in space?

5. How does atomic arrangement determine the hardness of a metal?

6. How is the arrangement of atomic structure studied and recorded?

7. How are the major types of crystals formed?

8. What is liquidus?

9. What is a eutectic point?

10. How are equilibrium diagrams derived?

11. What is the importance of the equilibrium diagram?

12. What is a solid solution in a metal?

13. Why are small equiaxed crystals the hardest type of crystals?

14. How do sharp edges affect the strength of a metal part?

SUGGESTED FURTHER READINGS

Bolz, Roger W.: "Production Processes," Penton Publishing Company, Cleveland, 1963.

Cooley, R. H.: "Complete Metalworking Manual," Arco Publishing Company, New York, 1963.

Lindberg, Roy A.: "Processes and Materials of Manufacture," Allyn and Bacon, Inc., Boston, 1964.

Monodolfo, L. F.: "Engineering Metallurgy," McGraw-Hill Book Company, New York, 1955.

Rogers, Bruce A.: "The Nature of Metals," The Iowa State University Press, The American Society of Metals, Ames, Iowa, 1964.

Weeks, William R.: *Metallurgy*, American Technical Society, Chicago, 1956.

28 PROPERTIES OF METALS

The properties of metals are divided into three major categories: the chemical properties, the physical properties, and the mechanical properties. The chemical properties determine whether or not the metallic elements will combine with each other and possible reactions and interactions of the metals in combination. The physical properties of metals are those that are inherent in the metal without any external force being applied to the metal, such as its electrical conductivity, thermal conductivity, melting point, and density. The mechanical properties of metals are those that indicate the strength of the metal or the reaction of the metal to an applied external force.

CHEMICAL PROPERTIES

The chemical properties of metals are not of direct concern to a welder, although they do concern the researcher in the welding field. Associated with the chemical properties of a metal are the chemical symbol, the crystalline form, the atomic weight, the atomic number, and the level of electron energy, or the quantum numbers. Metals can be separated into the rare-earth metals, such as uranium; the transition metals, such as nickel and iron; the noble metals, such as silver, copper, and gold; the semiconductor metals, such as tin and germanium; and the semimetals, such as bismuth and antimony. The semimetals have complex crystallographic structures, as in the salt-forming halogens, such as fluorine. The symbols for all elements, metallic or nonmetallic, are assigned arbitrarily; however, the crystalline structure, the atomic number, and the atomic weight have been determined as a result of patient and methodical research. All elements are currently placed on a periodic table, which arranges the elements in the order of their atomic numbers. The first table was derived by D. I. Mendeleev in 1869, and since then many elements have been added to his original table. The first element on the table is hydrogen, which has an atomic number of 1 and an atomic weight of 1.008. The atomic number of an element is determined by the number of protons in the nucleus. This number is the same as the number of electrons in the shells that surround the nucleus. The atomic number describes the chemical properties of the atom. The atomic weight of an element is determined by the total number of protons and neutrons within the nucleus. The atomic weight of an element is derived by comparing its weight to the atomic weight of

CHART 28–1 Chemical properties of selected metals

ELEMENT	SYMBOL	CRYSTALLINE FORM AT ROOM TEMPERATURE	ATOMIC NUMBER	ATOMIC WEIGHT
Aluminum	Al	Fcc	13	26.98
Copper	Cu	Fcc	29	63.54
Gold	Au	Fcc	79	197.00
Iron	Fe	Bcc	26	55.85
Lead	Pb	Fcc	82	207.21
Nickel	Ni	Fcc	28	58.69
Silver	Ag	Fcc	47	107.88
Tin	Sn	Tet	50	118.70
Tungsten	W	Bcc	74	183.92
Zinc	Zn	Hcp	30	65.38

carbon. The atomic weight of carbon has been established as 12.01 amu (atomic mass units). This figure, then, is used as a basis for establishing the atomic weight for all elements (Chart 28-1).

PHYSICAL PROPERTIES   The physical properties of a metal (or an element) are those that are inherent to the metal, regardless of any force applied to the metal. Some physical properties of a metal include melting point, boiling point, density, thermal conductivity, electrical resistivity, coefficient of thermal expansion, volumetric shrinkage, magnetic properties, and specific heat. Other physical properties of metals are its resistance to corrosion, color, nuclear radiation characteristics, and reflectivity (Chart 28-2).

CHART 28-2   Physical properties of selected metals

ELEMENT	MELTING POINT, °F	BOILING POINT, °F	DENSITY AT 68°F, PSI	THERMAL CONDUCTIVITY, AT 68°F, CAL/ (CM²)(CM)(SEC)(°C)	ELECTRICAL RESISTIVITY, $\mu \Omega$/CM	THERMAL EXPANSION COEF., AT 68°F, μIN./(IN.)(°F)	VOL. SHRINKAGE, %
Aluminum	1220	4442	0.0975	0.53	2.7	13.1	−6.4
Copper	1981	4703	0.324	0.94	1.7	9.2	−4.2
Gold	1945	5380	0.698	0.71	2.4	7.9	−5.2
Iron	2798	5430	0.284	0.18	9.7	6.5	−2.8
Lead	621	3137	0.4097	0.08	20.6	16.3	−3.5
Nickel	2647	4950	0.3220	0.22	6.8	7.4	NA
Silver	1761	4010	0.379	1.00	1.6	10.9	−5.0
Tin	449	4120	0.2637	0.15	11.0	13.0	−2.8
Tungsten	6170	10706	0.697	0.40	5.6	2.6	NA
Zinc	787	1663	0.258	0.27	5.9	22.0	−4.2

The physical properties most important for the welder to know are the melting point, the density, the thermal conductivity, the electrical resistivity, the coefficient of thermal expansion, and the volumetric shrinkage. The physical properties of metals are not affected by any outside pressure or force. They remain a fairly constant, accurate characteristic. For example, the melting point of aluminum remains at 1220°F regardless of what type of heat energy is used to melt it. The same is true of its electrical resistance properties or any of its other physical properties. The density of a metal is determined by using the weight of 1 in.^{3} of that metal. Specific gravity is the weight of a particular metal when it is compared to the weight of an equal volume of water; or, specific gravity is the ratio between the density of metal and the density of water. Knowing the density of a metal is important in the arc welding process because the welder must know and understand what will happen when two dissimilar metals are mixed. If one metal has a higher density than another, then the lighter, or the less dense, metal will possibly float upon the surface of the heavier metal, not mixing homogeneously.

Thermal conductivity is the rate at which heat is transmitted by conduction through a piece of metal. The amount of thermal conductivity in a metal determines the amount of heat needed in order to weld a particular metal. For example, copper has a high thermal conductivity, requiring that a large heat source be used in order to bring copper-based metals to a molten stage. The thermal conductivity of a metal is expressed as compared to silver, which has been given a rating of 1.00. Three theories have been developed to explain the phenomena of conductivity: the electron transfer theory, the molecular transfer theory, and the space lattice vibration theory. Thermal conductivity is always high in pure metals so that alloying a metal will lower the conductivity of that metal. Also, thermal conductivity is inversely proportional to temperature so that as the temperature of a metal increases, its thermal conductivity decreases.

The electrical conductivity of a metal is also affected by the temperature of the metal. The lower the temperature of a metal, the greater the electrical conductivity of the metal. Metals at extremely low temperatures, or metals at approximately 0°K, become superconductors, because the valence electrons flow much easier in cold metal. Electrons, once they are accelerated in a particular direction by an emf in a conductor have been found to flow constantly if no resistance to the flow is offered; it is a force that will stay in motion until acted upon by another force. This other force is called electrical resistivity. Electrical conduction is based on electron flow. As the metal atoms lose their valence electrons, the atoms change into positive metal ions. These metal ions interfere with electron flow and cause a decrease in electron motion, thus causing a resistance in the conductor. Although there is an electron flow and an ion resistance to an electron flow, there is no change in the mass in the base metal. The flow can be measured by the standard measurement for electrical resistivity which is measured in microhms per centimeter.

Knowing the coefficient of thermal expansion for a metal is extremely important for welders because this information indicates the amount of expansion or contraction to expect in a metal when heat is applied to it or extracted from it. The hotter the metal, the more it expands at a given rate. Also, metal contraction, when a metal is cooled or going through a cooling process, is indicated by the coefficient of thermal expansion. Many engineering books indicate a linear coefficient of thermal expansion which is more informative than a volume of thermal expansion. For example, aluminum has a thermal expansion of 13.1, which means that aluminum expands at the rate of 13.1 millionths of an inch for each inch in length for each degree in Fahrenheit above 68°F. Therefore, a 1-in. piece of aluminum raised to 70°F is 26 millionths of an inch longer at 70°F than it is at 68°F. The expansion and contraction of a metal can be determined, then, by the amount of heat applied to the metal. The thermal expansion of the metal is responsible for many of the locked-in stresses in the base metal after welding, stresses that result in distortion of the base metal. As the linear coefficient of thermal expansion increases, the more care a welder must take in welding that particular metal.

The corrosion resistance properties of metals have never been clearly defined or identified numerically; however, materials that are extremely susceptible to corrosion are the most active, such as magnesium, zinc, and aluminum. Those least susceptible are the least active, such as platinum. Ferrous metals corrode readily and nickel-based alloys have a higher corrosion resistance than silver, gold, and platinum. Platinum offers the most resistance to corrosion, or can be said to be the least active. All the noble metals are anodic and provide protection for cathodic metals, exhibiting a galvanic protection. If iron or other cathodic metal is submerged in a salt solution, for example, it must be protected from the solution in order to maintain itself, but noble metals do not need this protection.

MECHANICAL PROPERTIES The mechanical properties of a metal usually indicate its strength, hardness, ductility, or toughness. The mechanical properties of metal share a direct relationship to crystalline structure. As long as the metal has a polycrystalline structure, the mechanical properties of that particular metal are fairly reliable. However, if a metal is anistropic, the reliability of the mechanical properties of the metal depends upon the direction of the primary axis of the crystalline structures. On the other hand, metals that are isotropic in nature are those that do not depend upon a single crystal for strength, or the rotation of a single crystal for strength; but rely instead on a multi-axis polycrystalline structure. In most metals, there is no preferred orientation of the primary axis. Generally speaking, all transition metals are isotropic, even though individual grains within the structure may be anistropic.

STRENGTH Robert Hooke discovered that if the strains placed on a metal are strong enough, a simple relationship between the strains or the force applied to the metal and the corresponding stress within the metal exists. This stress-strain relationship is directly associated with the interatomic spacing of the atoms within metal (Figure 28–1). Hooke's law states that in an elastic material the stress is proportional to the strain. In other words, when a metal is placed under tension, it increases in length as, in a proportionate ratio, it decreases in width (Figure 28–1a). If the metal is elastic, the metal resumes its original shape when the tension is removed (Figure 28–1b). Also, the

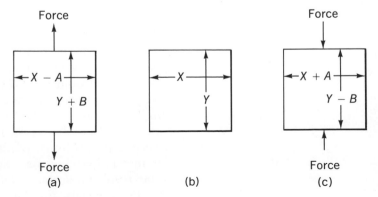

Fig. 28–1 Interatomic spacing—Hooke's law. (a) Under tension. (b) At rest. (c) Under compression.

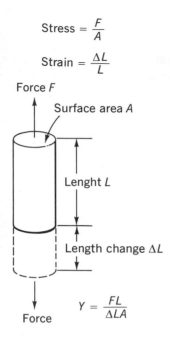

Stress = $\dfrac{F}{A}$

Strain = $\dfrac{\Delta L}{L}$

Force F

Surface area A

Lenght L

Length change ΔL

Force

$Y = \dfrac{FL}{\Delta L A}$

Fig. 28−2 Modulus of elasticity.

reverse is true, according to Hooke's observations. When a metal is compressed under a force, the length of the metal is reduced, and the width of the metal increases. Again, if the metal is elastic, it resumes its original shape.

Metals such as steels are said to be elastic because they return to their original size and shape when external stresses are removed. Materials that retain the deformation caused by the external forces are called plastic. Such metals are the alkali metals. For a certain range of stress, a metal may be elastic; but, if that range of force or limit of force is exceeded, the metal will become plastic. A relationship, known as Young's modulus or the modulus of elasticity, exists between stress or force exerted upon a metal and the associated strain. Young's modulus recognizes the following relationship: Y is equal to the tensile stresses divided by the longitudinal strain, or the modulus of elasticity is equal to the stress divided by the strain, where Y is the modulus of elasticity (Figure 28−2). Stress is defined as the force per unit area causing the strain, or force over area A is equal to stress. Strain is the deformation or elongation produced per unit length of the overall length, or stress is equal to the change in the delta L divided by L. The equation can then be rewritten for the modulus of elasticity as: Y is equal to force times length, divided by the change in length times the area.

The type of elasticity most often considered is that which deals with the change of length, either under a tensile loading or a compression loading. The coefficient of expansion or modulus is independent of the dimensions of the material used and will be the same for all materials. Therefore, a small sample of a metal will yield the same results as a large sample of that metal. To verify this, use the above formula and assume that a sample of metal has twice the length, represented by $2L$, and twice the cross-sectional area, $2A$. The same force F is used. The resultant stress would be F divided by $2A$ and the resulting strain would be the delta L or the change in L divided by $2L$. The 2s would divide out and the formula would still be the same. The stress and the strain are changed in the same proportions, yielding a constant for that particular metal. Tests can be made on a convenient sample and the results will show the same value, or the same modulus of elasticity, for that particular metal.

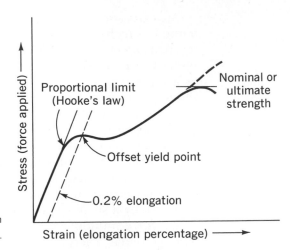

Fig. 28-3 Stress/strain diagram (Young's modulus).

Young's modulus refers to an engineering diagram. The major items that are indicated through the use of stress-strain diagrams are the proportional limit, or the elastic limit, for a particular metal; the offset yield point, which is the 2 percent elongation to the plastic range of the metal; the ultimate strength of the metal; and the breaking strength of that metal (Figure 28-3). Young's modulus of elasticity is expressed in pounds per square inch; or, in the metric system, it is expressed in dynes per square centimeter. At the same time that Young's modulus is established during the testing of a material, another ratio, called Poisson's ratio, becomes apparent. Poisson's ratio is reflected by the lateral growth of a metal as it is being subjected to applied stress. It is the ratio between the strain and the lateral growth of the test specimen. Poisson's ratio is indicated by the symbol *Mu*, and expresses the relationship between the elongation of the metal under tension, or *Y* plus some distance, *B*, and the reduction in area, or *X* minus *A* (Chart 28-3).

CHART 28-3 Elastic constant of selected metals

METAL	YOUNG'S MODULUS, Y OR E IN PSI	POISSON'S RATIO, MU
Aluminum	9,000,000	0.33
Copper	16,000,000	0.33
Iron	28,500,000	0.29
Tungsten	59,000,000	0.28
Steel	29,000,000	0.28

An extensometer is used for measuring the small extensions which occur when a metal rod is placed under tensile loads. One type of extensometer uses two pairs of oppositely pointed screws pressed into the test specimen at a known distance. A microscope mounted on a lever system measures the elongation of this known distance between the two sets of screws.

The extension, compression, or bending of a bar will accumulate potential energy due to the elastic properties and resultant molecular atomic action within the metal. If a bar has a length L and a cross-sectional area A and a force is exerted on the bar that either stretches it or compresses it by a small length, the potential energy is the average force multiplied by the displacement in length. The potential energy stored can be expressed by the equation:

$$\text{Potential energy} = \frac{(\text{force}^2)(\text{length})}{2(\text{area})(\text{Young's modulus})}$$

$$PE = \frac{F^2 L}{2AY}$$

An example of this storage of potential energy is a springboard depressed by a diver to the greatest bend of the board. As the diver springs downward, he loses potential energy and the board stores an equal amount of potential energy in order to comply with the law of conservation of energy and in order for the board to resume its original position.

The numerical value of Young's modulus for elastic material is large because it represents the forces that are needed to stretch a material to twice its original length. For example, for mild steel the modulus of elasticity is 29,000,000 lb/in.^{2}. The elastic limit is 35,000 psi and the breaking strength is 60,000 psi.

The modulus of elasticity is an important property of metal; however, the modulus of elasticity changes as the temperature of the metal increases. The stress-strain relationship decreases with increasing temperatures, a condition present in the welding situation. The elastic strain or the strain-elastic limit is greater when temperatures are higher, such as in the welding condition; however, the original stress-strain relationships are assumed once again when the metal cools.

HARDNESS   Hardness is defined in metals as a resistance to the penetration of the metallic surface. Hardness can be thought of as the resistance of a metal to scratching, denting, drilling, filing, or other deformation. The hardness of metal has been placed on arbitrary scales. The first hardness scale compared metals to organic materials and was called Moh's scale. Moh's scale of hardness is commonly used in determining the hardness of rocks and minerals. Hardness is recorded in terms of a scale in which no. 1 is talc, no. 2 is gypsum, no. 7 is quartz, no. 8 is topaz, no. 9 is sapphire, and no. 10 is diamond. The hardness of a metal may be changed by heat treating. For example, a chisel may be made out of a medium-carbon steel and then heat treated. In the heat-treated condition, the chisel can cut the very metal from which it was made. The amount of hardness in a metal also indicates its brittleness. Generally, as the hardness of a metal increases, the brittleness increases proportionately. The hardness of the metal, however, does not relate directly to the hardenability of the metal.

The hardenability of a metal is determined by a Jominey test to determine how well a metal hardens from the center of the metal to the interface of the metal. Many metals can be hardened only on the surface, while the interior remains unhardened. These metals would then be said to have low hardenability. A metal that is capable of being hardened throughout its structure is said to have high hardenability.

DUCTILITY   Ductile metals can be formed without failure, meaning without breaking or cracking. They can be drawn through a die, shaped by cold working, or have their shape changed without early failure in the metal. Ductility is determined by a standard stress-strain curve and is represented by the percentage of elongation of a metal coupled with its reduction in area. This information is derived from the standard tensile tests for metals. Usually, ductile metals have an fcc structure. Metals that do not have a great amount of ductility have bcc structures, and brittle metals have hcp or tetragonal space lattice systems.

MACHINABILITY The machinability of a metal is indicated by percentages. All machinable metals are compared to a basic standard, and the comparison yields a percentage rating which indicates the ease of cutting a particular metal. The standard metal used for the 100 percent machinability rating is a steel, coded by the American Iron and Steel Institute Index as B1112 steel. The Society of Automotive Engineers have coded this steel as no. 1112. Machinability ratings are expressed for all other metals as a percentage, which is based on a comparison to this free-cutting steel. For example, aluminum has a machinability rating that ranges between 300 to 2000 percent greater than free-cutting steel, and the rating for nickel steels ranges from 40 to 50 percent when compared with the free-cutting steel. Carbon steels generally range from 40 to 60 percent, and cast iron from 50 to 80 percent.

TOUGHNESS Toughness is an important mechanical property that indicates the resistance of a metal to impact and is sometimes called impact toughness. Impact toughness measures the amount of force that is required to fracture a test specimen completely with one rapid application of force or a sharp blow. The impact toughness of a metal is expressed in foot pound-force. The most common tool used for testing impact toughness is a hammer.

FATIGUE Metallic fatigue is perhaps the most important factor determined through the testing of the mechanical properties of a metal. Young's modulus, Hooke's law, standard stress-strain diagrams, and the tensile, bending, and compression tests yield the results that indicate when a particular metal will fail or pull apart. However, the metals that are in the services of man are exposed to more than one type of force and are subjected to reverse or alternating stresses. These stresses eventually cause failure, which is indicated

No force

Test specimen

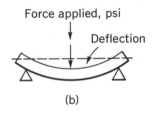

Supports

(a)

Force applied, psi

Deflection

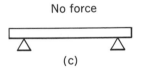

(b)

No force

(c)

Fig. 28–4 Fatigue cycle.

by cracks forming at stress levels. These levels would not be noticeable if force were applied only in one direction.

If a metal is subjected to completely reversed stresses of sufficient magnitude, a crack eventually appears that will grow until sudden failure of the cross section occurs. Physically, a minute crack originates at a point of high stress. This point may be a surface scratch, an oxident occlusion, or some other imperfection in the metal. The crack enlarges during continued repetitive loading until there is insufficient material left to support the load. Fatigue failures are generally determined by subjecting a metal to a given stress or force that is repeatedly applied to a particular portion of the metal (Figure 28–4). The basic fatigue cycle includes supporting a test specimen from one end or two, and applying a force in pounds per square inch to a location that will deflect the test specimen. Then, the force is released and the test specimen is allowed to return to its original position. Fatigue cycle testing is done within the elastic limit of a metal. If enough force is applied to overcome the elastic limit of the metal, permanent deflection will result. The fatigue cycles range from 2 or 3 cycles for a brittle metal to as high as 20,000,000 or 30,000,000 cycles for an extremely elastic metal. The number of cycles until ultimate failure occurs is referred to as the endurance limit of the metal.

Most of the mechanical properties of a metal are directly concerned with the mechanism of failure or the mechanism of slip. The mechanism of failure or slip occurs as a result of two possibilities. If a force applied to a metal is within the elastic range, the crystals or dendrites will not be affected by the force; therefore, the crystals

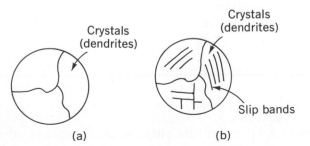

Fig. 28–5 Mechanism of slip. (a) Elastic range. (b) Plastic range.

Crystals (dendrites)

Crystals (dendrites)

Slip bands

(a)   (b)

will maintain their original shape and form. However, if enough force is applied, the crystals can no longer maintain their form and are said to be plastic (Figure 28-5). The mechanism of slip and the ultimate failure of a metal are based on what happens within each individual crystal and the location of each crystal relative to the next crystal. When the force is applied, the primary and secondary axes will be disrupted or broken, or the crystal may move in relationship to the next crystal. When a crystal moves, it creates a small void between the crystals, and fragments a small portion of each adjoining crystal. These fragments are called crystallites. When crystallites are formed, slip bands, which indicate the disruption of the primary and secondary axes of a crystal, can be observed by microscopic examination. These slip bands are indicators that a metal has been permanently deformed and that it is in the first stage of failure. The formation of crystallites is the second stage of failure; and the last stage of failure is the ultimate separation of the metal. An example of the formation of crystallites can be observed by twisting a common paper clip until it fails. First to appear are minute ridges, resulting from the twisting motion, which are minute dislocations and deformations of the metal. As the twisting of the metal continues, the formations of these voids become larger and larger until the ultimate failure of the paper clip. The mechanism of slip moves atoms through complete interatomic spacings, which distorts the lattice orientation of the metal slightly. Another mechanism of failure is twinning. Twinning occurs when the atoms are not moved one or more complete spaces in atomic bonding but are moved just a portion of the way. The effects of slipping can be removed through a metallurgical polishing of the specimen, while the mechanism of twinning cannot be removed by polishing the surface of the metal (Figure 28-6).

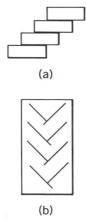

(a)

(b)

Fig. 28-6 Slipping and twinning. (a) Slipping. (b) Twinning.

QUESTIONS

1. Into what three categories can the properties of metals be classified? What do the categories indicate about the properties of a metal?

2. Which of the three is the most important in the study and practice of welding? Why?

3. What is an atomic number?

4. Why is the knowledge of the coefficient of thermal expansion necessary to the welder?

5. What does Fe stand for? How are the symbolic notations for metals determined?

6. What is volumetric shrinkage?

7. What is a valence electron?

8. What is Hooke's law? Young's modulus? Poisson's ratio? Of what importance is a knowledge of these concepts to a welder?

9. How is the toughness of a metal measured?

10. What is a fatigue cycle?

11. What is the difference between slipping and twinning?

SUGGESTED FURTHER READINGS

Almen, John O., and Paul H. Black: "Residual Stresses and Fatigue in Metals," McGraw-Hill Book Company, New York, 1963.

Lindsay, Robert Bruce: "Theoretical Physics," pp. 311–312, D. Van Nostrand Reinhold Company, New York, 1954.

Lubahn, J. D., and R. P. Frelgar: "Plasticity and Creep of Metals," John Wiley & Sons, Inc., New York, 1961.

Mendenhall, C. E., et al.: "College Physics," D. C. Heath & Company, Lexington, Mass., 1956.

Nadai, A.: "Theory of Flow and Fracture of Solids," vol. 1, McGraw-Hill Book Company, New York, 1950.

Shanley, F. R.: "Mechanics of Materials," McGraw-Hill Book Company, New York, 1967.

Stanford, E. G., et al. (eds.): "Progress in Applied Materials Research," Heywood Books, London, 1967.

Westergaard, Harold M.: "Theory of Elasticity and Plasticity," John Wiley & Sons, Inc., New York, 1952.

Wulff, John, et al. (eds.): "Mechanical Behavior," John Wiley & Sons, Inc., New York, 1965.

29 HEAT TREATING

Heat treating is the heating and cooling of a metal in its solid state under controlled conditions in order to improve its mechanical properties. The purposes of heat treating are manifold. For example, heat treating is used to increase hardness and strength; to remove internal stresses; to increase ductility, toughness, and softness; to refine the grain structure; and to obtain a variety of other physical and mechanical properties.

Heat treating can be grouped under two main headings: surface hardening and true hardening. The surface hardening of metal may be thought of loosely as changing only the exterior or the properties of the immediate surface of the metal. In true heat treating, or the interior heat treating of metal, the properties of the metal have been altered completely throughout the metal.

SURFACE HARDENING   Surface hardening is sometimes called case hardening because it forms a hardened case around the softer, underlying metal. The major types of surface hardening or case hardening are carburizing, cyaniding, nitriding, chromizing, siliconizing, induction hardening, and flame hardening.

Many steel objects require a hard outer surface to resist wear and a tough inner core to absorb shocks, such as in a hammer or a crank shaft. The treatment used to obtain these conditions is generally classified as surface hardening. Surface hardening may be accomplished by two different methods. First, the composition of the outer surface in a low-carbon steel can be changed, a process called case hardening; and second, the surface of a medium- or high-carbon steel can be heat treated, as in flame hardening.

Carburizing   The oldest known method of producing a hard surface on a low-carbon steel is carburizing. In carburizing, a steel of a low carbon content, usually below 0.25 percent, is heated to a red heat while in contact with some carbon material, which can be in the form of a solid, a liquid, or a gas. Iron has an attraction for carbon at a temperature close to and above its transformation temperature. As iron changes from a bcc to an fcc structure, carbon is allowed to precipitate through the iron because of the fcc arrangement. The carbon enters the surface of the metal to form a solid solution with the iron, which converts the outer surface to a high-carbon steel. As the operation continues, the carbon is gradually diffused into the interior of the part. The depth of the case depends upon the length of time and the temperature used in the treatment. The depth of the case also depends upon the type of carbon rich material in which the object is suspended for heating (Figure 29-1).

Carbon-rich material

Object
to be case hardened

Fig. 29-1   Carburizing.

For better control and to produce almost any desired hardness depth or carbon content in the case, gas carburizing is used. In this process, the metal is heated to approximately 100° above its upper transformation temperature in an atmosphere rich in one of the hydrocarbon gases, such as carbon monoxide, natural gas, propane, or methane. The steel used in this process is usually low-carbon steel, approximately 0.15 percent carbon. In the course of the gas carburizing process, the outside layer of metal absorbs some of the carbon in the gases and is converted into a high-carbon steel with a carbon content that ranges from 0.90 to 1.20 points of carbon. If the metal receives proper heat treatment, it will obtain an extremely hard surface with a soft, ductile, and tough center core. The case thickness achieved when using the gas carburizing process ranges from 0.01 to 0.06 in. in depth for treatments that last from 1 to 8 hr.

Since there is some grain growth in the steel during the carburizing treatment, the metal should be heated until the core of the metal is at its transformation temperature and then cooled; thereby refining the core structure. The steel should then be reheated to a point above the transformation range and quenched rapidly to produce a hard, fine structure.

Nitriding Nitriding uses the nitrogen in ammonia gas as a hardening agent. The part to be nitrided is placed in a special furnace and is heated to a temperature between 900 and 1000°F and held there for a period of time. The amount of time varies between 10 to 72 hr. A nitrided case that averages a thickness between 0.012 and 0.018 in. usually requires between 35 and 72 hr in the furnace. The furnace has a special atmosphere of ammonia gas. The nitrogen from the ammonia combines with the steel to form an extremely hard case that can range between 0.005 and 0.02 in. in thickness.

Nitrogen has a greater hardening ability with certain alloys and ferrous metals. Special alloy steels have been developed for nitriding. For example, aluminum, in amounts ranging from 1 to 1½ percent of the alloy, has proven especially suitable in steel because it combines with the ammonia gas to form a stable, hard compound. The carbon content of the base metal is usually less than 0.35 percent. Other alloying elements that are used are aluminum plus chromium or molybdenum. However, these special nitriding alloys are more expensive.

Nitriding, unlike carburizing, does not require quenching to obtain hardness. Nitriding also does not require as high a temperature as does carburizing. Because of these two factors, it does not distort, crack, or change any condition of the metal, and because of the relatively low temperatures used, there is little or no loss in core toughness. For these reasons, and because nitriding is a finishing process, the part can be machined to an exact size before it is nitrided and placed into immediate service after the treatment. Generally, dies for die casting and die-casting injection chambers for metal die casting machines are nitrided. Of all the gas hardening processes, nitriding also results in the hardest cases; however, the disadvantages of the process are that the aluminum-alloy steels are expensive and the time for the nitriding treatment is longer than for the carburizing procedures.

Cyaniding Cyaniding is a case-hardening process that utilizes the absorption of carbon and nitrogen. The part to be cyanide case hardened is immersed in a bath of molten sodium cyanide, which is a salt containing both carbon and nitrogen, at a temperature of approximately 1550°F. A cyaniding treatment, lasting from 30 to 45 min., yields a 2 percent carbon content increase of the case with a case thickness of about 0.005 in. Cyaniding is useful when a thin hard case, between 0.001 and 0.015 in., is required in a short period of time. The cyanide casing operation is the fastest type of casing that can be applied to a mild steel core; however, the disadvantages are that the cyanide salts are extremely toxic and the process is messy. Special protective equipment must be worn by personnel operating in the vicinity of the cyanide salts because of its extremely poisonous nature. Parts also are limited to a size that can be handled by one or two men.

Chromizing Chromizing differs from the more standard case hardening procedures in that a chromium carbide is diffused into the ferrous metal, converting the surface of the metal into a stainless steel case. This stainless steel case has a high hardness and a low coefficient of friction. Chromizing is applied to hydraulic rams, pump shafts, die tools, and dies, such as drop-forging dies. Chromizing is used to improve the corrosion resistance and the heat resistance of a metal. Chromizing is not limited to ferrous metals but may be applied to many nonferrous metals, such as cobalt, molybdenum, nickel, and tungsten.

The major use for chromizing, however, is in the case hardening of ferrous metals that contain above 0.60 percent, or 60 points, of carbon. The steel must contain this percentage of carbon in order for the chromium carbides to precipitate through the carbon. The weld resistance of the metal will be increased because this process turns the surface of the metal into a stainless steel. The process is usually done at an elevated temperature ranging from 1650 to 2000°F. The application of the chromium at this temperature resembles to some extent the application of the ammonia gas in nitriding. The chromium is in a chromium-rich gas that is exposed to the metal while the metal is at an elevated temperature. In the nitriding process, the ammonia breaks down into nitrogen, but in chromizing, the chromium-rich gas is injected onto the surface of medium- or high-carbon steel changing it through diffusion to a stainless, case-hardened steel.

Ihrigizing Ihrigizing, or siliconizing, is the diffusion of silicon into the surface of a ferrous-based metal. This silicon creates a case that ranges in depth from 0.005 to 0.1 in. thick, depending upon the carbon content of the ferrous material and upon the length of time that the material is in contact with a silicon carbide and chlorine gas mixture. Siliconizing, or ihrigizing, lends itself to automation readily. A common practice when siliconizing is to place the silicon carbide and chlorine gas in a liquid mixture and to bring this mixture to a temperature between 1700 and 1850°F. The mixture is contained in a tank with appropriate heaters to maintain this temperature. The items to be case hardened are placed on a conveyor track that simply runs through this tank of liquid silicon, carbide, and chlorine gas. The depth of the case of the part is controlled by the length of time the part is in the hot solution (Figure 29–2). Often this conveyor track, prior to ihrigizing, goes through a cleaning tank which cleanses the part before the case is applied. The conveyor also can go through a rinse tank that will cleanse the part again after the case has been applied. A further change in the basic system can be made by using two tanks. One tank is filled with silicon carbide floating in a liquid and the other tank, which is heated, contains chlorine.

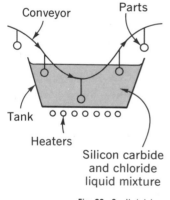

Fig. 29–2 Ihrigizing.

The part is moved from one tank to the next. Silicon cases range in hardness from 80 to 85 on the Rockwell B scale. Even though the silicon cases are not too hard, they are difficult to machine because of their extremely low coefficient of friction. Silicon cases have high non-galling properties, making the case useful for mated moving parts, such as valve guides, cylinder liners, and pump shafts.

Flame hardening   This is a surface hardening method that uses an oxyacetylene flame to heat treat the surface of the metal. Flame hardening can be performed on only medium- or high-carbon steel. When flame hardening is applied to steels with over 70 points of carbon, extra care must be used in order to prevent surface cracking of the high-carbon steel. Flame hardening is based on the rapid heating of the outer surface of a ferrous metal to or above its transformation temperature. The minimum distance that the oxyacetylene flame must be held from the base metal is approximately $9/16$ in. If the oxyacetylene flame is closer than $9/16$ in., the base metal will be deformed by the flame. The metal is moved rapidly under the flame, allowing the flame to heat the base metal only on the surface. This surface heating creates two heat-affected zones, a primary heat-affected zone, where the transformation of the metal has taken place; and a secondary heat zone, where grain growth has been developed, or where the grains have been enlarged or decreased by the application of heat to the base metal. Immediately after it is heated, the metal is subjected to a quenching spray, generally water, that hardens the metal area that has undergone transformation. The depth of hardness depends entirely upon the hardenability of the material being treated since no other elements are being added or diffused, as in case hardening. By proper control, the interior of the metal is not affected by this process. Often an average application will be the heat treating of the complete piece to a certain specified softness or toughness. The exterior then may be flame hardened so that the finished piece will resemble an item that has been case hardened (Figure 29–3). A further adaptation of the flame harden-

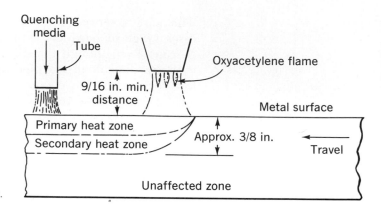

Fig. 29–3 Flame-hardening effects.

ing process is flame spin hardening in which the object to be flame hardened is rotated at a high speed and the flame is allowed to play over the complete piece, hardening the complete outer layer of the piece. The length of time that the flame is allowed to play over the ferrous metal is usually around 10 sec.

Induction hardening This method is similar to flame hardening, with the exception that the heat is generated in the metal by an induced electrical alternating current. The only metals that can be induction hardened are those that are conductors or semiconductors. The ACHF obtained from a pulsating magnetic field about a wire produces the heat in induction hardening. The heat results from molecular agitation induced by the electricity. The high-frequency, low-voltage, high-amperage current produces a great number of eddy currents which are primarily responsible for the heating of the metal, although hysteresis is another source of heating (Figure 29–4).

An inductor block, similar to a primary coil in a transformer, is placed around the part to be hardened. This coil does not touch the metal. A high-frequency current is passed through the block or the coil and induces a sympathetic current in the surface of the metal, which creates heat, a process called hysteresis. As the temperature of the metal reaches the transformation range, the power is turned off, the heat source is removed, and the area is quenched, usually by a spray. The important aspect of induction heating is its rapid action. For example, it

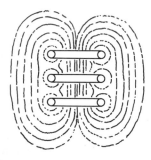

Fig. 29–4 Induction heating-magnetic lines of force (flux).

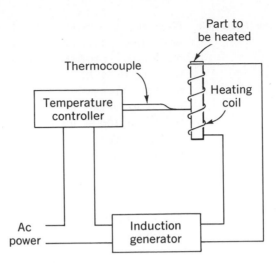

Fig. 29-5 Induction heating.

Fig. 29-6 Induction coil design. *Note:* Coils induce current only into their immediate area and the closer they are to each other, the more concentrated is the induced current.

requires only a few seconds to heat steel to a depth of $\frac{1}{8}$ in. (Figures 29-5 and 29-6).

The heat produced by induction is the result of both current and frequency. Higher currents produce stronger magnetic fields, while higher frequencies produce more pulsations of the field within a given time. A given degree of heating can be obtained either by using high current at low frequencies or low current at high frequencies. However, high-frequency induction heating of metals requires a device that can convert 60-Hz power to a high frequency of several hundred or more cycles per second. The frequency most used for metal treating applications is 450 kHz/sec.

When induction heating had its origin, frequencies from 1000 to 2000 Hz were produced by motor generators. Frequencies from 10 to hundreds of cycles can now be obtained by spark-gap oscillators. The more reliable radio-frequency oscillators have replaced the more troublesome spark-gap oscillators, and now there is a vacuum-tube oscillator that is capable of producing 450,000 or more cycles per second. Recently, special transformer arrangements, called frequency multipliers, have successfully produced frequencies of 180 to 540 Hz. Because of different designs, and because of the availability of a wide range of high-frequency power, induction can be used for brazing, welding, soldering, heat treating, and many other material-processing methods.

HEAT TREATING   Heat treating gives a metal tremendous advantages as a structural material because heat treating changes the mechanical properties of metal. The purpose of heat treating is to harden or soften a metal. The first step toward either softening, also called annealing, or hardening, is to elevate the temperature of the metal, generally a ferrous metal, to a transformation state or the point at which the atoms in the structure either transform the structure from one form to another, such as in allotropism, or the point at which the crystals in the solution begin to realign themselves. Almost all heat treating is done with ferrous metals and most of the heat treating is done with medium-carbon, high-carbon, low-alloy, or high-alloy steel. Medium-carbon, high-carbon, and low-alloy steels are all heat treated in the same manner, but a different heat treating method is used with high-alloy steels.

The reason for softening or annealing a metal is to increase its machinability rating so that such operations as rolling, drawing, and forging can be done more easily. Annealing itself is no more than heating a metal until it is above the transformation range and then cooling the metal slowly to stop hard types of crystalline structures from forming in the metal. Heat treating for hardening metals involves heating a metal to a high temperature and then cooling it at a faster rate than when annealing. The comparatively faster cooling allows a different type of crystalline structure to be formed in the metal, which is hard and many times internally strained because of the crystal types. All heat treating, whether it softens or hardens metal, affects the microstructure of the metal. The types of crystals that are formed as a result of the cooling or quenching of the metal determine the hardness or the softness of the metal. The study of the behavior of steel at different stages of heat treating is accomplished by a specific diagram, called variously an S curve; a time-temperature transformation diagram (TTT); an isothermal transformation diagram; or, more precisely, a graph of isothermal transformation on cooling (IT) (Figure 29–7). Isothermal transformation diagrams are derived by cutting large numbers of small samples from a bar of steel. These pieces have homogeneous structures. These small samples are individually hung on wires and placed

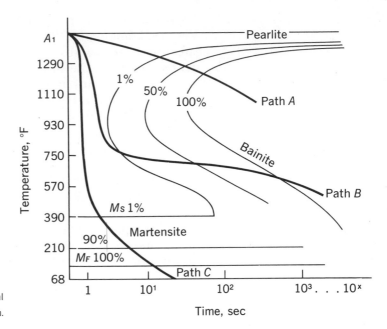

Fig. 29–7 Subcritical isothermal transformation on cooling diagram.

in a furnace where they are heated to the proper temperature for complete austenitizing, or until they have gone completely through transformation. After all the samples have gone through transformation, they are placed immediately into a second furnace, or a holding furnace, which has been preheated to a set temperature, which is below the transformation temperature. These samples are then removed individually and cooled in a progressive cycle in an ice-brine or ice-water quenching medium. The cooling is controlled precisely and the length of time from their removal from the furnace and their introduction into the quenching medium is controlled to a geometric progression of time, such as 1 sec, 5 sec, 10 sec, 20 sec, 40 sec, 80 sec, and so on. Each sample is then tested for hardness and also checked microscopically for the type of crystalline structures that are present. From this information the isothermal diagram is derived. Therefore, each steel will have one IT diagram. With this information, then, the results of any heat-treating operation can be predicted with a given steel of known transformation qualities.

All steels, whether they are to be softened or hardened, should be brought into or above the transformation range of that metal. After this step has been done, the hardening or the softening of the metal is accomplished by the amount of heat used and the rate that the heat is extracted from the metal. The subcritical isothermal transformation on cooling diagram indicates how quickly heat should be withdrawn (Figure 29-7). If the heat is withdrawn slowly, pearlite is formed. Pearlite is the softest crystalline structure found in ferrous metal. If the heat is withdrawn from the metal rapidly, then the hardest crystalline structure is formed, which is martensite. A knowledge of each crystalline type is mandatory in order to determine what heat treating should be done and what method of heat treating should be used.

Metal, upon reaching an austenitic temperature, transforms into austenite with three major grain types; pearlite, bainite, or martensite. The development of these three crystalline types depends upon the temperature loss imposed on the austenite in the ferrous metal. The first decomposing of austenite yields a pearlite structure. Pearlite is the softest crystalline form in ferrous metal and is formed by retarding the cooling rate significantly. If the cooling rate were accelerated, bainite structure would be formed from the decomposing austenite. If the austenite were quenched as quickly as possible, the austenite would completely decompose into martensite, the hardest form of ferrous metal. All of the groups are comprised of a mixture of cementite and ferrite. Cementite is a compound of iron and carbon that contains 6.68 percent carbon and 93.32 percent iron. The amount of cementite separating the ferrite determines the hardness and strength of the crystal, which in turn determines the strength of the complete structure. Cementite and ferrite appear in pearlite as plates (Figure 29-8). The cementite plates are extremely hard; ferrite plates are soft. A number of these plates make up one crystal. The formation of lamellar pearlite can be coarse, medium, or fine (Figure 29-9). The coarser the pearlite, the softer the pearlite. Fine pearlite is formed close to the temperature at which upper or feathery bainite is formed. Pearlite is formed by heating a steel to an austenitic temperature and then quenching it, approximately along path *A* in Figure 29-7. Feather bainite results at that point at which the cementite disperses. This feathery bainite consists of the broken-up cementite plates which

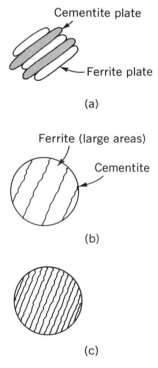

Cementite plate

Ferrite plate

(a)

Ferrite (large areas)

Cementite

(b)

(c)

Fig. 29-8 Pearlite. (a) Lamellar pearlite. (b) Coarse lamellar pearlite. (c) Fine lammelar pearlite.

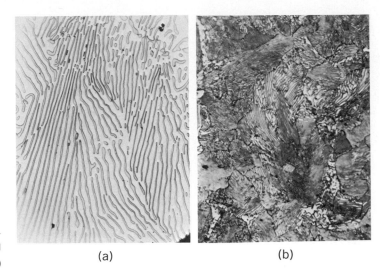

Fig. 29-9 Pearlite. (a) Coarse—2500×. (b) Fine—2500×. (Courtesy of U.S. Steel Corporation.)

(a)

(b)

have assumed a new axis and are no longer lamellar in structure. Bainite is created in a metal by decomposing the austenite at a fixed rate. The nose, or the knee, of the IT diagram is ignored and the austenite is cooled approximately on path *B* in Figure 29-7. Whether path *B* is close to either the pearlite ranges or the martensite ranges determines whether or not feathery bainite or acicular bainite is formed from the austenite (Figure 29-10).

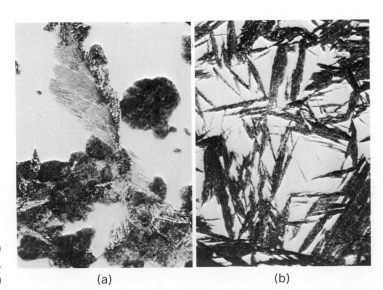

Fig. 29-10 Bainite. (a) Feathery (upper) —2500×. (b) Acicular (lower)—2500×. (Courtesy of U.S. Steel Corporation.)

(a)

(b)

Acicular bainite occurs when the cementite plates that have been broken up in the upper bainite ranges have been reduced in size but increased in strength to produce a stronger structure. The cementite plates are now apparent as small, needlelike plates. The cementite can be thought of as having fine grains. These fine grains are the result of the precipitation of the cementite from <u>austenite</u> by diffusion, the physical movement of the atoms. The cementite requires a concentration of carbon atoms to form the cementite lamina, and time is the basic requirement for forming this concentration of carbon atoms.

Martensite, the hardest substance found in a hardened steel, is formed by extracting the heat from a ferrous metal at an extremely fast rate (path C in Figure 29-7). Martensite begins forming at an M_s line, or at the start of the martensite line. Approximately 1 percent of martensite will be formed at this temperature, which is approximately 390°F. If the temperature of the steel continues to fall at a fast rate, it is possible to form a structure of nearly 100 percent martensite, which will be very brittle (Figure 29-11). One hundred percent martensite is difficult to obtain because there is usually an amount of retained austenite that remains untransformed. It will transform eventually, but uncontrollably, into untempered martensite. The martensite left in this condition will cause the ferrous metal to crack because of the extreme stresses

Fig. 29-11 Martensite. (a) 1 percent —500×. (b) 90 percent—500×. (Courtesy of U.S. Steel Corporation.)

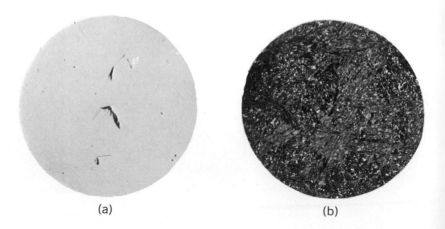

(a)   (b)

set up by the martensite crystals. The martensite crystals will have enough energy so that the standard bcc system is distorted and internally distressed. This stress must be relieved if the martensite is to be maintained.

Pearlite is formed at about 930°F (Figure 29–7). Bainite results from the decomposing of austenite between the temperatures of 930° and 390°F. Martensite is decomposed from austenite between 390°F and ambient, or room, temperature.

QUENCHING Cooling has the effect of freezing the mixture of austenite, cementite, and ferrite. The resulting structure depends upon the rate of temperature loss of the metal. If austenite is allowed to cool slowly, the grain formation will be uniform and the austenite will decompose into cementite and ferrite layers, called pearlite. If the austenite is quenched rapidly, the austenite will not have a chance to separate into other components and the iron carbide, or cementite, will be dispersed at a different level. The rate at which metal is cooled is dependent upon the type of coolant used, and whether the coolant is agitated or not and heated or not.

The type of coolant used and the temperature of the coolant determine the rate of temperature loss. It is not always advisable to cool a piece of metal rapidly since the resulting large temperature differences between the internal and external areas of the metal set up stresses which cause cracks to appear at the surface. An example of this situation would be cooling a large piece of metal in water at 70°F. Water cools the surface of the metal rapidly while the internal area is just as hot as before quenching. When a medium-carbon or a high-carbon steel is heated to its austenitic temperature and cooled quickly in cool water, martensite forms immediately upon cooling on the exteriors of the metal. The interior of the metal, however, is still austenite. When martensite is derived from the decomposing of austenite, the martensite expands. However, when the exterior of the metal has already had a hard surface of martensite formed completely around it, there is no room for the decomposing of austenite into ferrite and for the core of the metal to expand. Therefore, stresses are set up and cracks result.

The major media used for quenching are water, brine, oil, air, and metal. Water is generally used at room temperature. Brine is a solution of salt and water. Brine has the fastest cooling rate of all because the salt in the water disperses heat faster than pure water. Oil and low-temperature metal in a liquid state are used for types of martensitic tempering; and air, either still air or forced air, is used for cooling many metals.

Oil has a more even temperature gradient than that of water. The more the quenching medium can control the temperature loss of the metal, the more efficient it is. The major quenchers used commercially are oil, liquid, molten salts, and molten metals.

The use of quenchants is dependent upon the hardenability of the steel. The maximum hardenability of steel and the attainable level of hardness of steel are two different things. Steel can reach maximum hardness only if the proper temperature, proper time held at that temperature, and the proper rate of cooling are used. To some extent, the maximum hardness is determined by the chemical structure of the metal and the type of coolant used. A steel with low hardenability will not need a coolant that cools uniformly. Water, a rapid coolant, could be used for this type of steel. Oil, on the other hand, is used on metals with a high hardenability rating because the steel must be cooled at an even rate. Also, molten salt and molten low-temperature metals are used with metals that have a high hardenability rating.

Several recent quenching developments like new cycling and new quenchants will soon have widespread applications in the heat-treating industries. A few years ago, an aerospace firm developed what is now called cryogenic quenching. Cryogenic quenching is a process by which a liquified gas serves as the quenchant. The process is capable of producing sheet metal parts with physical properties comparable to water-quenched parts but without any distortion. In cryogenic quenching, liquid nitrogen replaces the usual water-quenching agent. The heat transfer takes place by radiation and conduction through the nitrogen vapor blankets, resulting in a more uniform removal of heat which reduces distortion.

Another relatively new method of quenching alleviates

distortion in thin parts. This method uses a solution containing a 40 percent polyalkylenelycol and 60 percent water. This solution reacts fast enough to impart a good corrosion resistance to heat-treated parts, even to nonferrous parts such as aluminum, and reduces distortion, as does cryogenic quenching.

Cracking and warping of steel during the quenching operation can be avoided by a method of interrupted quenching. In this operation, the steel is heated to the transformation range and is then cooled to below 1000°F but above 400°F. The traditional heat-treating operation is then performed on the metal. The final heat treatment in interrupted quenching depends upon the type of steel and the mechanical properties that are desired.

ANNEALING Annealing is the process of softening the metal for further working. Annealing is accomplished by heating the metal to approximately 50°F above the upper critical temperature for hypoeutectoid steel and approximately 50°F above the lower critical temperature for hypereutectic steels. The metal is held at these heating zones until the carbide disperses in the solid solution and austenite is completely formed. The metal is then cooled slowly; usually the temperature loss is between 15 and 30°F/hr until the metal is at room temperature. When ferrous metal is allowed to cool down to room temperature in the furnace at this extremely slow rate, the process is called full annealing. However, the process can be speeded up by using isothermal annealing. Isothermal annealing occurs when the steel is allowed to cool down so that either medium or coarse pearlite is formed. The steel is held at this temperature for an appropriate length of time so that coarse or medium pearlite is formed completely throughout the volume of the metal, and then it is cooled to room temperature in still air. The microstructure of fully annealed hypereutectic steel consists of coarse lamellar pearlite with a network of cementite surround it. This cementite which surrounds the pearlite in a hypereutectic fully annealed steel is brittle, generating planes of weakness. This network or coring of cementite also creates a thick, hard grain boundary which causes poor machinability. For these reasons, hypereutectic steels are annealed by using iso-thermal annealing so that cementite coring of the grains does not occur.

Normalizing, which is another type of annealing, consists of heating the steel to approximately 100°F above the critical temperature of the steel and holding the steel at this transformation temperature until transformation is complete. The steel is removed from the furnace and allowed to cool in still room-temperature air. Normalizing is employed after steel has been cold worked or has been unevenly heated or cooled, as when it is cast or welded. Normalizing removes the built-up stresses that are present in the metal as a result of these operations, giving the metal ductility and toughness. Normalizing results in the production of a finer pearlitic structure than is obtained by full annealing, and it results in a steel that has greater hardness and strength than is obtained by full annealing.

Spheroidizing, which is also incorporated in the annealing processes, is used to improve the machinability of steel. This process is usually applied to high-carbon steels with 0.75 percent or greater amounts of carbon. Spheroidizing produces the softest, most ductile, hypereutectoid steels possible but at the expense of strength. In spheroidizing, the steel is heated for a prolonged time below its transformation range, usually from 1250°F to the lower transformation range of 1333°F, until the carbon-containing cementite gathers into globular form. This transformation of cementite from a plate or needlelike structure into a round ball shape makes it easier for a cutting tool to shear through it.

Stress-relief annealing is also a process that is done below the subcritical annealing temperatures. Stress annealing is similar to spheroidizing with the exception that the temperature used ranges from 1000 to 1200°F. Stress relieving simply unlocks the built-in stresses within the grain structure, removing the residual stresses caused by welding, machining, or some other form of working the metal. The stresses are simply dissipated by the extremely slow temperature loss of the metal. Stress-relieving, or process annealing, is accomplished by heating the metal in a furnace to an appropriate temperature and then turning the furnace off, leaving the metal in the furnace to cool off at a slow rate until the metal reaches room temperature.

HARDENING AND TEMPERING   The process of hardening is the process of reheating the metal to a set temperature, which is slightly above the transformation range, allowing austenite to form. The metal is then cooled at a controlled rate to form one of the three major types of crystalline structures; either pearlite, bainite, or martensite. The formation of pearlite is a soft structure, generally known as annealed structure. Hardening, on the other hand, involves the bainite and martensite structures. The three basic hardening procedures for the development of martensite, tempered martensite, and bainite are conventional hardening and tempering (Figure 29-12), martempering (Figure 29-13), and austempering (Figure 29-14). Austempering is sometimes known as bainite tempering because bainite is formed in the metal.

Conventional hardening and martemperature hardening are related, with the exception that the quenching medium is controlled more closely in martempering than

Fig. 29-12 Conventional hardening and cooling diagram (IT diagram).

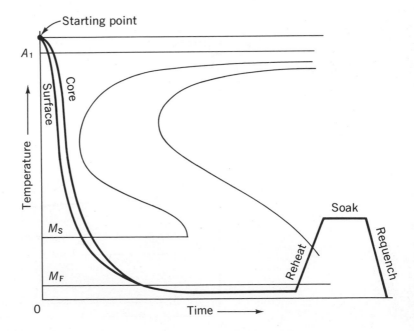

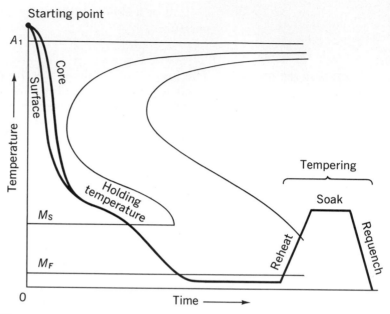

Fig. 29–13 Martempering.

Fig. 29–14 Austempering (bainite tempering).

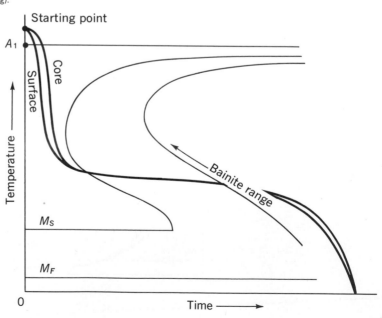

in conventional hardening and tempering. If the ferrous metal is brought up to an austenitic temperature and cooled in a cold-water solution, the difference between core temperature and surface temperature creates heat-treating cracks in the metal, and there is a high probability of distortion in the metal to be treated. Marquenching or martempering is one means that stops this distortion and keeps the heat-treating cracks from forming. Martempering is similar to conventional hardening, except that in quenching the temperature of the metal is reduced to approximately 400°F, or slightly above the beginning of the martensite line, the M_s Line. This temperature is held until the core temperature and surface temperature are the same, and then the metal is cooled to room temperature. Martempering of steel involves quenching the steel, which is at the austenitizing temperature, in hot oil, molten salt, or molten metal that is at a temperature at the upper part of, or above, the M_s temperature range.

The steel is kept in the quenching medium until the temperature throughout the steel is uniform, and then it is cooled at a moderate rate, usually in air. The formation of martensite occurs fairly uniformly throughout the workpiece during the cooling in the room-temperature air, avoiding the formation of excessive amounts of residual stresses. Martempering is known also as marquenching, but it is not a replacement for the tempering process, just for the hardening process. After the martempered parts have cooled to room temperature, they are tempered as though they had been conventionally quenched. Because the final phase of cooling during martempering is relatively slow, light and heavy sections transform from the surface to the center in about the same time, thus minimizing or eliminating the distortion that results from unequal transformation rates present in the conventional hardening of metal.

The success of martempering depends upon close control of variables in the entire process. It is important that the original structure of the material being austenitized be uniform. Also, the use of a protective atmosphere in austenitizing is required because oxide or scale will act as a barrier to uniform quenching in hot oil, salt, or metal. The process variables that must be controlled include the austenitizing temperature, the temperature of the martempering bath, the time in the martempering bath, the salt contamination of the bath, and the agitation and rate of cooling of the martempering bath.

In most instances, austenitizing temperatures for martempering will be the same as those used for conventional hardening in oil. Occasionally, however, medium-carbon steels are austenitized at higher temperatures prior to martempering to increase the marquenched hardness. The austenitizing temperature is important because it controls the grain size and the amount of cementite in solution; and it does have some effect upon the beginning martensite temperature. The main factor in determining the M_s temperature, or the temperature of the martempering bath, is the alloying contents of the metal. When the steel is cooled quickly from the austenitic state, the carbon simply does not have time to disperse. It is literally frozen in or near its original position in the austenite, preventing the segregation that is required to make cementite or pearlite, resulting instead in martensite. The formulas that have been developed to calculate the M_s temperature for various alloys are as follows: The M_s temperature in Fahrenheit is equal to 1000 minus 650 times the percentage of carbon; minus 70 times the percentage of manganese; minus 35 times the percentage of nickel; minus 70 times the percentage of chromium; and minus 50 times the percentage of molybdenum in the steel.

Time in the martempering bath depends upon the thickness of the metal, the temperature, the quenching medium, and the degree of agitation in the quenching medium. Because the object of martempering is to develop martensitic structure with low thermal and transformational stresses, there is no need to hold the steel in the martempering bath for extended periods of time. Excessive holding time lowers the final hardness because it permits transformation to structures other than martensite. In addition, stabilization of austenite may occur in moderately alloyed steels that are held for an extended period of time at the martempering temperature. The martempering time for temperature equalization is about 4 to 5 times the time required for quenching in anhydrous salt in the same temperature.

The agitation of the martempering salt or oil comsiderably increases the hardness obtainable for a given sectional thickness in comparison to that hardness obtainable in still quenching. In some instances, the rapid cooling produced by the most vigorous agitation increases distortion; thus mild agitation is often used to obtain minimum distortion at the sacrifice of some hardness.

Cooling from the martempering bath ordinarily is done in still air to avoid large differences of temperature between the surface and the interior of the steel. Forced air-cooling with a fan is occasionally used if the cross-sectional area is more than ¾ in. thick. Cooling steel in cool oil or water after removing it from the martempering bath is considered undesirable because thermal gradients cause unequal stress patterns to be formed, increasing the distortion, which would destroy the objective for using the martempering procedure.

After the part has been either conventionally hardened or martempered hardened, the piece must be tempered, or some of the hardness must be withdrawn from the piece or else it will crack during the storage period. The storage period can be overnight with martempered items and a matter of just a few hours with conventionally hardened items. The extreme hardness is withdrawn by the tempering process. The tempering operation is a process of reheating the metal to a set temperature below the transformation range for a period of time until the desired mechanical properties are attained (Chart 29-1).

CHART 29-1 Tempering temperatures

TEMPERATURE, °F	TEMPER COLOR	APPLICATION
590	Pale blue	Hammers, screwdrivers
545	Violet	Axes, chisels, knives
525	Purple	Scribes, scratch awls, center punches
490	Yellow brown	Wood chisels, hammer faces, shearing blades
465	Straw	Dies, taps, punches
425	Light straw	Drills, reamers, milling cutters
380	Light yellow	Lathe tools, shaper tools

AUSTEMPERING Austempering converts austenite into a hard structure called bainite. In austempering, the steel is quenched in molten salts at a temperature that ranges from 450 to 800°F. This temperature produces a structure having the desired degree of toughness and ductility. When a constant temperature is held for sufficient time to complete the transformation from austenite, the resulting structure is called bainite. Bainite is relatively tougher than tempered martensite. Cracking and warping are less likely to occur because the temperature drop in austempering is less severe than in ordinary quenching or martempering. A disadvantage of austempering is that only metal with thin cross-sectional areas can be austempered, as for example, knife blades, razor blades, and wires. Steels containing 0.6 percent or more carbon are easily austempered. Austempering is a function of the quenching bath temperature and the length of time that the austenitic metal is in the bath. Austempering baths are generally molten salt mixtures that contain sodium carbonate and barium chloride. This salt bath is generally held at a specified temperature by a thermostatically controlled heater. For example, if a particular steel were to be quenched at 650°F, the bath temperature would be 650°F. The steel would be submerged in the molten salt and would remain in this molten salt quenching bath for a specified number of seconds. The metal would then be removed from the salt bath and the part would be allowed to cool in still air. The amount of time and the temperature of the bath are important because the metal has to be in the bath long enough for the bainite to form completely and the temperature must be hot enough to control the type of bainite formed. Austempering is capable of creating a tougher article than one that has been either conventionally hardened and tempered or martempered.

QUESTIONS

1. What are the differences between hardening and tempering?

2. What are the major types of case hardening?

3. How does flame hardening differ from induction hardening?

4. How does an induction coil induce heat into a metal?

5. What is ihrigizing?

6. What is hysteresis?

7. How does martempering differ from austempering and conventional tempering?

8. What are the limitations of austempering? Conventional tempering?

9. How is the IT diagram used?

10. How is a bainite crystalline structure derived? Martensite? Pearlite?

11. What are the major quenching media?

12. How does the quenching medium help determine the hardness of a heat-treated metal?

13. What are the differences between annealing and normalizing?

14. To what practical uses can a welder apply his knowledge of heat treating?

SUGGESTED FURTHER READINGS

Ashburn, Anderson: '68 Outlook: Heat Treating, *American Machinist*, vol. 112, pp. 141–142, January 15, 1968.

Begeman, L. Myron, and B. H. Amstead: "Manufacturing Process," John Wiley & Sons, Inc., New York, 1963.

Beingessner, James A.: The Changing Character of Commercial Heat Treating, *Welding Engineer*, vol. 44, pp. 53–57, July 1968.

Bingham, John F.: Guidelines for Heat Treating High-Strength Fastener, *Metal Progress*, vol. 93, pp. 99–102, May 1968.

Braymer, Daniel T.: Program Control for Heat Treating of Metals Forecast, *Electrical World*, vol. 196, p. 60, June 24, 1968.

Clayton, Kirk L.: Heat Treating Gears and Shafts by Induction, Cutting Costs with Modern Induction Heat Treating, *Metal Progress*, vol. 94, pp. 68–69, 74, June 1968.

Enos, George: "Elements of Heat Treatments," John Wiley & Sons, Inc., New York, 1960.

Grossman, Marcus: *Principles of Heat Treatment*, Haddon Craftsman, Inc., Pittsburgh, Penn., 1955.

Keyser, A. Carl: "Materials of Engineering," Prentice-Hall, Inc., Englewood Cliffs, N.J., 1959.

Koebel, Norbert K.: Vacuum Heat Treating: Process with a Bright Future, *Metal Progress*, vol. 94, p. 93, September 1968.

Kohn, Max, and Martin J. Starfield: "Materials and Processes," The MacMillan Company, New York, 1952.

Making, Shaping, and Treating of Steel, U.S. Steel Corporation, Pittsburgh, Penn., 1957.

Mondolfo, L. F., and Otto Zmeskal: "Engineering Metallurgy," McGraw-Hill Book Company, New York, 1955.

Rusinoff, S. E.: *Manufacturing Processes*, American Technical Society, Chicago, 1962.

Scott, James K.: Organic Quenchant Aids, Heat Treatment of Dip Brazed Aluminum Parts, *Metal Progress*, vol. 95, pp. 80–82, March 1969.

30 DESTRUCTIVE TESTING

Destructive testing uses various types of mechanical tests to derive information concerning the mechanical properties of materials. The mechanical tests either totally destroy the sample or impair the surface of the sample. The most common destructive mechanical tests are the tensile test, compression test, bending test, tortion test, fatigue test, and hardness test.

So far as anyone knows, the origin of mechanical testing is obscure. Most of the tests have been developed primarily in relation to the more recent uses of metal. Existing records of any precise mechanical test date back no more than three or four centuries, but the earlier civilizations, Rome, Greece, and Egypt, had some knowledge of testing for mechanical strength. The more recent developments in mechanical testing can be regarded as beginning in the late seventeenth and early eighteenth centuries. Over the next 100 years, the most important advances in the science of mechanical testing were first made in France and later in other European countries, a result of the establishment of engineering schools in European universities. Today, mechanical testing supplies all the necessary information to infer the safety and strength of all metals. The destructive tests supply a great amount of the quantitative information that is used for comparison of mechanical properties.

TENSILE TESTING   The tensile-testing machine consists of a mechanism for exerting a pull on a test piece coupled to a device which measures the load or stress. The stressing mechanism may be hydraulic, consisting of a cylinder with a piston carrying one of the test pieces; or it may consist of a screw driven in an axial direction by the rotation of a nut; or it may be a simple fulcrum and beam device.

The simplest measuring device employed is a pivoted steelyard to which the load is transmitted through the test piece by a knife edge a short distance from the fulcrum. The load is then balanced by the steelyard to a reference position between the weight stops. The weights may be moved along by a drive screw mechanism or by hand (Figure 30–1). An arrangement of this form constitutes a single-lever testing machine, a type which has been much used in the past, particularly when the maximum load requirements are not great. A single-lever testing machine

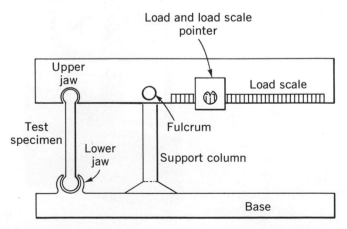

Fig. 30–1 Pivoted steelyard.

of 100-ton capacity was in regular use at the National Physical Laboratory for many years. A single-lever machine is still adaptable to tensile tests that are performed at high temperatures. The single-lever principle is also employed extensively today for long-term tensile loading of materials at high temperatures in which the load remains unaltered for the duration of the test. This type of test determines the creep of the metal at elevated temperatures.

The principle of measuring loads by steelyard and jockey weights can be extended to higher applied loads by transmitting to the steelyard through an auxiliary lever the load of the test beams. An example of a multilever machine of this kind is the 300-ton Karkaldy machine which can accommodate test pieces up to 20 ft in length.

More popular nowadays is the self-indicating type of machine in which the load is applied to the test piece by a hydraulic force and the pressure in the system is balanced against the deflection from the vertical pendulum by a subsidiary hydraulic ram. The displacement of the pendulum varies with pressure so that the movement of the load applied to the test piece can be utilized directly to operate a dial indicator from which instantaneous values of load can be read (Figure 30-2).

The universal testing machine can be used for tensile testing. Tests for tensile strength, percentage of elongation, elastic limit, yield point, and reduction of area can be

Fig. 30-2 Universal tester. (Courtesy of Ametek/Testing Equipment.)

made. The test specimen size is an important aspect of tensile testing. The tensile strength of many metals is high; for example, mild steel has a tensile strength of 62,000 psi. A machine to test this would have to be large; however, the mechanical properties remain true for the whole piece even if only a portion of this area is tested and then the results multiplied by the size ratio. For example, a 505 bar (no. 1 in Chart 30-1) has a cross-sectional area

CHART 30-1 Tensile test specimen dimensions

SPECIMEN NUMBER	DIMENSIONS, IN.						
	A	B	C	D	E	F	G
1	0.500 : 0.01	2	$2\frac{1}{2}$	$\frac{3}{4}$ D	$4\frac{1}{4}$	$\frac{3}{4}$	$\frac{3}{8}$
2	0.437 : 0.01	1.75	2	$\frac{5}{8}$ D	4	$\frac{3}{4}$	$\frac{3}{8}$
3	0.357 : 0.007	1.4	$1\frac{3}{4}$	$\frac{1}{2}$ D	$3\frac{1}{2}$	$\frac{5}{8}$	$\frac{3}{8}$
4	0.252 : 0.005	1.0	$1\frac{1}{4}$	$\frac{3}{8}$ D	$2\frac{1}{2}$	$\frac{1}{2}$	$\frac{1}{4}$
5	0.126 : 0.003	0.5	$\frac{3}{4}$	$\frac{1}{4}$ D	$1\frac{3}{4}$	$\frac{3}{8}$	$\frac{1}{8}$

of 0.2 in.²; therefore, the testing may be done on this bar and then the results multiplied by 5 to determine the tensile strength for 1 in.² of metal. The tensile strengths that are used for the testing of a weld come in 2 full sample sections and a reduced sample sections tensile test. The reduced section tensile test reduces the sample to approximately 0.2 in.². The full section tests are based on 1 in.² of metal (Figure 30-3).

Before the test piece is inserted into the jaws that clamp it into the machine, the gage length is marked along the parallel length. The gage length for a standard 505 tensile test bar is 2 in. The gage length is marked at each end and the test piece is mounted in the machine. The load is applied and steadily increased until yield occurs. One method of observing the yield is to hold one point of a pair of dividers at one of the gage length marks and lightly scribe across the other end. When a sudden elongation occurs, it will be both seen and felt. At the same time, the operator of the universal testing machine can observe when the load indicator stops and falls back. The load at this point should be recorded before forward motion is arrested.

When yielding is complete, the load starts to rise. The beginning of this yielding is the elastic limit. The metal is now in the plastic state and the deformation of the test

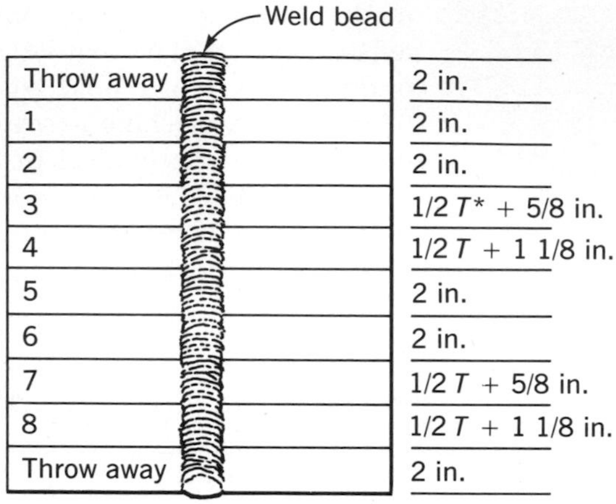

Weld bead

Throw away		2 in.
1		2 in.
2		2 in.
3		1/2 T* + 5/8 in.
4		1/2 T + 1 1/8 in.
5		2 in.
6		2 in.
7		1/2 T + 5/8 in.
8		1/2 T + 1 1/8 in.
Throw away		2 in.

1 and 5	Full-section tensile test
2 and 6	Reduced-section tensile test
3 and 7	Nick-break test
4 and 8	Free bend test

Fig. 30–3 Standard weld specimen (coupon).

*T = plate thickness, in.

piece is permanent. The rate of straining will not need to be increased and by the time the maximum load is reached, the gage length will have undergone a uniform extension if the material being tested has any ductility. If the material is brittle, the piece will not elongate but will break or fracture. The maximum load should be noted and the breaking load should also be noted. The maximum load is the ultimate strength of the metal. The test piece is then brought back together while still clamped into the machine, and the gage length is remeasured. The gage length of the test piece is then compared to the original gage length of the test piece. For example, if the test piece were 2 in. prior to testing and after testing the gage length were 2.185 in., the elongation would be 9 percent.

The test pieces are then removed from the universal testing machine and each end of the fractured piece is measured with the micrometer. An average of these two measurements is computed and this measurement is compared to the original diameter of the test bar. If the original size of the diameter of the test bar were 0.505 in. and the diameter of the bar after fracture were 0.405 in., the reduction of area would be 0.100 or about 19.8 percent (Figure 30–4).

COMPRESSION TESTING A great many of the machines used for tensile testing are universal testing machines, which can be employed for compression testing as well as tensile testing. Special fittings or shackles are part of the original equipment of the universal testing machine. The compression load is applied axially if the parts to be compression tested have parallel ends. If the ends are not parallel, the piece may slip from between the platens and cause damage. Specially designed machines for compression testing are commercially available for carrying out the testing of concrete. Test pieces are made from rock and samples of aggregate. Such machines are frequently hand-operated, hydraulic machines with a pressure gage as the load indicator and with a large load capacity. Some small machines are capable of applying 100 tons of pressure, but they are usually limited to small test pieces. Though the compression test ranks low on the list of routine acceptance tests for metals, there are fields in which it can be used in order to obtain useful data, such as in plastics and ceramics. The compression test is not used for most metals because it is not as reliable as the tensile test and the reduction of area test as an indicator of ductility. A test piece being tested for

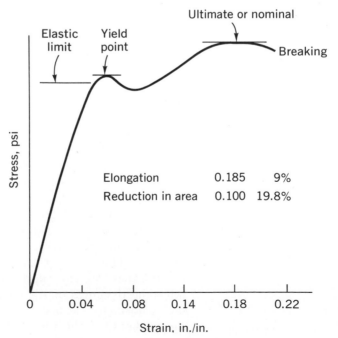

Fig. 30–4 Tensile test 0.505 bar.

tensile strength which is in tension may break after a 20 percent reduction in area; whereas, an identical test piece might be compressed to 80 percent reduction without failure. This, then, is not an indicator of the mechanical property of ductility for a metal.

BENDING TEST Bend testing can be accomplished with two types of tests, the free bend test and the guided bend test. Bend tests of ductile metals are usually conducted on a qualitative basis with no attention paid to the initial stages of bending or the amount of force employed in bending. Bending is continued far beyond the limit of elastic properties of metal and into the plastic zone. The failure of the weld bead in the free bend test means that the bead does not have strength that is equal to the base metal. For example, in mild steel the tensile strength of the base metal is approximately 60,000 psi. If the weld were to rupture during the free bend test, the logical conclusion is that the weld bead is not strong enough for that particular base metal. A few cracks and checks may become apparent during and after the free bend test has been made. Corner cracks are generally thought to be negligible, as long as they do not extend more than $\frac{1}{16}$ in. into the base metal. Any surface defect, any crack or depression, or a slag inclusion, may not be over $\frac{1}{16}$ in. or the weld bead is considered a failure.

The elongation of the piece may also be measured by placing gage lines on each side of the weld bead. This distance is measured to the nearest hundredths of an inch, 0.001 in. After the bend test, the gage lines are then remeasured and a percentage of elongation can be equated. If, by some means, the elongation of the test specimen cannot be established, the elongation can be calculated by the formula $100T$ divided by $2R$, where T is the thickness of the specimen and R is the radius of the piece.

The test specimens for the free bend test range between 6 and 21 in. in length. The thickness ranges from $\frac{1}{4}$ to $2\frac{1}{2}$ in. The width of test specimens is dependent upon the thickness of the piece; generally it is equal to $1\frac{1}{2}$ times the thickness of the specimens. The fixture for applying force to the test specimens is determined also by the thickness of the piece. If the piece is small, between

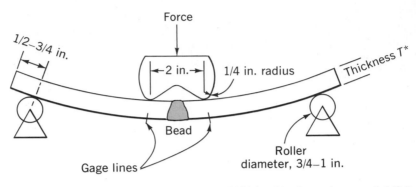

Force

1/2–3/4 in.

←2 in.→ 1/4 in. radius

Thickness T*

Bead

Roller
diameter, 3/4–1 in.

Gage lines

*Width of test specimen = 1 1/2 T

(a)

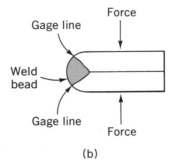

Force

Gage line

Weld
bead

Gage line

Force

(b)

Fig. 30–5 Free bend test. (a) Initial
bend. (b) Close bend.

¼ and ½ in., the gage lines should be approximately
1¼ in.; however, for all other tests, the gage length
is usually 2 in. (Figure 30–5).

The guided bend test is designed to show surface imperfections near and in the weld bead. The guided bend tester
is designed around a certain thickness of plate. For example, in Figure 30–6, the numbers that are shown are multiplied by the thickness of the plate. The mandrel is 4
times the thickness of the plate in width and 18 times the
thickness of the plate in length. The radius of the corners
is 2 times the thickness of the plate in diameter plus ⅛
in. For example, if the radius on the mandrel were designed for ½-in. plate, the radius would be 2 in. The weld
test specimens are indicated in Figure 30–3, no. 4 and no.
8. Number 4 specimen is tested with the face of the bead
out or the face of the bead toward the mandrel, and specimen no. 8 is tested with the root of the bead toward the
mandrel. After the piece is tested, the outside of the bend

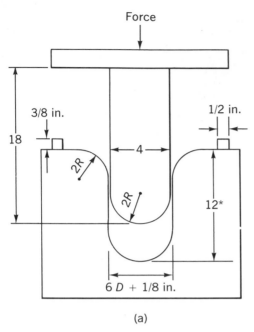

(a)

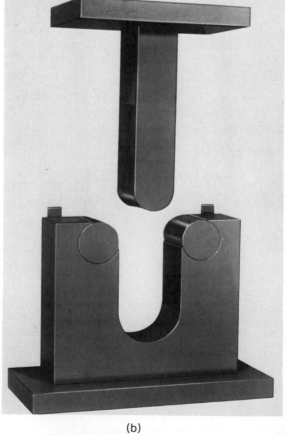

Fig. 30–6 Guided bend tester. (a) Dimensions. *Whole numbers are multiplying factors for the plate thickness (12 = 12 × specimen thickness). (b) Commercial test device. (Photograph courtesy of Ametek/Testing Equipment.)

(b)

is checked for surface discontinuances. If an occlusion or tear is greater than $1/16$ in., the specimen is thought to have failed the test. The test will also show if there is any lack of penetration where the weld bead fuses to the workpiece.

NICK-BREAK TEST

The nick-break test is designed to show if interior inclusions, such as gas pockets and slag inclusions, exist and to show the degree of porosity in the weld bead (Figure 30–7). The nick-break test is a simple test in which the force may be applied by a press or the sharp blow of a hammer. The intensity or the swiftness of the force to break the piece is not important. A visual inspection of the piece is the test. Also, during the visual inspection, the inclusions or discontinuances in the weld metal should not be over $1/8$ in. in length or the piece will have failed the test. The standard nick-break test specimens are indicated in Figure 30–3, nos. 3 and 7. The width of the specimen is dependent upon the thickness of the metal. The width of the specimen is usually $1 1/2$ times the thickness, plus $1/2$ in. The slots for the nick-break test are put in with a saw to a depth of $1/4$ in. on each side. One of the nick-break test specimen pieces is broken from one side and the other nick-break specimen is broken from the other side. An adaptation of the standard nick-break test is the fillet nick-break test (Figure 30–8). The standard length for a specimen in the fillet nick-break test is from 4 to 6 in. The test is applied by simply applying force in the back of the fillet, crushing the angle flat. One requirement of the fillet nick-break test is that there be no tack welds on the other side of the fillet weld. One small tack weld will have enough holding power to invalidate the test.

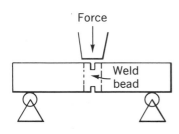

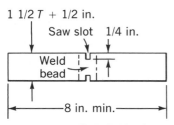

Fig. 30–7 Nick-break test.

IMPACT TESTING

Impact testing determines the relative toughness of a material. Toughness is defined as the resistance of a metal to fracture after plastic deformation has begun. This plastic deformation is begun and finished by the swing of a weighted pendulum. The weighted pendulum swings down, striking the test piece as it swings through its arc. The energy that is required to fracture the test piece is recorded in foot pound-force on the scale. The tougher

Force

Fillet
weld

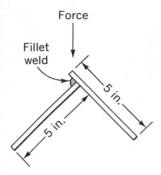

Fig. 30-8   Fillet nick-break test.

the material, the higher the energy absorbed by the test piece. Unalloyed metals are rated in the order of their toughness as follows: copper, nickel, iron, magnesium, zinc, aluminum, lead, tin, cobalt, bismuth. Bismuth is rated no. 10, indicating that it can absorb a small amount of energy before plastic deformation and fracture, while copper is highly ductile and can withstand a great deal of plastic deformation.

Two major tests for determining impact toughness are the Izod test and the Charpy test. Both of these methods use the same type of machine and both yield a quantitative value in foot pound-force. The major difference is in the test specimen and the striking area of the pendulum on the test specimen. The Izod test specimen is generally 2.952 in. in length with a small V notch in the test bar (Figure 30-9). The Charpy method of impact toughness testing uses two different test specimens, the V bar and the keyhole bar (Figure 30-10). Both test specimens, for the Charpy method and the Izod method of impact toughness testing,

Fig. 30-9   Impact testing, Izod method.

use a square bar 3.94 in. square, or 10 mm square (Figure 30-11).

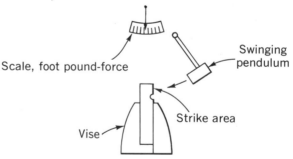

Scale, foot pound-force

Swinging pendulum

Strike area

Vise

IZOD test specimen

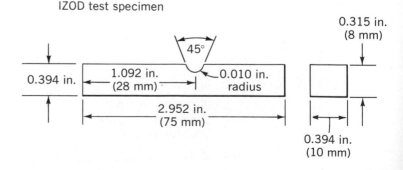

HARDNESS TESTING

Scale,
foot pound-force

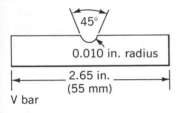

Force of
pendulum path

Charpy test specimens

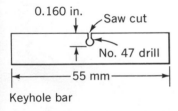

V bar

Keyhole bar

Fig. 30–10 Impact testing, Charpy method.

Hardness is considered a mechanical property that is sometimes thought to be derived by destructive testing. Sometimes it is thought to be tested by a nondestructive testing device. Either way, testing does deform the surface of the metal slightly. Hardness is the ability of a metal to resist penetration, to resist abrasive wear, or to resist the absorption of energy under impact loads. These definitions can be thought of as penetration, wear hardness, and rebound hardness.

The Rockwell hardness tester and the Brinell hardness tester are two means to determine the resistance of a metal to penetration. The scleroscope is an instrument to test the rebound hardness of a metal. The hardness scale for any of the hardness testing methods can be computed into another scale: a Rockwell B scale of 100 is equal to a Rockwell C scale of 24, which is equal to a Brinell scale of 245 and a scleroscope scale of 34. There are standard tables available from the manufacturers of hardness testing equipment that compute these comparisons (Chart 30–2).

Rockwell hardness tester The Rockwell hardness tester is a machine that measures hardness by determining the depth of penetration of a penetrator into the specimen under certain fixed conditions of testing. The penetrator may be either a steel ball (the B scale), or a diamond penetrator (the C scale). The hardness number is related to the depth of indentation. The higher the number on the scale, the harder the material. The machine is operated by placing the specimen on the platen of the machine. A minor load of 10 kg is first applied, which causes an initial

CHART 30–2 Hardness comparison

ROCKWELL B	VICKERS PYRAMID	ROCKWELL C	BRINELL	SCLEROSCOPE
	944	69	755	98
	737	60	631	84
	503	50	497	68
	394	40	380	53
	294	30	288	41
100	248	24	245	34
98	229	20	224	32
89	179	10	179	25
78	143	0	143	21
⋮	⋮	—	⋮	⋮

penetration. This is the beginning point of the hardness test. The minor load is applied to force the penetrator through any scale or other surface hindrance that may be present on the material. After the minor load has been applied, the read-out dial is zeroed, and then the major load is applied. The major load is 60 kg or 100 kg when the B scale or the steel ball is used as a penetrator, and 150 kg when the diamond penetrator is used. The ¹⁄₁₆-in. diameter ball is used for material that is medium hard and the ¹⁄₈-in. penetrator may be used for soft metals, such as lead, bismuth, antimony, and tin. The diamond penetrator is used for the hard metals, ranging from iron in hardness on up. The ¹⁄₁₆-in. diameter ball is used for the copper-based metals and other medium-hard metals. After the major load is applied and then removed, the reading is taken while the minor load is still being applied (Figure 30–12).

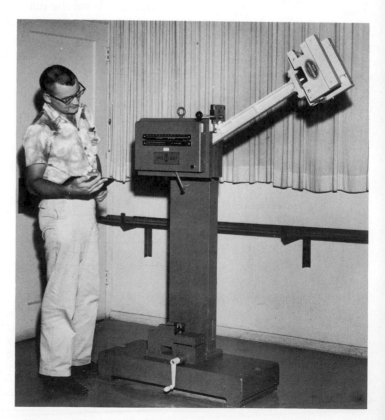

Fig. 30–11 Impact tester. (Courtesy of Tinius Olsen Testing Machine Company.)

The letters B and C were adopted to differentiate between the two Rockwell methods. The letter B applies to tests with a ball penetrator of 1/16-in. diameter and 100-kg major load. A kilogram is approximately 2.2 lb. A Rockwell C test is designed for testing hard materials with a diamond penetrator having a slightly rounded point and a major load of 150 kg. There are other special tests that can be accomplished with the Rockwell test machine, including the A test, the D test, and the E test, depending upon the type and the thickness of the material being tested.

The advantages of the Rockwell tests are that they can be performed rapidly; and, if the specimen is prepared carefully, the accuracy is consistent. The accuracy of the Rockwell tester is thought to be greater than that of the Brinell system in the higher hardness ranges. Another important consideration is that the hardness numbers can be read directly from the indicator dial. The hardness number is not read with the aid of a microscope and for inexperienced operators, the results can be more easily obtained.

Fig. 30-12 Rockwell hardness tester.
(Courtesy of Ametek/Testing Equipment.)

Brinell test J. A. Brinell, a Swedish engineer, in 1900, published information about a method for determining the relative hardness of steel which has come into extensive use. The Brinell hardness test uses a hardened steel ball 10 mm in diameter, 0.3937 in., which is forced, under a pressure of 3000 kg for hard metals and 5000 kg for soft metals, into a flat surface on the test specimen. The ball is allowed to remain there for a minimum of 10 sec for iron and ferrous materials or for at least 30 sec for nonferrous metals. Then the hardened steel ball is removed, leaving a slight spherical impression. The diameter of this impression is then measured with a special microscope. The hardness of the specimen is defined as the quotient of the pressure divided by the area of the surface of the impression, which is assumed to be spherical. From this measurement, the Brinell hardness number is calculated. See Figure 30-13 and the following formula.

$$B = \frac{P}{\frac{\pi D}{2}\left(D - \sqrt{D^2 - d^2}\right)}$$

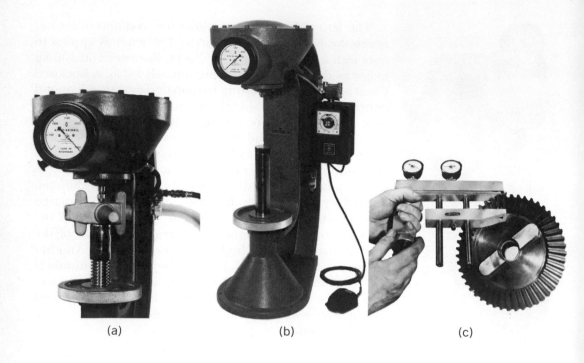

(a)   (b)   (c)

where B = Brinell hardness number
P = 3000 or 500 kg load
D = diameter of the ball
d = diameter of the impression
π = <u>3.141</u>   592   653   589   793

The Vickers pyramid testing machine is a British instrument that employs the same principle as the Brinell hardness tester and the Rockwell testing machines. The penetrator is a diamond in the form of a pyramid. After the indentation has been made, a microscope attachment measures the distance across the diagonal indentation. One advantage of this machine is that it is suitable for any degree of hardness. A Vickers pyramid number is comparable to the Rockwell, Brinell, and scleroscope systems as well (Chart 30-2).

Scleroscope The scleroscope is an instrument invented by A. F. Shore for determining the rebound hardness of metals. It consists of a vertical glass tube in which a hammer slides freely (Figure 30-14). This hammer is pointed at the lower end and is allowed to fall about 10 in. onto the sample being tested. The distance the hammer rebounds back up the vertical glass tube, or slide, is a measure of the rebound hardness of the specimen. A scale on the tube is divided into 140 equal parts. The hardness is expressed as the number on the scale to which the hammer rebounds. Measured in this way, the hardness of different substances are as follows: glass, 130; porcelain, 120; the hardest steel, 110; mild steel, 25 to 30; grey iron, 39; babbit, 4 to 10; brass, 12; zinc, 8; copper, 6; and lead, 2. The relationship between the scleroscope number and the ultimate strength is given by the following equation: ultimate strength is equal to 4.1 times the scleroscope number minus 15.

A simple estimate of ultimate strength can also be derived from a Brinell hardness number. A Brinell hardness number can be divided by 2, then multiplied by 1000, which yields a tensile strength in pounds per square inch from the Brinell number. For example, if a Brinell number of 140 were observed on the Brinell hardness tester, the ultimate strength could be calculated by dividing 140 by 2, which equals 70, representing 70,000 psi tensile strength for the material.

Fig. 30-14 Scleroscope hardness tester. (a) Air operated. (b) Standard. (Courtesy of Shore Instrument and Manufacturing Company.)

(a) (b)

QUESTIONS

1. What does a destructive tensile test reveal about a metal?

2. What are the major types of tensile testing machines?

3. Why are reduced section test specimens used?

4. What are four common tests used for testing weld coupons?

5. What are the advantages of the free bend tests over the guided bend tests?

6. What is the difference between the Izod and the Charpy impact tests?

7. What is hardness?

8. Why are there different methods to determine the hardness of a metal?

9. What are the different Rockwell scales? Why are they different?

10. What are the principles of the Rockwell hardness tester? Brinell hardness tester? Scleroscope hardness tester?

11. What is destructive testing of materials?

SUGGESTED FURTHER READINGS

Breneman, J. W.: "Strength of Materials," McGraw-Hill Book Company, New York, 1952.

Hull, F. C., and H. R. Welton: Work Hardened Surfaces of Fatigue Specimens, *Metal Progress*, vol. 48, pp. 1287–1288, December 1945.

Kennedy, A. J.: "The Processes of Creep and Fatigue in Metals," John Wiley & Sons, Inc., New York, 1963.

Odquist, Folke K: "The Mathematical Theory of Creep and Creep Rupture," The Clarendon Press, Oxford, 1966.

O'Neill, H.: "The Hardness of Metals and Its Measurement," Sherwood Press, Inc., Cleveland, 1934.

Peek, R. L., Jr., and W. E. Ingerson: Analysis of Rockwell Hardness Data, *Am. Soc. Testing Mater., Spec. Tech. Publ.*, no. 106, June 26–30, 1939.

Polokowski, N. H., and E. J. Ripling: "Strength and Structure of Engineering Materials," Prentice-Hall, Inc., Englewood Cliffs, N.J., 1966.

Shames, Irving H.: "Mechanics of Deformable Solids," Prentice-Hall, Inc., Englewood Cliffs, N.J., 1964.

Shanley, F. R.: "Mechanics of Materials," McGraw-Hill Book Company, New York, 1967.

Smith, R. L., and G. E. Sandland: The Diamond Pyramid for Hardness Testing, *J. Iron Steel Institute*, London, 1925.

Timoshenko, S., and D. H. Young: "Elements of Strength of Materials," Van Nostrand Reinhold Company, New York, 1962.

Van Vlack, Lawrence: "Elements of Materials Science," Addison-Wesley Publishing Company, Inc., Reading, Mass., 1967.

31 NONDESTRUCTIVE TESTING

Today, in advanced technology, fields of construction and manufacturing build many things that need 100 percent reliability. Because of this demand for high reliability, the field of nondestructive testing has originated. Nondestructive testing is a means to define and locate flaws within a material or a product without destroying or defacing the product. The major ways to locate flaws, defects, cracks, and discontinuances are by liquid penetrant testing, magnetic particle testing, radiographic testing, and ultrasonic testing. Many other means of testing materials nondestructively are employed, but these are the four major types used in welding. Of course, there is always the original type of nondestructive testing available to the welder—a visual inspection of the weld. Visual inspection involves measuring the weld and noticing whether the weld surface has any porosity holes or slag inclusion or if the weld itself is undercut or overlapped.

ULTRASONIC TESTING

The elastic property of metal can be described as displacement of metal particles or molecules in an object that has been struck by another object. If the surface of a metal is displaced, the displacement occurs inward, moving the molecules of the metal inward. As long as they are still elastic, they return to their original position after the force of the blow has been dissipated. A further application of this concept is that energy is transmitted through a solid portion of material by a series of small displacements or molecular movements. Each molecule displaces the next one in line with it, until a total displacement has occurred. The first displacement is caused by the initial impact of a force, and the second major displacement is caused by the return of the molecules to their original position. A total displacement of these molecules can be expressed as a cycle made up of the two opposing movements. The length of time it takes the cycle to progress one complete movement is the frequency of the cycle. For example, if six cycles are occurring in 1 sec, the frequency would be 6 Hz (cycles per second). Any void within a metal disrupts this cyclic movement of molecules and this disruption could be portrayed on some type of readout device. Ultrasonics, used in the testing of materials, uses waves that reach a frequency range of 200,000 to 25,000,000 Hz. Several types of waves that pass through solid matter can be used. Longitudinal or compression waves are those in which the particle vibrations continue in the same direction as the sound waves. Particle vibrations of sheer waves or transverse waves are at right angles to the wave motion. With some degree of success, sheer waves can be produced along a free surface so that they ripple across the surface, checking for deflection just below the surface. These sound waves have properties similar to those exhibited by light waves. In other words, the sound waves can be reflected, focused, or refracted. The major characteristic of sound waves is that they can pass through a solid object with little absorption.

Generation of wave lengths through a solid is usually accomplished with a device called a transducer, which changes electrical energy into sound energy, or sound waves, by utilizing piezoelectric principles. For example, a paddle pushed through the water by some mechanical force produces a wave in water. The same is true of waves of sound. Transducers operate at frequencies well above 20,000 Hz. There are two types: the sending transducer and the receiving transducer. The sending transducer converts mechanical energy into sound energy, and the receiving transducer converts the sound energy back into electric energy. These are the input and the output of the energy source for the ultrasonic testing device. Many times the same transducer has been used to generate the mechanical energy and also act as a receiver, recovering the mechanical energy or the sound waves.

When sound waves vibrate through the medium to be tested, the process of attenuation occurs. Attenuation is the loss in a wave form resulting from the density of the material and can be thought of as internal friction in the object being tested. Therefore, a compensation is needed for the difference in the measurements from the input pulse and the output pulse. The lost sound waves usually emerge as heat, which changes the density of the material. This factor, however, is small and is usually ignored.

In ultrasonic testing, transducers are placed on the surface of the article that is going to be tested. A couplant is used to help make contact between the transducer and the surface of the material. A couplant is an oily, glycerine-base substance which performs two major functions, removing the air between the transducer and the test specimen and providing a medium for the transfer of the sound vibrations.

A problem of major importance in ultrasonic testing is the divergence of the beam, or the sound wave, as it travels through the material. As the wave travels, it diverges or spreads and becomes wider; as it becomes wider it also becomes weaker. There are two possible ways to control beam divergence; one way is to increase the size of the transducer but this way has a physical limitation. The other method is to increase the frequency of the cycles. Increasing the frequency does not have any theoretical or physical limitations, the reason why most ultrasonic testers operate with a fairly small transducer and a high frequency.

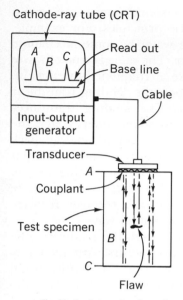

Cathode-ray tube (CRT)

Read out

Base line

Cable

Input-output generator

Transducer

Couplant

Test specimen

Flaw

Fig. 31−1 Pulse echo ultrasonic system.

Three types of ultrasonic test devices are available on the commercial market: the pulse echo system, the through transmission system, and the resonance system. The pulse echo system is the most widely used (Figure 31−1). It generates an ultrasonic wave that passes through the material from the transducer. The wave is returned through the same transducer. The signal is then displayed upon the cathode-ray tube (CRT). The pulse echo system has evenly timed pulses of ultrasonic sound waves which are transmitted into the material being tested. The pulses reflect from the discontinuances in their path or from any boundary of the material being tested. As the waves pass through the top surface of the material being tested, there will be a pip on the CRT, which will indicate the beginning of the test specimen. If the test specimen has a flaw within it, the wave will bounce off, indicating the distance from the surface of this flaw, as in *B* in Figure 31−1. The specimen will also be tested as far as the thickness indicated by *C* in Figure 31−1. The pulse echo ultrasonic system can be thought of also as a measurement device, or a one-sided micrometer, with which the thickness of materials can also be measured. The pulse echo system relies entirely upon reflected energy reaching the receiver-sender-transducer for any information as to the thickness of the test specimen, any flaw, or any discontinuances within the test specimen.

Through transmission system The through transmission system requires the use of two transducers, one used for the transmitting or sending of the sound waves, and the other used for the receiving of the sound waves. As in the pulse echo system, short pulses of ultrasonic sound are transmitted into and through the material. In a few applications, a continuous sound wave could be but very seldom is used. Unlike the pulse echo system, the echoes returning to the sending or transmitting transducer are ignored. The receiving transducer, however, which is aligned with the sending transducer, conveys the information in the form of a sound wave, and then in the form of electrical energy, to the CRT. The soundness or quality of the material being tested is measured in terms of the energy lost by the sound beam as it travels through the material, which limits the use of this system. It can only

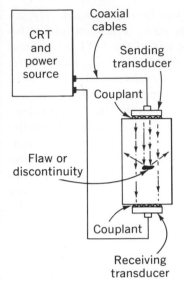

Fig. 31-2 Through transmission ultrasonic system.

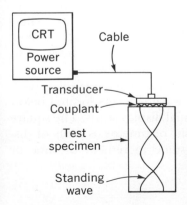

Fig. 31-3 Resonance ultrasonic system.

be applied to items whose sides are relative or parallel to each other (Figure 31-2).

Ultrasonic resonance system The resonance system makes use of resonance phenomena to measure the thickness of material and to determine the quality of bonded materials. It is also used, but to a lesser degree, to detect gross discontinuities. The resonance system transmits ultrasonic sound waves into the material, making this system similar to the pulse echo system, except that the waves are always continuous longitudinal waves (Figure 31-3). Wave frequency is varied until standing waves are set up in the system, causing the item to resonate or to vibrate at a greater amplitude. A difference in vibration resonance is then sensed by the generator indicator instrument, the transducer, and this information is displayed on the CRT screen. A change in the resonant frequency, which cannot be accounted for as a change in the material thickness, is usually a discontinuance or a flaw in the test material itself.

Two methods are used in all ultrasonic testing: the contact testing method and the immersion testing method. The contact method involves placing the transducer directly on the test material. The transducer is coupled to this material through a thin layer of fluid known as couplant. Contact testing is most frequently used in field and shop locations, for the equipment is usually portable and can be moved easily, and the only requirement is that there be a 110-volt, ac power supply. In immersion testing both the test material and the transducer are immersed in a liquid couplant, sending the ultrasonic waves through the liquid into the test specimen. The transducer in this case does not touch the object but is simply immersed in the couplant fluid, which is usually water. The one major requirement for immersion testing is that the back surface of the test specimen be supported so that total contact is avoided.

Three basic types of visual display units show the qualities or soundness of an object being tested: the A scan, the B scan, and the C scan (Figure 31-4).

The A scan display is a presentation of time versus the amplitude on the CRT. The time versus amplitude is the most efficient way of revealing the existence of discontinuities. From the location of the pips, the relative depth and

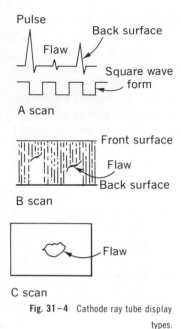

Pulse

Flaw

Back surface

Square wave form

A scan

Front surface

Flaw

Back surface

B scan

Flaw

C scan

Fig. 31–4 Cathode ray tube display types.

size of a flaw in the material can be determined. The interpretation of the A scan is easy. The first pulse, indication, or pip appears at the left of the CRT screen and the back surface reflection is on the right-hand side of the screen. The discontinuities are shown somewhere between the initial pulse, the front surface, and the back surface reflection pip. The height of the pip on the screen represents the amplitude of the wave reflection coming from the material being tested. Comparing the height of the flaw pip with known discontinuities pips from a material can establish the size of the discontinuity. The distance from the pulse to the flaw pip and the distance from the flaw pip to the back surface reflection are proportional to the elapsed time it takes the pulse to travel through the material. However, during the initial pulse, the reflection of the sound wave may be blocked by a discontinuity directly under the transducer, which can cause a dead zone at the front edge of the test specimen. Therefore, the flaw pip should be determined from the reflected back surface of the specimen.

The B scan unit is used mostly in the medical applications of ultrasonic testing and is generally not used for nondestructive testing; but when used, it is capable of yielding a cross-sectional view of the material being tested. The B scan shows the reflection of the front and back surfaces of the test material and any flaws that lie between these two surfaces.

The C scan display unit gives a plain view similar to an x-ray picture. The external details of the test specimen are projected onto a plane that is the CRT screen. The reflection from the discontinuity is the only one that is presented. The presentation shows the shape of a discontinuity as viewed from the surface of the test material.

The only other presentation that may be displayed on the CRT screen is the square wave markers located below the horizontal sweep on all visual display scans. The square wave markers are used as units of time or as units of distance to relate the scan to some increment of time or distance.

RADIOGRAPHIC TESTING

In the radiographic process, a ray is emitted from a controllable source, and this ray penetrates a test specimen

X-ray tube

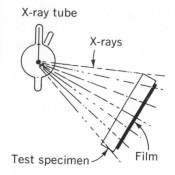

Fig. 31-5 Basic x-ray radiographic process.

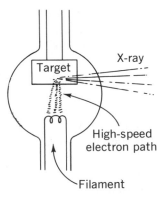

Fig. 31-6 X-ray generation.

film (Figure 31-5). The errors or flaws in the test specimen are superimposed upon the film, providing a visual record of the analysis for future reference. X-ray analysis can be used to reveal the internal structure of most materials, revealing errors in fabrication and assembly. Also gas bubbles, fractures, and other weld imperfections and base-metal imperfections can be identified in the x-ray analysis. X-ray radiographic testing is most effective in discerning or identifying small defects. It is not too practical for complex shapes, and the shapes of the items to be x-rayed should be of fairly simple design. X-ray is also a relatively expensive nondestructive testing method because of the high cost of equipment and the extreme safety precautions that must be taken to insure operator safety.

Continuous x-rays are produced when a high-energy electron collides with the nucleus of an atom (Figure 31-6). An x-ray source must be able to produce free electrons and direct them at a target with sufficient velocity so that they reach the nucleus of the constituent atoms of the target. Also, the source must provide for a uniform distribution of the resulting beam of x-rays into a desired pattern. A filament, which is similar to an incandescent light, can be placed near a highly charged cathode, causing the electrons freed by the heating of the filament to be repelled toward and attracted by an anode. The collision of the electrons with the anode produces x-rays of a continuous nature, if the potential difference between the anode and the cathode is high. The shape of the anode target determines the pattern of the emitted x-rays. The intensity of the x-rays produced is directly proportional to the number of electrons freed at the filament. Freeing electrons increases the filament temperature, which is proportional to the current passing through it. Therefore, the logical criterion for rating x-ray machines is the number of milliamperes the filament draws. The intensity of the x-ray varies according to the inverse of the square of the distance from the source. Doubling the distance between the filament and the target quarters the intensity of the x-ray beam, a factor true with any source of electromagnetic radiation.

Film Radiographic film is composed of a thin transparent plastic base on one or both sides of which a thin emulsion of gelatin has been placed. This gelatin usually has a thickness of 0.001 in. Suspended in this gelatin emulsion are tiny grains of silver bromide, which, when exposed to x-rays, gamma rays, or some physical rays, can be chemically developed into a metallic silver. This development is proportional to the exposure of the ray source. After this film has been exposed and developed, greater penetration of the x-rays through an object shows up as denser areas on the film. Some of the most important qualities to consider when choosing film include its speed, its sensitivity, its ability to reproduce fine detail, and the grain size of the silver bromide after the film has been developed. All these factors must be considered when selecting the film for a particular job.

After the film is exposed, its development is usually accomplished by some automated process, except in limited applications. The basic developing process involves the five following steps: First is developing, placing the film into a solution to convert the exposed silver bromide process at some desired stage. Then the fixer removes the unused silver bromide from the undeveloped areas. The wash removes all the previously used chemicals, leaving the film in an inert state; and, last, the film is dried, which is done to facilitate viewing and storage.

A characteristic of x-rays is that they are invisible and incapable of detection by any of the human senses. They can also ionize matter, or film, and they are absorbed by matter in direct proportion to the density and thickness of the matter and in direct proportion to the energy of the x-rays. Other characteristics are that they can travel in straight lines at the theoretical speed of light, and they have no electrical charge nor do they have any mass. They can be thought of to some extent as a photon or as light energy. As light energy, the energy of x-rays is directly proportional to the wave length. Because of some of these factors, x-rays are exceedingly hazardous. Because x-rays cannot be detected by any human sense, exposure can often lead to overexposure, causing permanent damage to the human body. The accumulative effect of x-rays upon living tissue has far-reaching consequences. The maximum permissible dose of x-ray exposure per day and the lifetime accumulative exposures per year are prescribed by the National Bureau of Standards. The material that is used to protect the operator from the x-rays is generally lead.

Gamma rays Gamma rays are also used as a power source in the exact manner that x-rays are used. The materials that are used for the radioactive isotopes are cesium 137, cobalt 60, iridium 192, samorium 153, and other radioactive elements. Radium was the first radioactive isotope to be used for the production of gamma rays in a testing device. Gamma rays penetrate thick sections of metal far more easily than x-rays. Gamma rays are used when x-rays of high power must be used, for the gamma rays or the radioisotope producing gamma rays does not require an external power source and is small. Gamma ray exposure time is generally longer than the exposure time for the corresponding x-ray radiographs. However, radiographs that use gamma rays as a wave ray power supply lack the definition or sharpness of character that x-ray produces. Generally, a person who is skilled in the use of x-ray equipment is also skilled in the use of gamma ray radiography.

Gamma rays do not require any external power source. The gamma rays given off by the radioactive isotopes are independent of any electrical source. The x-ray, however, is dependent directly upon electrical power sources. The power requirements for x-ray machines range from 50 to 24,000 kV (kilovolts). A 24,000-kV x-ray machine is capable of photographing approximately 20 in. of steel.

LIQUID PENETRANT TESTING Liquid penetrant testing, one of the simplest nondestructive tests that can be applied to a metal, is a chemical means to detect the surface openings and porosity of material. Liquid penetrant testing is a method to find cracks, seams, porosity, folds, inclusions, and any other surface defect that may occur in a metal. It is not limited to ferrous metal but can be applied successfully to nonferrous, nonmagnetic metals. Liquid penetrant testing is an excellent means for testing for porosity and weld inclusions in stainless steel, magnesium, aluminum, brass, and many more nonferrous metals. Liquid penetrant testing requires six basic steps. Each must be carefully performed in order to achieve the proper results. The six basic steps include surface preparation, application of penetrant, removal of excess penetrant, developing, inspection, and cleaning.

Surface preparation The preparation of the metallic surface is one of the most important steps in the liquid penetrant testing procedure. If the metal to be tested inhibits the penetrant from entering the metal, the metal cannot be tested. Such inhibiting contaminants as dirt, grease, scale, acids, chromates, and many other types of contaminants must be removed from the surface of the metal. Of the many possible cleaning methods, solvent cleaning is most popular. The solvent that cleans the metal must be capable of removing the many contaminants from the components to be tested, and it must also be volatile so that it will evaporate out of the defects in the metal. If the solvent that is used to clean the metal is not one that vaporizes readily, the solvent will dilute or prevent the liquid penetrant from entering into the surface flaws. Solvents that are currently used for cleaning metallic surfaces are acetone, isopropyl alcohol, methylene chloride, and perchlorethylene.

Penetrant application All liquids can be considered penetrants because their penetrating capability is a natural force resulting from capillary action. However, liquid penetrants used for tests have a greater number of demand on them that must coincide with their natural capillary force. These demands are a higher fluidity, a lower viscosity, and a higher wetability. Besides these qualities, a liquid penetrant must also be able to carry a dye that will mark the metal. The dye that is floated within the liquid penetrant solution can be either a fluorescent or a visible dye. The dye is applied either by dipping, immersing, pouring, spraying, or brushing. After the liquid penetrant with the dye has been applied to the surface, there is a waiting period, the length of which depends on the type of penetrant used and the type of metal being tested. After this waiting period, the excess penetrant must be removed.

Removal The excess penetrant must still be liquid during the removal process. Extreme care should be taken not to remove the penetrant from the possible defective openings in the metal. If the penetrant has dried, then the cycle must be restarted by recleaning the metal and reapplying the penetrant. The three types of penetrants are categorized by the solutions used to remove them. They are the water washable, the solvent removable, and the postemulsification.

Developing After the excess penetrant has been removed, the developer is applied in order to make the discontinuities readily visible. The two major types of developers are the wet developer and the dry. Wet developers can be either solvent base or water base; generally both are applied with a pressurized spray. The dry developer, which performs the same general functions as the wet developer, differs only in the fact that it is not carried in a liquid and is applied in a powder form. The wet developer can be applied while the surface is still damp to the touch, but the dry developer must be applied only to surfaces that are dry.

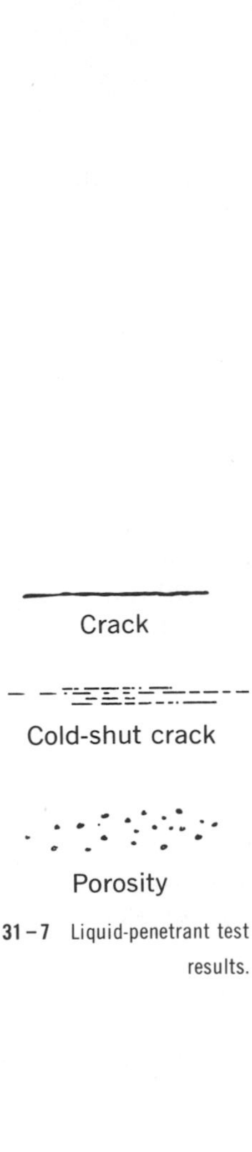

Crack

Cold-shut crack

Porosity

Fig. 31–7 Liquid-penetrant test results.

Inspection Inspection is the next step in the liquid penetrant test. The inspection can be done with either a normal light, with visible type dyes, or with a black light, with fluorescent dye penetrants. With either method, two types of indications are revealed: true indications or false indications. True indications are caused by penetrants bleeding out from actual discontinuities in the metal (Figure 31–7). The standard true flaws that are indicated are the cracks; the cold-shut cracks; fatigue cracks, which resemble to a great extent the cold-shut cracks; pits; and porosities. The large crack is represented by a line of some width. The large crack, after the developer is applied, becomes apparent quickly. The cold-shut crack, which is an undersurface crack that is bleeding through the surface, is represented by a line of dots and requires a few minutes after the developer has been added to come to the surface. Porosity is noticeable quickly in that an indication of porosity comes to the surface almost immediately, as do indications of the large crack. Porosity is generally indicated by dots that identify the area of porosity.

The false or nonrelevant indications are not caused by flaws at the surface of the metal. The major reasons for false indications are failure to follow the correct liquid penetrant application or rough, irregular surfaces of the test metal.

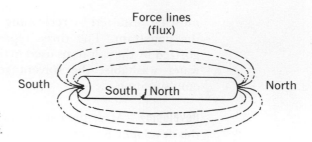

Fig. 31-8 Surface defects—magnetic lines of flux.

MAGNETIC PARTICLE TESTING

Magnetic particle inspection is based upon the magnetic lines of flux or force lines that can be generated within the test specimen. These force lines are parallel to each other while traveling through the test specimen if there are no surface or subsurface defects in the piece (Figures 31-8 and 31-9). In the bottom half of Figure 31-8, the force lines are going through the piece without any deviation; however, if there is a surface defect, the magnetic lines of force, or the flux lines, will create a small north-south hole area at the location of the surface defect, causing the iron powder that is used in magnetic particle inspection to form around the surface defect. This surface defect is called a magnetic leakage, or an indication.

There are two types of power supplies that can be used to develop the magnetic lines of force in the test specimen. They are a permanent magnet or an electromagnet. No matter which power source is used, the magnetic lines of force will be the same. However, the electromagnet is capable of producing a stronger force field than the permanent magnet. After the magnetic force lines have been generated in the test specimen, a finely grained iron powder is sprayed or blown over the surface of the test specimen. This powder may be either suspended in a liquid or it may be dry. The liquid applications are either blown on with some kind of spray device or the test specimen is immersed completely in a bath that has the particles suspended in a light or medium oil. If the test specimen is too large to be immersed, portable prods can be used to establish the magnetic field in the test specimen. Another way to establish a magnetic field in a large test specimen is to wrap the specimen with arc welding cables.

Magnetic particle inspection can also indicate subsurface defects if these defects are close to the surface. These

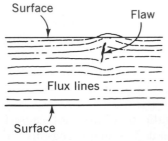

Fig. 31-9 Subsurface defects—magnetic lines of flux.

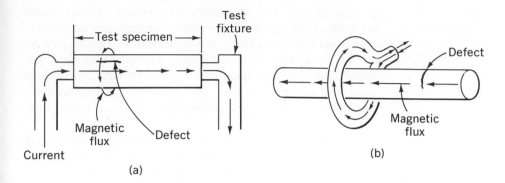

Fig. 31-10 Magnetization. (a) Longitudinal (parallel defects). (b) Circular (perpendicular defects).

subsurface defects disrupt the magnetic lines of force and bend them around the defect (Figure 31-10). A subsurface defect is indicated by the magnetic powder collecting around the broken lines of force at the surface of the test specimen.

Two types of surface defects can be detected by magnetic particle inspection: defects that are parallel to the interface of the workpiece and defects that are perpendicular to the interface or sides of the test specimen. Longitudinal, or parallel, defects are indicated because the electrical current flow sets up magnetic flux lines at a 90° angle to the current flow, following the left-hand rule of current flow. These magnetic lines of force must cross the defect in a perpendicular line in order to derive the full benefit from magnetic particle inspection. If the defect is parallel to the lines of force, there is a possibility that the defect would not be noticeable and there would be no indication of a surface or a subsurface flaw in the test specimen (Figure 31-10a).

In order to check a test specimen fully, both longitudinal and perpendicular indications must be checked. The perpendicular indications are inspected by using a circular coil which has current flowing through it in one direction. This coil will set up magnetic flux lines that are traveling through the test specimen longitudinally. Any defect that is perpendicular to these magnetic lines of flux will show an indication. The two tests together, the longitudinal test and the circular test, complete the testing of metal by magnetic particle inspection.

The power requirements for electromagnets are either dc or rectified ac current, which will produce a current flowing in only one direction. The power requirements are high amperage and low voltage. The major items that are inspected in welding are penetration, fusion, cracks in the weld, and slag inclusions. The particles that are applied to the test specimen can be of different colors according to the color of the specimen. The magnetic particle colors come in yellow, red, green, brown, and other colors.

For most of the applications, the test specimen need not be demagnetized; however, if the test specimen needs to be restored to its nonmagnetic or paramagnetic state, it can be passed through a high field intensity ac coil, which will sufficiently demagnetize the part so that it will resume its original magnetic state.

QUESTIONS

1. What defects can ultrasonic testing discover?

2. How do ultrasonic waves indicate where these defects are in material?

3. How is electrical energy converted into sound energy?

4. What types of wave forms are commercially available with ultrasonic equipment?

5. What does CRT stand for?

6. How are x-rays liberated?

7. What are the safety restrictions when using x-ray testing equipment?

8. What is the effect of the exposure of film to x-rays?

9. How does the shape of the x-ray target affect the shape of the ray?

10. What other power sources are capable of replacing the x-ray power source?

11. How is the test by liquid penetrants conducted?

12. Why are clean surfaces important when using liquid penetrants to test metals for surface and subsurface defects?

13. How does the right-hand rule of current flow apply to the magnetic particle testing method?

14. How are the iron powder particles applied to the test specimen in magnetic particle testing?

SUGGESTED FURTHER READINGS

Bergmann, Ludwig: "Ultrasonics," John Wiley & Sons, Inc., New York, 1938.

Fearon, J. H., and E. G. Stanford: "Progress in Non-Destructive Testing," The MacMillan Company, New York, 1959.

Lambel, J. H.: "Principles and Practices of Non-Destructive Testing," John Wiley & Sons, Inc., New York, 1962.

Lytel, Allen: "Industrial X-Ray Handbook," Howard W. Sams & Co., Inc., New York, 1962.

McDonald, Pat: Ultrasonics—New Tool for Industry, *Radio Electronics,* vol. 39, pp. 58-59, February 1968.

McMasters, Robert C.: "Non-Destructive Testing Handbook," The Ronald Press Co., New York, 1963.

Rockley, J. C.: "An Introduction to Industrial Radiology," Butterworths and Company, London, 1964.

Rossi, Boniface E.: "Welding Engineering," McGraw-Hill Book Company, New York, 1954.

INDEX